2011

Edexcel
Biology AS

Student Workbook

BIOZONE

Edexcel *Biology AS* 2011

Student Workbook

First edition 2010

ISBN 978-1-877462-54-2

Copyright © **2010** Richard Allan
Published by **BIOZONE International Ltd**

Printed by REPLIKA PRESS PVT LTD using paper
produced from renewable and waste materials

Acc No.	237975	
Class No.	570: ALL	
Date on System	16/2/2012	Initial MS

About the Writing Team

Lissa Bainbridge-Smith worked in industry in a research and development capacity for eight years before joining Biozone in 2006. Lissa has an M.Sc from Waikato University.

Kent Pryor has a BSc from Massey University majoring in zoology and ecology. He was a secondary school teacher in biology and chemistry for 9 years before joining Biozone as an author in 2009.

Tracey Greenwood joined the staff of Biozone at the beginning of 1993. She has a Ph.D in biology, specialising in lake ecology, and taught undergraduate and graduate biology at the University of Waikato for four years.

Richard Allan has had 11 years experience teaching senior biology at Hillcrest High School in Hamilton, New Zealand. He attained a Masters degree in biology at Waikato University, New Zealand.

Purchases of this workbook may be made direct from the publisher:

www.biozone.co.uk

UNITED KINGDOM:

BIOZONE Learning Media (UK) Ltd.
Bretby Business Park, Ashby Road, Bretby,
Burton upon Trent, DE15 0YZ, **UK**
Telephone: 01283-553-257
FAX: 01283-553-258
E-mail: sales@biozone.co.uk

AUSTRALIA:

BIOZONE Learning Media Australia
P.O. Box 2841, Burleigh BC,
QLD 4220, **Australia**
Telephone: +61 7-5535-4896
FAX: +61 7-5508-2432
E-mail: sales@biozone.com.au

NEW ZEALAND:

BIOZONE International Ltd.
P.O. Box 13-034, Hamilton 3251, **New Zealand**
Telephone: +64 7-856-8104
FAX: +64 7-856-9243
E-mail: sales@biozone.co.nz

Preface to the 2011 Edition

This is the first year Biozone has offered a workbook specifically designed to meet the needs of students enrolled in the Edexcel Biology course. This workbook has been designed so that key biological concepts are covered within the contexts prescribed in the course specification. This approach means the Edexcel workbook is ideal for students undertaking the concept based approach (as prescribed by the Edexcel scheme), and also for those following the context based Salters-Nuffield scheme. The specific nature of this workbook allows both students and educators to easily locate material relevant to their course. This workbook has a strong focus on scientific literacy and learning within relevant contexts. The chapters are organised into blocks of related information aligned to the Edexcel units and topics.

▶ Concept maps introduce each unit of the workbook, integrating the content across chapters to encourage linking of ideas.

▶ An easy-to-use chapter introduction comprising succinct learning objectives, a list of key terms, and a short summary of key concepts.

▶ An emphasis on acquiring skills in scientific literacy. Each chapter includes a comprehension and/or literacy activity, and the appendix includes references for works cited throughout the text.

▶ *Web links* and *Related Activities* support the material provided on each activity page.

A Note to the Teacher

This workbook is a student-centred resource, and benefits students by facilitating independent learning and critical thinking. This workbook is just that; a place for your answers notes, asides, and corrections. It is not a textbook and annual revisions are our commitment to providing a current, flexible, and engaging resource. The low price is a reflection of this commitment. Please do not photocopy the activities. If you think it is worth using, then we recommend that the students themselves own this resource and keep it for their own use. I thank you for your support.

Richard Allan

Acknowledgements

We thank all those who have contributed to this edition:

• Donna Allan, Sue FitzGerald, Gwen Gilbert, and Mary McDougall, for efficient handling of the office • Will Robinson, Gemma Conn for their artistic contributions to this edition, • Pip Sullivan and Tim Lind for their contributions • Raewyn Poole, University of Waikato, for information provided in her MSc thesis: Culture and transformation of *Acacia* • TechPool Studios, for their clipart collection of human anatomy: Copyright ©1994, TechPool Studios Corp. USA (some of these images were modified by R. Allan and T. Greenwood) • Totem Graphics, for their clipart • Corel Corporation, for vector clipart from the Corel MEGAGALLERY collection • 3D artwork created using Poser IV, Curious Labs and Bryce.

Photo Credits

Royalty free images, purchased by Biozone International Ltd, are used throughout this workbook and have been obtained from the following sources: Corel Corporation from their Professional Photos CD-ROM collection; IMSI (Intl Microcomputer Software Inc.) images from IMSI's MasterClips® and MasterPhotos™ Collection, 1895 Francisco Blvd. East, San Rafael, CA 94901-5506, USA; ©1996 Digital Stock, Medicine and Health Care collection; © 2005 JupiterImages Corporation www.clipart.com; ©Hemera Technologies Inc, 1997-2001; ©Click Art, ©T/Maker Company; ©1994., ©Digital Vision; Gazelle Technologies Inc.; PhotoDisc®, Inc. USA, www.photodisc.com.

The writing team would like to thank the following individuals and institutions who kindly provided photographs: • Alan Sheldon Sheldon's Nature Photography, Wisconsin for the photo of the lizard without its tail • International plant nutrition institute (IPNI) for the competition photos on plant mineral deficiencies (S. Srinivasan, Bobby Golden, T. Tindall, T. Wyciskalla, Gunter *et.al*) • Dan Butler for the photo of his injured finger • Dartmouth College for the freeze fracture image of a plasma membrane • Dept. of Natural Resources, Illinois, for the photograph of the threatened prairie chicken • Ed Uthman for the image of the nine week human embryo • Dr. Nita Scobie, Cytogenetics Department, Waikato Hospital for the chromosome photos • Stephen Moore for his photos of aquatic invertebrates • David Wells at Agresearch for photos on cloning • PEIR digital library for the image of the Atherosclerotic plaque • www.coastalplanning.net for the image of a marine quadrat • Andrew Dunn, *www.andrewdunnphoto.com* • Alison Roberts for the photo of the plasmodesmata • California Academy of Sciences for the photo of the ground finch *'Britain's Biodiversity' are as follows*: **IMSI**: acorn barnacle; **EII**: hedgehog, hermit crab, red elf cup fungus, Duke of Burgundy fritillary, woodmouse, oak, field vole; **Corel**: badger, falcon, bluebell woodland, nuthatch, European otter, red fox, puffin, common toad.

We also acknowledge the photographers that have made their images available through **Wikimedia Commons** under Creative Commons Licences 2.5. or 3.0: • Duncan Wright • gone, gone, gone • Georgetown University Hospital • Indian Nomad • Janet Stephens • Jacoplane • kaylaya WCC • Lucien Monfils • ms-donna • RDC/Welcome Trust • SHAC • Trounce • University of Hawaii • Yelyos

Contributors identified by coded credits are:
BF: Brian Finerran (Uni. of Canterbury), **BH**: Brendan Hicks (Uni. of Waikato), **BOB**: Barry O'Brien (Uni. of Waikato), **CDC**: Centers for Disease Control and Prevention, Atlanta, USA, **DS**: Digital Stock, **EII**: Education Interactive Imaging, **EW**: Environment Waikato, **FRI**: Forestry Research Institute, **IF**: I. Flux (DoC), **IMSI**: International Microcomputer Software Inc, **NIH**: National Institute for Health, **RA**: Richard Allan, **RCN**: Ralph Cocklin, **TG**: Tracey Greenwood, **VM**: Villa Maria Wines, **WMU**: Waikato Microscope Unit

Cover Photographs

Main photograph:
The white bellied spider monkey (*Ateles belzebuth*) is found in Colombia, Brazil, Ecuador, Peru, and Venezuela. Its status is endangered, mainly as a result of habitat loss from deforestation. This species is highly social, living in groups of 20 to 40 members, and live 30-40 years
PHOTO: Pete Oxford

Background photograph:
Autumn leaves, Image ©2005 JupiterImages Corporation www.clipart.com

Contents

UNIT 3: Practical Biology and Research Skills

Practical Biology and Research Skills

UNIT 1: Lifestyle, Transport, Genes and Health

Topic 1: Lifestyle, Health and Risk

Biological Molecules

Healthy Lifestyle, Healthy Heart

Topic 2: Genes and Health

Membranes and Exchange Surfaces

Proteins, Genes and Health

CODES: **Activity** is marked: • to be done; ✓ when completed   † Denotes support for recommended core practical

CONTENTS (continued)

CODES: **Activity** is marked: ▪ to be done;   ✔ when completed   † Denotes support for recommended core practical

Getting The Most From This Resource

This workbook is designed as a resource to increase your understanding and enjoyment of biology. This workbook is suitable for students in their AS year of Edexcel Biology. The course guide on page 10 (or on the Teacher Resource CD-ROM) indicates where the material for your course is covered in the workbook. It is hoped that this resource will reinforce and extend the ideas developed by your teacher. It is **not a textbook**; its aim is to complement the texts written for your course. The workbook includes the following useful features:

Features of the Concept Map

A summary of the curricular emphasis in each major section of the workbook. This panel also helps you to understand how the concepts within the unit link together

Chapter panels identify and summarise the material covered within each chapter.

Encouraging Key Competencies

Thinking - bringing ideas together
Relating to others - communicating
Using language, symbols, and text
Managing self - independence
Participating and contributing

Each section of the workbook emphasises skills and knowledge to be gained.

A summary of why this material is important and where it fits into your understanding of your course content.

Features of the Chapter Topic Page

The Edexcel unit and topic to which this chapter applies.

The important key ideas in this chapter. You should have a thorough understanding of the concepts summarised here.

The page numbers for the activities covering the material in this subsection of objectives.

Denotes recommended core practical activity.

The objectives provide a point by point summary of what you should have achieved by the end of the chapter.

A list of key terms used in the chapter. These terms appear in the chapter's vocab activity and can be used to create a glossary for revision purposes. The list represents the minimum literacy requirement for the chapter.

Periodicals of interest are identified by title on a tab on the activity page to which they are relevant. The full citation appears in the **Appendix** on the page indicated.

You can use the check boxes to mark objectives to be completed (a **dot** to be done; a **tick** when completed).

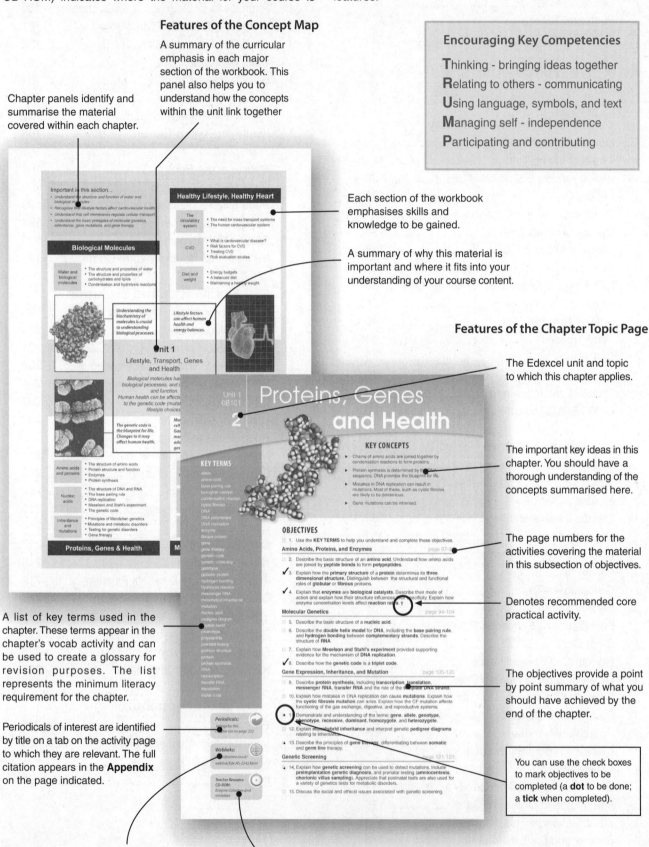

The Weblinks on many of the activities can be accessed through the web links page at: *www.biozone.co.uk/weblink/Edx-AS-2542.html* See page 11 for more details.

Extra resources for this chapter are available on the Teacher Resource CD-ROM (for separate purchase).

Using the Activities

The activities make up most of the content of this book. Your teacher may use the activity pages to introduce a topic for the first time, or you may use them to revise ideas already covered by other means. They are excellent for use in the classroom, as homework exercises and topic revision, and for self-directed study and personal reference.

Perforations allow easy removal so that pages can be submitted for grading or kept in a separate folder of related work.

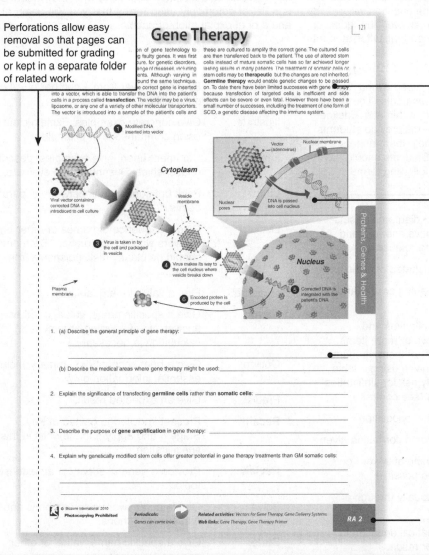

Gene Therapy 121

Introductory paragraph:
The introductory paragraph provides essential background and provides the focus of the page. Note words that appear in bold, as they are 'key words' worthy of including in a glossary of terms for the topic.

Easy to understand diagrams:
The main ideas of the topic are represented and explained by clear, informative diagrams.

Write-on format:
Your understanding of the main ideas of the topic is tested by asking questions and providing spaces for your answers. Where indicated by the space available, your answers should be concise. Questions requiring more explanation or discussion are spaced accordingly. Answer the questions adequately according to the questioning term used (see the next page).

A tab system at the base of each activity page is a relatively new feature in Biozone's workbooks. With it, we have tagged valuable resources to the activity to which they apply. Use the guide below to help you use the tab system most effectively.

Using page tabs more effectively

Periodicals:

Genes can come true,

Related activities: Vectors for Gene Therapy, Gene Delivery Systems

Web links: Gene Therapy, Gene Therapy Primer

RA 2

Students (and teachers) who would like to know more about this topic area are encouraged to locate the periodical cited on the Periodicals tab.
Articles of interest directly relevant to the topic content are cited. The full citation appears in the Appendix as indicated at the beginning of the topic chapter.

Related activities
Other activities in the workbook cover related topics or may help answer the questions on the page. In most cases, extra information for activities that are coded R can be found on the pages indicated here.

Web links
This citation indicates a valuable video clip or animation that can be accessed from the web links page specifically for this workbook.
www.biozone.co.uk/weblink/Edx-AS-2542.html

INTERPRETING THE ACTIVITY CODING SYSTEM
Type of Activity
D = includes some data handling or interpretation
P = includes a paper practical
R = *may* require extra reading (e.g. text or other activity)
A = includes application of knowledge to solve a problem
E = extension material

Level of Activity
1 = generally simpler, including mostly describe questions
2 = more challenging, including explain questions
3 = challenging content and/or questions, including discuss

Explanation of Terms

Questions come in a variety of forms. Whether you are studying for an exam, or writing an essay, it is important to understand exactly what the question is asking. A question has two parts to it: one part of the question will provide you with information, the second part of the question will provide you with instructions as to how to answer the question. Following these instructions is most important. Often students in exams know the material but fail to follow instructions and therefore do not answer the question appropriately. Examiners often use certain key words to introduce questions. Look out for them and be absolutely clear as to what they mean. Below is a list of commonly used terms that you will come across and a brief explanation of each.

Commonly used Terms in Biology

The following terms are frequently used when asking questions in examinations and assessments. The terms given in bold are key terms defined within the Edexcel biology specification. The additional terms provide students with an expanded key term bank to help them understand and answer questions appropriately. Students should have a clear understanding of each of the following terms.

Advantage: Compare two process, diagrams or sets of data. Give comparative answers and clearly state the feature referred to.

Analyse: Interpret data to reach stated conclusions.

Appreciate: To understand the meaning or relevance of a particular situation.

Compare: Give an account of similarities and differences between two or more items, referring to both (or all) of them throughout. Comparisons can be given using a table. Comparisons generally ask for similarities more than differences (see contrast).

Contrast: Show differences. Set in opposition.

Deduce: Reach a conclusion from information given.

Define: Give the precise meaning of a word or phrase as concisely as possible.

Demonstrate: Show the effects of (usually through a practical experiment).

Derive: Manipulate a mathematical equation to give a new equation or result.

Describe: Give a detailed account, including all the relevant information.

Design: Produce a plan, object, simulation or model.

Determine: Find the only possible answer.

Discuss: Give an account including, where possible, a range of arguments, assessments of the relative importance of various factors, or comparison of alternative hypotheses.

Distinguish: Give the difference(s) between two or more different items.

Draw: Represent by means of pencil lines. Add labels unless told not to do so.

Estimate: Find an approximate value for an unknown quantity, based on the information provided and application of scientific knowledge.

Evaluate: Assess the implications and limitations.

Explain: Give a clear account including causes, reasons, or mechanisms.

Give: Give a specific name, value, or other answer. No supporting argument or calculation is necessary.

Identify: Find an answer from a number of possibilities.

Illustrate: Give concrete examples. Explain clearly by using comparisons or examples.

Interpret: Comment upon, give examples, describe relationships. Describe, then evaluate.

Link: Point out connections between separate points.

List: Give a sequence of names or other brief answers with no elaboration. Each one should be clearly distinguishable from the others.

Measure: Find a value for a quantity.

Name: Give a specific name, value, or other answer. No supporting argument or calculation is necessary.

Outline: Give a brief account or summary. Include essential information only.

Predict: Give an expected result.

Recall: Present knowledge gained at Key Stage 4 and apply it to the units in this specification.

Review: Provide a general survey of an extensive topic.

Solve: Obtain an answer using algebraic and/or numerical methods.

State: Give a specific name, value, or other answer. No supporting argument or calculation is necessary.

Suggest: Propose a hypothesis or other possible explanation.

Summarise: Give a brief, condensed account. Include conclusions and avoid unnecessary details.

Understand: Describe and explain the underlying principles and apply the knowledge to situations.

In Conclusion

Students should familiarise themselves with this list of terms and, where necessary throughout the course, they should refer back to them when answering questions. The list of terms mentioned above is not exhaustive and students should compare this list with past examination papers / essays etc. and add any new terms (and their meaning) to the list above. The aim is to become familiar with interpreting the question and answering it appropriately.

Resources Information

Your set textbook should be a starting point for information about the content of your course. There are also many other resources available, including journals, magazines, supplementary texts, dictionaries, computer software, and the internet. Your teacher will have some prescribed resources for your use, but a few of the readily available periodicals are listed here for quick reference. The titles of relevant articles are listed with the activity to which they relate and are cited in the appendix. Please note that listing any product in this workbook does not, in any way, denote Biozone's endorsement of that product and Biozone does not have any business affiliation with the publishers listed herein.

Supplementary Texts

For further details of text content, or to make purchases, link to the relevant publisher via Biozone's resources hub or by typing: **www.biozone.co.uk > resources > supplementary**

Adds, J., E. Larkcom & R. Miller, 2004.
Exchange and Transport, Energy and Ecosystems, revised edition 240 pp.
Publisher: Nelson Thornes
ISBN: 0-7487-7487-4
Covers exchange processes, transport systems, adaptation, and sexual reproduction. Practical activities are included in several of the chapters.

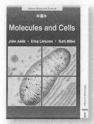

Adds, J., E. Larkcom & R. Miller, 2003.
Molecules and Cells, revised edition 112 pp.
ISBN: 0-7487-7484-X
Includes coverage of the basic types of biological molecules, with extra detail on the structure and function of nucleic acids and enzymes, cellular organisation, and cell division. Practical activities are provided for most chapters.

Cadogan, A. and Ingram, M., 2002
Maths for Advanced Biology
Publisher: NelsonThornes
ISBN: 0-7487-6506-9
Comments: *Covers the maths requirements of AS/A2 biology. Includes worked examples.*

Indge, B., 2003
Data and Data Handling for AS and A Level Biology, 128 pp.
Publisher: Hodder Arnold H&S
ISBN: 1340856475
Comments: *Examples and practice exercises to improve skills in data interpretation and analysis.*

Clegg, C.J., 1999.
Genetics and Evolution, 96 pp.
ISBN: 0-7195-7552-4
Concise but thorough coverage of molecular genetics, genetic engineering, inheritance, and evolution. An historical perspective is included by way of introduction, and a glossary and a list of abbreviations used are included.

Fullick, A., 1998
Human Health and Disease, 162 pp.
Publisher: Heinemann Educational Publishers
ISBN: 0435570919
Comments: *An excellent supplement for courses with modules in human health and disease. Includes infectious and non-infectious disease.*

Periodicals, Magazines and Journals

Details of the periodicals referenced in this workbook are listed below. For enquiries and further details regarding subscriptions, link to the relevant publisher via Biozone's resources hub or by going to: **www.biozone.co.uk > Resources > Journals**

Biological Sciences Review (Biol. Sci. Rev.) *An excellent quarterly publication for teachers and students. The content is current and the language is accessible.* Subscriptions available from Philip Allan Publishers, Market Place, Deddington, Oxfordshire OX 15 OSE.
Tel. 01869 338652 **Fax**: 01869 338803
E-mail: sales@philipallan.co.uk

New Scientist: *Published weekly and found in many libraries. It often summarizes the findings published in other journals. Articles range from news releases to features.*
Subscription enquiries:
Tel. (UK and international): +44 (0)1444 475636. (US & Canada) 1 888 822 3242.
E-mail: ns.subs@qss-uk.com

Scientific American: *A monthly magazine containing mostly specialist feature articles. Articles range in level of reading difficulty and assumed knowledge.*
Subscription enquiries:
Tel. (US & Canada) 800-333-1199.
Tel. (outside North America): 515-247-7631
Web: www.sciam.com

Biology Dictionaries

Access to a good biology dictionary is of great value when dealing with the technical terms used in biology. Below are some biology dictionaries that you may wish to locate or purchase. They can usually be obtained directly from the publisher or they are all available (at the time of printing) from www.amazon.com. For further details of text content, or to make purchases, link to the relevant publisher via Biozone's resources hub or by typing: **www.biozone.co.uk > Resources > Dictionaries**

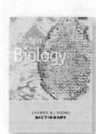

Clamp, A. **AS/A-Level Biology. Essential Word Dictionary**, 2000, 161 pp. Philip Allan Updates.
ISBN: 0-86003-372-4.
Carefully selected essential words for AS and A2. Concise definitions are supported by further explanation and illustrations where required.

Edexcel Course Guideline

Candidates taking Edexcel AS Biology are required to complete Unit 1 (6BIO1), Unit 2 (6BIO2) and Practical Unit 3 (6BIO3). Candidates taking Edexcel A2 Biology will first need to complete the requirements for the AS course as well as Unit 4(6BIO4), Unit 5 (6BIO5), and Practical Unit 6 (6BIO6). The content of the specification can be approached from the concept-led approach (Edexcel), or a context-led approach (Salter-Nuffield). A single common assessment schedule (see below) applies, although the material in this workbook has been structured on the former. An asterisk on the scheme denotes a recommended core practical.

AS Specification

Topics in Edexcel AS workbook (unless indicated)

Unit 1 (6BI01) : Lifestyle, Transport, Genes and Health

Topic 1: Lifestyle, Health and Risk

Water as a solvent. Carbohydrate and lipid structure and function. Condensation and hydrolysis.	Biological Molecules
Limitations to diffusion. Mass transport systems. Mammalian heart. Cardiac cycle. Blood vessels.	Healthy Lifestyle, Healthy Heart
*Describe how the effect of caffeine on heart rate can be studied in *Daphnia*. Include ethical issues.	Practical Biology & Research Skills
Blood clotting and CVD.	
CVD, risk factors, prevention,treatment.	
Blood cholesterol levels, HDL:LDL ratio.	Healthy Lifestyle, Healthy Heart
*Describe how to measure vitamin C level in food.	
Energy budgets and its effect on weight.	
Illness data and mortality rates. Risk perception.	

Topic 2: Genes and Health

Membrane structure. Osmosis. Transport processes.	Membranes & Exhange Surfaces
*Practical investigation of membrane structure.	
Gas exchange surfaces. Mammalian lung.	
Amino acids, proteins, enzymes.	
*Investigation of enzyme reaction rates.	Practical Biology & Research Skills
Nucleotides, nucleic acids, DNA replication, genetic code, genes, protein synthesis.	
Mutations: causes and consequences of CF.	Proteins, Genes & Health
Monohybrid inheritance, pedigree analysis.	
Gene therapy. Somatic versus germ line therapy.	
Genetic profiling and screening.	
Prenatal testing: amniocentesis, CVS. Ethics.	

Unit 2 (6B102): Development, Plants and the Environment

Topic 3: The Voice of the Genome

Prokaryotic v eukaryotic cells. Ultrastructure of eukaryotic cells. Function of rER and Golgi. Tissues.	Cells & Microscopy
Mitosis for growth and asexual reproduction.	
*Identify the stages of mitosis.	
Meiosis as a source of variation.	
Mammalian gametes. Mammal and plant fertilisation.	
Stem cell research and definitions.	
*Describe how to demonstrate totipotency by using plant tissue culture techniques.	Variation & Heredity
Cell specialisation through gene expression.	
Effects of environment on genotype expression.	
Polygenic inheritance.	

Topic 4: Biodiversity and Natural Resources

Compare plant and animal cell ultrastructure.	Cells & Microscopy
Starch and cellulose. Cellulose microfibrils.	
Compare and identify sclerenchyma and xylem.	
*Describe how to measure tensile strength.	
Importance of water and inorganic ions to plants.	Plants as Resources
*Describe how to practically investigate mineral deficiencies and antimicrobial properties in plants.	
Drug testing protocols.	
Biodiversity, endemism. Measuring biodiversity.	
Niche. Adaptation to niche.	
Natural selection, adaptation, and evolution.	Biodiversity & Evolution
Taxonomy methods.	
Conservation methods.	

Unit 3 (6B103): Practical Biology & Research Skills

Includes skilful and safe use of apparatus, precise measurements and observations, proper data presentation and analysis. Prepare a report based upon a site visit or specific issue.	Practical Biology & Research Skills

A2 Specification

Topics in Edexcel A2 workbook (unless indicated)

Unit 4 (6B104): The Natural Environment & Species Survival

Topic 5: On the Wild Side

Chloroplast. Biochemistry of photosynthesis.	
Productivity, plant respiration. Energy transfer.	
Carbon cycle. Reducing CO_2 levels.	
Distribution and control of organism numbers.	Energy & Ecosystems
*Describe how to reliably study the ecology of a habitat (biotic and abiotic measurements).	
Niche. Organism distribution and abundance.	
Ecological succession.	
Global warming, causes and effects.	
*Describe how to investigate effect of temperature on an organisms developmental rate.	
Analyse and interpret global warming evidence.	Global Warming & Evolution
Natural selection, gene mutation, and evolution.	
Reproductive isolation and speciation.	
Evidence for evolution.	

Topic 6: Infection, Immunity and Forensics

Genetic code. Protein synthesis, role of RNA.	
Post transcriptional modifications.	Genes, Proteins & Relationships
DNA profiling.	
*Describe how DNA can be amplified by PCR, and separated using gel electrophoresis.	
Compare the structure of bacteria and viruses.	
Microbes, decomposition and the carbon cycle.	
Pathogen infection routes. Barriers to infection.	
Bacterial infection: TB. Viral infection: HIV.	
Non-specific defences. Antigen and antibodies.	
B- and T-cells. Acquiring immunity.	The Fight Against Disease
Evolutionary race theory for HIV and TB.	
Antibiotics and their mode of action.	
*Describe testing antibiotic efficiency on bacteria.	
Hospital acquired infections. Control and treatment.	
Forensics to estimate time of death.	

Unit 5 (6B105): Energy, Exercise and Coordination

Topic 7: Run For Your Life

Muscle fibre structure. Fast vs slow twitch.	
Muscle contraction. Movement.	
Aerobic respiration.	
*Describe how to investigate respiration.	Muscles & Energy
Role of ATP. Glycolysis and cellular respiration.	
ATP synthesis. ETC. Chemiosmosis. Lactate.	
Cardiac muscle control. ECGs, CVD diagnosis.	
Cardiac output control and ventilation.	
*Describe testing effects of exercise on breathing.	
Negative feedback. Homeostasis. Thermoregulation.	Homeostasis & Exercise
Hormones as gene switches.	
Exercise: too much v too little. Medical technologies for sport injuries. Performance enhancement.	

Topic 8: Grey Matter

Plant photoreceptors. Environmental cues.	
Neruones. Synapses. Vision.	
Nervous v hormonal regulatory systems.	
Structure and function of the human brain.	
Imaging: MRI, fMRI, CT scans.	Sensing & Responding
Brain development theories. Visual window.	
*Describe how to investigate habituation.	
Ethics of using animals for research.	
Chemical imbalances. Drugs at synaptic connections. HGP and drug development.	Drug Development AS: Plants as Resources
Drugs produced by GMOs. Risks and benefits.	

Unit 6 (6B106): Practical Biology & Investigative Skills

Planning and implementing practicals: observations and recording, data interpretation and analysis, evaluation and communication of results.	Practical Biology & Investigative Skills

Using Biozone's Website

The current internet address (URL) for the web site is displayed here. You can type a new address directly into this space.

Use Google to search for web sites of interest. The more precise your search words are, the better the list of results. EXAMPLE: If you type in "biotechnology", your search will return an overwhelmingly large number of sites, many of which will not be useful to you. Be more specific, e.g. "biotechnology medicine DNA uses".

Find out about our superb **Presentation Media**. These slide shows are designed to provide in-depth, highly accessible illustrative material and notes on specific areas of biology.

News: Find out about product announcements, shipping dates, and workshops and trade displays by Biozone at teachers' conferences around the world.

Podcasts: Access the latest news as audio files (mp3) that may be downloaded or played directly off your computer.

RSS Newsfeeds: See breaking news and major new discoveries in biology directly from our web site.

Access the **BioLinks** database of web sites related to each major area of biology. It's a great way to quickly find out more on topics of interest.

Weblinks: www.biozone.co.uk/weblink/Edx-AS-2542.html

BOOKMARK WEBLINKS BY TYPING IN THE ADDRESS: IT IS NOT ACCESSIBLE DIRECTLY FROM BIOZONE'S WEBSITE

Throughout this workbook, some pages make reference to web links and periodicals that are particularly relevant to the activity on which they are cited. They provide great support to aid understanding of basic concepts:

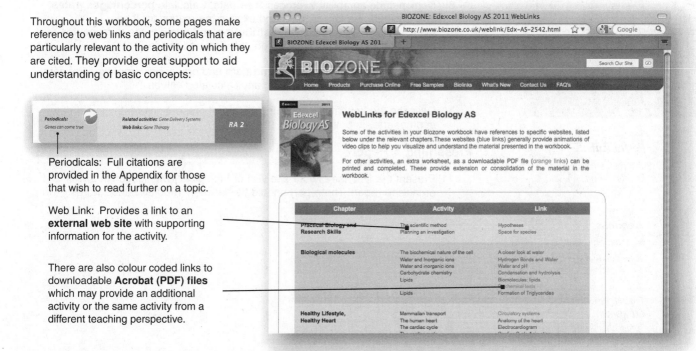

Periodicals: Full citations are provided in the Appendix for those that wish to read further on a topic.

Web Link: Provides a link to an **external web site** with supporting information for the activity.

There are also colour coded links to downloadable **Acrobat (PDF) files** which may provide an additional activity or the same activity from a different teaching perspective.

Practical Biology & Research Skills

KEY CONCEPTS

▶ The basis of all science is observation, hypothesis, and investigation.

▶ Scientists collect and analyse data to test their hypotheses.

▶ Data can be analysed and presented in various ways, including in graphs and tables.

▶ A scientific report summarises the results of a scientific investigation and makes the findings accessible to others.

KEY TERMS

accuracy
animal testing
bibliography
biological drawing
citation
control
controlled variable
data
dependent variable
ethical considerations
frequency
graph
histogram
hypothesis
independent variable
mean
measurement
median
mode
observation
percentage
precision
qualitative data
quantitative data
rate
raw data
report
sample
scientific method
table
trend (of data)
variable
X axis
Y axis

OBJECTIVES

☐ 1. Use the **KEY TERMS** to help you understand and complete these objectives.

Biological Investigations   pages 13-19

☐ 2. Describe and explain the basic principles of the **scientific method**.

☐ 3. Recognise the difference between dependent and independent variables, and how they are measured. Identify controlled variables and their significance.

☐ 4. Explain the difference between qualitative and quantitative data and give examples of their appropriate use.

☐ 5. Demonstrate an ability to **systematically record** data. Evaluate the accuracy and precision of any recording or measurements you make.

☐ 6. Explain systematic and random errors generated during investigations.

☐ 7. Demonstrate an ability to make accurate biological drawings.

Presenting Data   pages 20-30

☐ 8. Demonstrate an ability to process raw data. Calculate **percentages**, **rates**, and **frequencies** for raw data and explain the reason for these manipulations. Describe **data distribution** and relate these to measures of **central tendency**. Calculate **mean**, **median** and **mode**.

☐ 9. Describe the benefits of **tabulating data** and present different types of data appropriately in a table, including any calculated values.

☐ 10. Describe the benefits of **graphing** data and present different types of data appropriately in both graphs.

Presenting A Report   pages 31-35

☐ 11. Present the findings of your site visit or biological investigation in a well organised report. The report should be between 1500 and 2000 words. †

Periodicals:
listings for this
chapter are on page 228

Weblinks:
www.biozone.co.uk/
weblink/Edx-AS-2542.html

*Teacher Resource
CD-ROM:*
Spreadsheets and Statistics

The Scientific Method

Scientific knowledge grows through a process called the **scientific method**. This process involves observation and measurement, hypothesising and predicting, and planning and executing investigations designed to test formulated hypotheses. A scientific hypothesis is a tentative explanation for an observation, which is capable of being tested by experimentation. Hypotheses lead to predictions about the system involved and they are rejected or not on the basis of findings arising from the investigation.

Scientific hypotheses have specific characteristics (below) and may be modified as more information becomes available. For the purposes of investigation, hypotheses are often constructed in a form that allows them to be tested statistically. The **null hypothesis** (H_0) can be tested statistically, and may then be rejected in favour of accepting the alternative hypothesis (H_A). Scientific information is generated as scientists make discoveries through testing hypotheses.

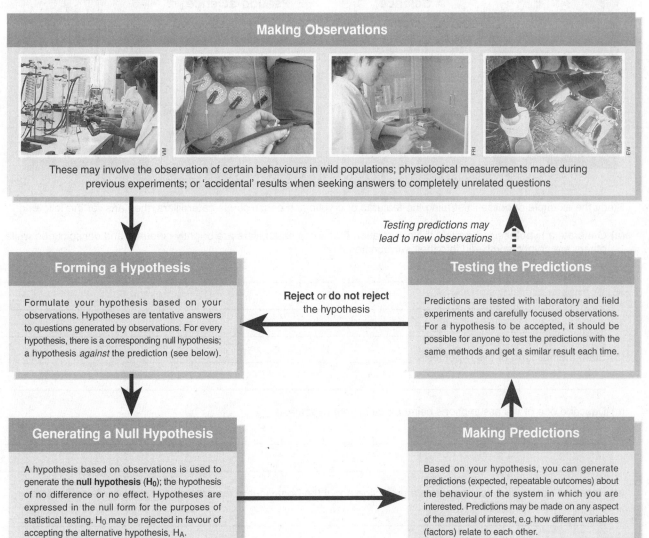

Making Observations

These may involve the observation of certain behaviours in wild populations; physiological measurements made during previous experiments; or 'accidental' results when seeking answers to completely unrelated questions

Testing predictions may lead to new observations

Forming a Hypothesis

Formulate your hypothesis based on your observations. Hypotheses are tentative answers to questions generated by observations. For every hypothesis, there is a corresponding null hypothesis; a hypothesis *against* the prediction (see below).

Testing the Predictions

Predictions are tested with laboratory and field experiments and carefully focused observations. For a hypothesis to be accepted, it should be possible for anyone to test the predictions with the same methods and get a similar result each time.

Reject or **do not reject** the hypothesis

Generating a Null Hypothesis

A hypothesis based on observations is used to generate the **null hypothesis** (H_0); the hypothesis of no difference or no effect. Hypotheses are expressed in the null form for the purposes of statistical testing. H_0 may be rejected in favour of accepting the alternative hypothesis, H_A.

Making Predictions

Based on your hypothesis, you can generate predictions (expected, repeatable outcomes) about the behaviour of the system in which you are interested. Predictions may be made on any aspect of the material of interest, e.g. how different variables (factors) relate to each other.

Practical Biology & Research Skills

Generating Predictions from Observations

Observation is a good way to develop the basis for formulating hypotheses and making predictions. An observation may generate a number of plausible hypotheses, and each hypothesis will lead to one or more predictions, which can be tested by further investigation.

Observation 1: Some caterpillar species are brightly coloured and appear to be conspicuous to predators such as insectivorous birds. It has also been observed that predators usually avoid them. These caterpillars are often found in groups, rather than as solitary animals.

Observation 2: Some caterpillar species are cryptic in appearance or behaviour. Their camouflage is so convincing that when alerted to danger, they are difficult to see against their background. Such caterpillars are usually found alone.

Periodicals:
The truth is out there

Web links: Hypotheses

The Differences between Science and Pseudo-Science

The scientific method is rigorous and based on observation, theory, experimentation, documentation, repeatability, and critical review. The underlying principles of science: to observe, question, theorise, and test are quite different from pseudo-science, which begins with a preconceived idea or theory and then looks for certain experimental results or evidence to support that idea, rejecting other contrary evidence in the process. Astrology, creationism, palmistry, and clairvoyancy are examples of this type of pseudo-science. Some of the differences between science and pseudo-science are outlined below:

Science
Ability & willingness to change
Absorbs all new discoveries
Ruthless peer review
Invites critical review
Ability to predict
Makes testable assertions
Verifiable
Few assumptions made

Pseudo-science
Fixed ideas
Selected favourable discoveries
No review
Hostile to criticism
Inability to predict
Makes non-testable assertions
Non-repeatable ("trust me")
Clings to cherished ideas

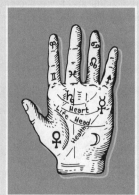

1. Study the example opposite illustrating the features of cryptic and conspicuous caterpillars, then answer the following:

   (a) Generate a hypothesis to explain the observation that some caterpillars are brightly coloured and conspicuous while others are cryptic and blend into their surroundings:

   Hypothesis: _____

   _____

   _____

   (b) State the null form of this hypothesis: _____

   _____

   (c) Describe one of the assumptions being made in your hypothesis: _____

   _____

   _____

   (d) Based on your hypothesis, generate a prediction about the behaviour of insectivorous birds towards caterpillars:

   _____

   _____

   (e) Describe a simple experiment to test your hypothesis and its prediction: _____

   _____

   _____

   _____

   _____

   _____

2. Explain why scientific hypotheses are usually expressed in the negative form (i.e. a **null hypothesis**):

   _____

   _____

   _____

   _____

Biological Drawings

Microscopes are a powerful tool for examining cells and cell structures. In order to make a permanent record of what is seen when examining a specimen, it is useful to make a drawing. It is important to draw **what is actually seen**. This will depend on the **resolution** of the microscope being used. Resolution refers to the ability of a microscope to separate small objects that are very close together. Making drawings from mounted specimens is a skill. Drawing forces you to observe closely and accurately. While photographs are limited to representing appearance at a single moment in time, drawings can be composites of the observer's cumulative experience, with many different specimens of the same material. The total picture of an object thus represented can often communicate information much more effectively than a photograph. Your attention to the outline of suggestions below will help you to make more effective drawings. If you are careful to follow the suggestions at the beginning, the techniques will soon become habitual.

1. **Drawing materials**: All drawings should be done with a clear pencil line on good quality paper. A sharp HB pencil is recommended. A soft rubber of good quality is essential. Diagrams in ballpoint or fountain pen are unacceptable because they cannot be corrected.

2. **Positioning**: Centre your diagram on the page. Do not draw it in a corner. This will leave plenty of room for the addition of labels once the diagram is completed.

3. **Size**: A drawing should be large enough to easily represent all the details you see without crowding. Rarely, if ever, are drawings too large, but they are often too small. Show only as much as is necessary for an understanding of the structure; a small section shown in detail will often suffice. It is time consuming and unnecessary, for example, to reproduce accurately the entire contents of a microscope field.

4. **Accuracy**: Your drawing should be a complete, accurate representation of the material you have observed, and should communicate your understanding of the material to anyone who looks at it. Avoid making "idealised" drawings; your drawing should be a picture of what you actually see, not what you imagine should be there. Proportions should be accurate. If necessary, measure the lengths of various

parts with a ruler. If viewing through a microscope, estimate them as a proportion of the field of view, then translate these proportions onto the page. When drawing shapes that indicate an outline, make sure the line is complete. Where two ends of a line do not meet (as in drawing a cell outline) then this would indicate that it has a hole in it.

5. **Technique**: Use only simple, narrow lines. Represent depth by stippling (dots close together). Indicate depth only when it is essential to your drawing (usually it is not). Do not use shading. Look at the specimen while you are drawing it.

6. **Labels**: Leave a good margin for labels. All parts of your diagram must be labelled accurately. Labelling lines should be drawn with a ruler and should not cross. Where possible, keep label lines vertical or horizontal. Label the drawing with:
    *   A title, which should identify the material (organism, tissues or cells).
    *   Magnification under which it was observed, or a scale to indicate the size of the object.
    *   Names of structures.
    *   In living materials, any movements you have seen.

Remember that drawings are intended as records for you, and as a means of encouraging close observation; artistic ability is not necessary. Before you turn in a drawing, ask yourself if you know what every line represents. If you do not, look more closely at the material. *Take into account the rules for biological drawings and draw what you see, not what you think you see!*

Examples of acceptable biological drawings: The diagrams below show two examples of biological drawings that are acceptable. The example on the left is of a whole organism and its size is indicated by a scale. The example on the right is of plant tissue – a group of cells that are essentially identical in the structure. It is not necessary to show many cells even though your view through the microscope may show them. As few as 2-4 will suffice to show their structure and how they are arranged. Scale is indicated by stating how many times larger it has been drawn. Do not confuse this with what magnification it was viewed at under the microscope. The abbreviation **T.S.** indicates that the specimen was a cross or transverse section.

Practical Biology & Research Skills

Cyclopoid copepod

Single eye
Antenna
Trunk
Egg sac
Thorax
Caudal rami
Setae
scale
0.2 mm

Collenchyma T.S. from Helianthus stem
Magnification x 450

Cytoplasm
Nucleus
Primary wall with secondary thickening
Chloroplast
Vacuole containing cell sap

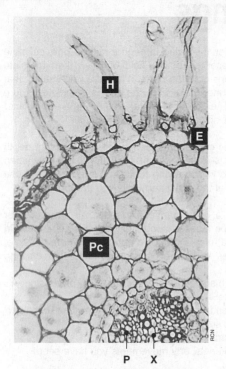

P   X

Specimen used for drawing

The photograph above is a light microscope view of a stained transverse section (cross section) of a root from a *Ranunculus* (buttercup) plant. It shows the arrangement of the different tissues in the root. The vascular bundle is at the centre of the root, with the larger, central xylem vessels (**X**) and smaller phloem vessels (**P**) grouped around them. The root hair cells (**H**) are arranged on the external surface and form part of the epidermal layer (**E**). Parenchyma cells (**Pc**) make up the bulk of the root's mass. The distance from point **X** to point **E** on the photograph (above) is about 0.15 mm (150 µm).

An Unacceptable Biological Drawing

The diagram below is an example of how *not* to produce a biological drawing; it is based on the photograph to the left. There are many aspects of the drawing that are unacceptable. The exercise below asks you to identify the errors in this student's attempt.

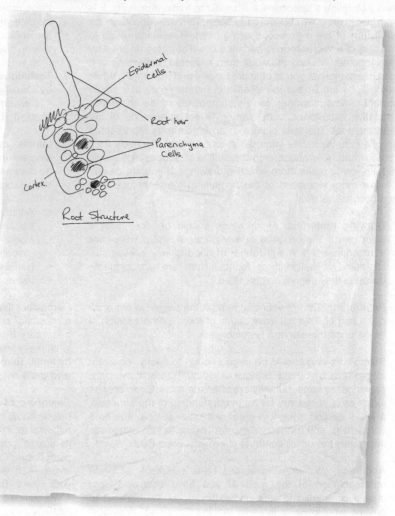

1. Identify and describe eight unacceptable features of the student's biological diagram above:

   (a) _____

   (b) _____

   (c) _____

   (d) _____

   (e) _____

   (f) _____

   (g) _____

   (h) _____

2. In the remaining space next to the 'poor example' (above) or on a blank piece of refill paper, attempt your own version of a biological drawing for the same material, based on the photograph above. Make a point of correcting all of the errors that you have identified in the sample student's attempt.

3. Explain why accurate biological drawings are more valuable to a scientific investigation than an 'artistic' approach:

   _____

   _____

   _____

Recording Results

Designing a table to record your results is part of planning your investigation. Once you have collected all your data, you will need to analyse and present it. To do this, it may be necessary to work a little with your data first, by calculating a mean or a rate. An example of a table for recording results is presented below. This example relates to the investigation described in the previous activity, but it represents a relatively standardised layout. The labels on the columns and rows are chosen to represent the design features of the investigation. The first column contains the entire range chosen for the independent variable. There are spaces for multiple sampling units, repeats (trials), and averages. A version of this table should be presented in your final report.

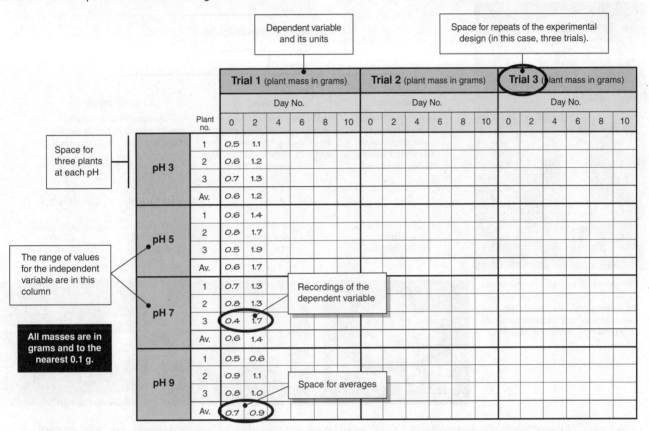

Dependent variable and its units

Space for repeats of the experimental design (in this case, three trials).

	Plant no.	Trial 1 (plant mass in grams)						Trial 2 (plant mass in grams)						Trial 3 (plant mass in grams)					
		Day No.						Day No.						Day No.					
		0	2	4	6	8	10	0	2	4	6	8	10	0	2	4	6	8	10
pH 3	1	0.5	1.1																
	2	0.6	1.2																
	3	0.7	1.3																
	Av.	0.6	1.2																
pH 5	1	0.6	1.4																
	2	0.8	1.7																
	3	0.5	1.9																
	Av.	0.6	1.7																
pH 7	1	0.7	1.3																
	2	0.8	1.3																
	3	0.4	1.7																
	Av.	0.6	1.4																
pH 9	1	0.5	0.6																
	2	0.9	1.1																
	3	0.8	1.0																
	Av.	0.7	0.9																

Space for three plants at each pH

The range of values for the independent variable are in this column

All masses are in grams and to the nearest 0.1 g.

Recordings of the dependent variable

Space for averages

1. In the space (below) design a table to collect data from the case study below. Include space for individual results and averages from the three set ups (use the table above as a guide).

Case Study
Carbon dioxide levels in a respiration chamber

A datalogger was used to monitor the concentrations of carbon dioxide (CO_2) in respiration chambers containing five green leaves from one plant species. The entire study was performed in conditions of full light (quantified) and involved three identical set-ups. The CO_2 concentrations were measured every minute, over a period of ten minutes, using a CO_2 sensor. A mean CO_2 concentration (for the three set-ups) was calculated. The study was carried out two more times, two days apart.

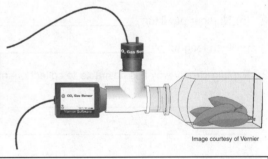

Image courtesy of Vernier

2. Next, the effect of various light intensities (low light, half-light, and full light) on CO_2 concentration was investigated. Describe how the results table for this investigation would differ from the one you have drawn above (for full light only):

_____

_____

Related activities: Manipulating Raw Data, Constructing Tables, Constructing Graphs

DA 2

Variables and Data

When planning any kind of biological investigation, it is important to consider the type of data that will be collected. It is best, whenever possible, to collect quantitative or numerical data, as these data lend themselves well to analysis and statistical testing. Recording data in a systematic way as you collect it, e.g. using a table or spreadsheet, is important, especially if data manipulation and transformation are required. It is also useful to calculate summary, descriptive statistics (e.g. mean, median) as you proceed. These will help you to recognise important trends and features in your data as they become apparent.

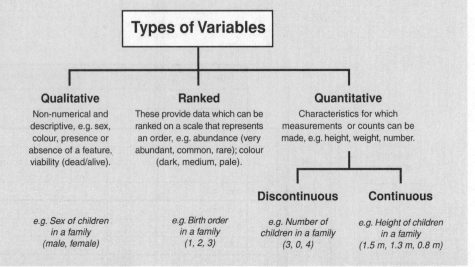

Types of Variables

Qualitative
Non-numerical and descriptive, e.g. sex, colour, presence or absence of a feature, viability (dead/alive).

Ranked
These provide data which can be ranked on a scale that represents an order, e.g. abundance (very abundant, common, rare); colour (dark, medium, pale).

Quantitative
Characteristics for which measurements or counts can be made, e.g. height, weight, number.

Discontinuous

Continuous

e.g. Sex of children in a family (male, female)

e.g. Birth order in a family (1, 2, 3)

e.g. Number of children in a family (3, 0, 4)

e.g. Height of children in a family (1.5 m, 1.3 m, 0.8 m)

The values for monitored or measured variables, collected during the course of the investigation, are called **data**. Like their corresponding variables, data may be quantitative, qualitative, or ranked.

A: Leaf shape

B: Number per litter

C: Fish length

1. For each of the photographic examples (A – C above), classify the variables as quantitative, ranked, or qualitative:

   (a) Leaf shape: _____

   (b) Number per litter: _____

   (c) Fish length: _____

2. Explain clearly why it is desirable to collect quantitative data where possible in biological studies: _____

   _____

   _____

   _____

   _____

3. Suggest how you might measure the colour of light (red, blue, green) quantitatively: _____

   _____

4. (a) Give an example of data that could not be collected in a quantitative manner, explaining your answer:

   _____

   _____

   (b) Sometimes, ranked data are given numerical values, e.g. rare = 1, occasional = 2, frequent = 3, common = 4, abundant = 5. Suggest why these data are sometimes called **semi-quantitative**:

   _____

   _____

Related activities: Describing Data

Periodicals:
Descriptive statistics

Manipulating Raw Data

The data collected by measuring or counting in the field or laboratory are called **raw data**. They often need to be changed (**transformed**) into a form that makes it easier to identify important features of the data (e.g. trends). Some basic calculations, such as totals (the sum of all data values for a variable), are made as a matter of course to compare replicates or as a prelude to other transformations. The calculation of **rate** (amount per unit time) is another example of a commonly performed calculation,

and is appropriate for many biological situations (e.g. measuring growth or weight loss or gain). For a line graph, with time as the independent variable plotted against the values of the biological response, the slope of the line is a measure of the rate. Biological investigations often compare the rates of events in different situations (e.g. the rate of photosynthesis in the light and in the dark). Other typical transformations include frequencies (number of times a value occurs) and percentages (fraction of 100).

Practical Biology & Research Skills

Tally Chart

Records the number of times a value occurs in a data set

HEIGHT / cm	TALLY	TOTAL
0-0.99	111	3
1-1.99	++++ 1	6
2-2.99	++++ ++++	10
3-3.99	++++ ++++ 11	12
4-4.00	111	3
5-5.99	11	2

- A useful first step in analysis; a neatly constructed tally chart doubles as a simple histogram.

- Cross out each value on the list as you tally it to prevent double entries. Check all values are crossed out at the end and that totals agree.

Example: Height of 6d old seedlings

Percentages

Expressed as a fraction of 100

Women	Body mass in kg	Lean body mass	% lean body mass
Athlete	50	38	76.0
Lean	56	41	73.2
Normal weight	65	46	70.8
Overweight	80	48	60.0
Obese	95	52	54.7

- Percentages provide a clear expression of what proportion of data fall into any particular category, e.g. for pie graphs.

- Allows meaningful comparison between different samples.

- Useful to monitor change (e.g. % increase from one year to the next).

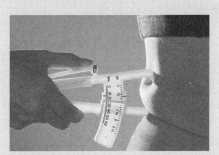

Example: Percentage of lean body mass in women

Rates

Expressed as a measure per unit time

Time / minutes	Cumulative sweat loss	Rate of sweat loss / mL min⁻¹
0	0	0
10	50	5
20	130	8
30	220	9
60	560	11.3

The Time column header and rate unit superscript: Rate of sweat loss / $mL\,min^{-1}$.

- Rates show how a variable changes over a standard time period (e.g. one second, one minute, or one hour).

- Rates allow meaningful comparison of data that may have been recorded over different time periods.

Example: Rate of sweat loss in exercise

---

1. Explain why you might perform basic data transformations: _____

_____

2. (a) Describe a transformation for data relating to the relative abundance of plant species in different habitats:

_____

(b) Explain your answer: _____

_____

3. Complete the transformations on the table (right). The first value is given for you.

TABLE: *Incidence of cyanogenic clover in different areas*

Working: 120 ÷ 158 = 0.76 = 76%

This is the number of cyanogenic clover out of the total.

Incidence of cyanogenic clover in different areas

Clover plant type	Frost free area		Frost prone area		Totals
	Number	%	Number	%	
Cyanogenic	120	76	22		
Acyanogenic	38		120		
Total	158				

Integrated Learning Resources Centre
Greenwich Community College
95 Plumstead Road

Related activities: Variables and Data

DA 2

Constructing Tables

Tables provide a convenient way to systematically record and condense a large amount of information for later presentation and analysis. The protocol for creating tables for recording data during the course of an investigation is provided elsewhere, but tables can also provide a useful summary in the results section of a finished report. They provide an accurate record of numerical values and allow you to organise your data in a way that allows you to clarify the relationships and trends that are apparent. Columns can be provided to display the results of any data transformations such as rates. Some basic descriptive statistics (such as mean or standard deviation) may also be included prior to the data being plotted. For complex data sets, graphs tend to be used in preference to tables, although the latter may be provided as an appendix.

Presenting Data in Tables

Tables should have an accurate, descriptive title. Number tables consecutively through the report.

Heading and subheadings identify each set of data and show units of measurement.

Independent variable in the left column.

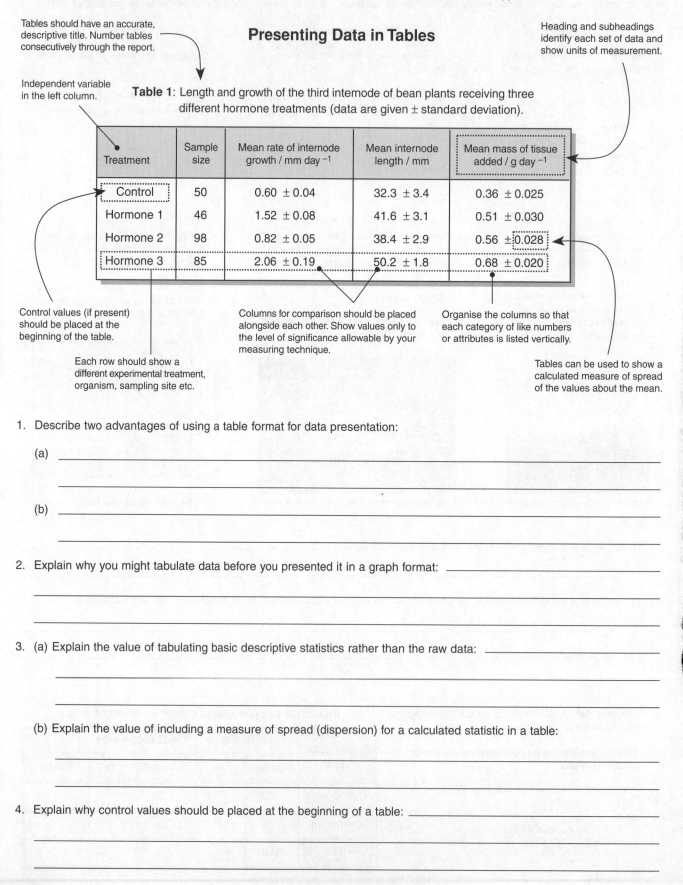

Table 1: Length and growth of the third internode of bean plants receiving three different hormone treatments (data are given ± standard deviation).

Treatment	Sample size	Mean rate of internode growth / mm day^{-1}	Mean internode length / mm	Mean mass of tissue added / g day^{-1}
Control	50	0.60 ± 0.04	32.3 ± 3.4	0.36 ± 0.025
Hormone 1	46	1.52 ± 0.08	41.6 ± 3.1	0.51 ± 0.030
Hormone 2	98	0.82 ± 0.05	38.4 ± 2.9	0.56 ± 0.028
Hormone 3	85	2.06 ± 0.19	50.2 ± 1.8	0.68 ± 0.020

Control values (if present) should be placed at the beginning of the table.

Each row should show a different experimental treatment, organism, sampling site etc.

Columns for comparison should be placed alongside each other. Show values only to the level of significance allowable by your measuring technique.

Organise the columns so that each category of like numbers or attributes is listed vertically.

Tables can be used to show a calculated measure of spread of the values about the mean.

1. Describe two advantages of using a table format for data presentation:

    (a) _____

    _____

    (b) _____

    _____

2. Explain why you might tabulate data before you presented it in a graph format: _____

    _____

    _____

3. (a) Explain the value of tabulating basic descriptive statistics rather than the raw data: _____

    _____

    _____

    (b) Explain the value of including a measure of spread (dispersion) for a calculated statistic in a table: _____

    _____

    _____

4. Explain why control values should be placed at the beginning of a table: _____

    _____

    _____

Constructing Graphs

Presenting results in a graph format provides a visual image of trends in data in a minimum of space. The choice between graphing or tabulation depends on the type and complexity of the data and the information that you are wanting to convey. Presenting graphs properly requires attention to a few basic details, including correct orientation and labelling of the axes, and accurate plotting of points. Common graphs include scatter plots and line graphs (for continuous data), and bar charts and histograms (for categorical data). Where there is an implied trend, a line of best fit can be drawn through the data points, as indicated in the figure below. Further guidelines for drawing graphs are provided on the following pages.

Presenting Data in Graph Format

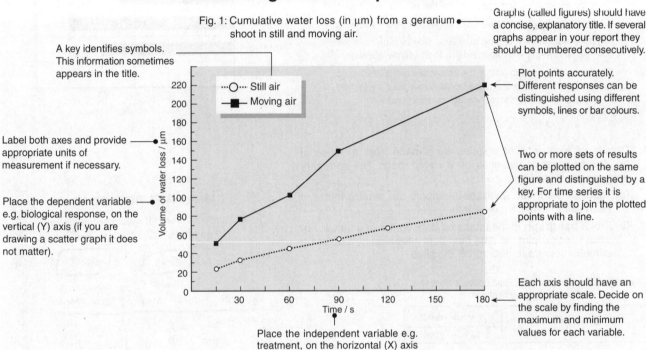

Fig. 1: Cumulative water loss (in μm) from a geranium shoot in still and moving air.

Graphs (called figures) should have a concise, explanatory title. If several graphs appear in your report they should be numbered consecutively.

A key identifies symbols. This information sometimes appears in the title.

Label both axes and provide appropriate units of measurement if necessary.

Place the dependent variable e.g. biological response, on the vertical (Y) axis (if you are drawing a scatter graph it does not matter).

Plot points accurately. Different responses can be distinguished using different symbols, lines or bar colours.

Two or more sets of results can be plotted on the same figure and distinguished by a key. For time series it is appropriate to join the plotted points with a line.

Each axis should have an appropriate scale. Decide on the scale by finding the maximum and minimum values for each variable.

Place the independent variable e.g. treatment, on the horizontal (X) axis

1. Describe an advantage of using a graph format for data presentation: _____

_____

2. (a) Explain the importance of using an appropriate scale on a graph: _____

_____

(b) Scales on X and Y axes may sometimes be "floating" (not meeting in the lower left corner), or they may be broken using a double slash and recontinued. Explain the purpose of these techniques:

_____

_____

3. (a) Explain what is wrong with the graph plotted to the right:

_____

_____

_____

_____

(b) Describe the graph's appearance if it were plotted correctly:

_____

_____

_____

_____

Fig. 1: Yeast growth against time

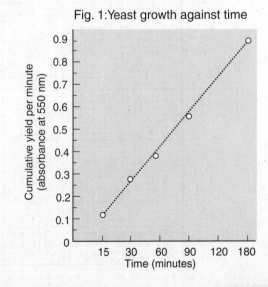

 © Biozone International 2010
Photocopying Prohibited

Periodicals:
Dealing with data

Related activities: Describing data

DA 2

Drawing Bar Graphs

Guidelines for Bar Graphs

Bar graphs are appropriate for data that are non-numerical and **discrete** for at least one variable, i.e. they are grouped into separate categories. There are no dependent or independent variables. Important features of this type of graph include:

- Data are collected for discontinuous, non-numerical categories (e.g. place, colour, and species), so the bars do not touch.

- Data values may be entered on or above the bars if you wish.

- Multiple sets of data can be displayed side by side for direct comparison (e.g. males and females in the same age group).

- Axes may be reversed so that the categories are on the x axis, i.e. the bars can be vertical or horizontal. When they are vertical, these graphs are sometimes called column graphs.

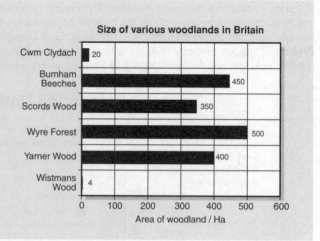

Size of various woodlands in Britain

Woodland	Area
Cwm Clydach	20
Burnham Beeches	450
Scords Wood	350
Wyre Forest	500
Yarner Wood	400
Wistmans Wood	4

Area of woodland / Ha

---

1. Counts of eight mollusc species were made from a series of quadrat samples at two sites on a rocky shore. The summary data are presented here.

   (a) Tabulate the mean (**average**) numbers per square metre at each site in Table 1 (below left).

   (b) Plot a **bar graph** of the tabulated data on the grid below. For each species, plot the data from both sites side by side using different colours to distinguish the sites.

Average abundance of 8 mollusc species from two sites along a rocky shore.

Species	Average/ no m^{-2}	
	Site 1	Site 2

Field data notebook

Total counts at site 1 (11 quadrats) and site 2 (10 quadrats). Quadrats 1 sq m.

Species	Site 1 Total	Site 1 Mean	Site 2 Total	Site 2 Mean
Ornate limpet	232	21	299	30
Radiate limpet	68	6	344	34
Limpet sp. A	420	38	0	0
Cats-eye	68	6	16	2
Top shell	16	2	43	4
Limpet sp. B	628	57	389	39
Limpet sp. C	0	0	22	2
Chiton	12	1	30	3

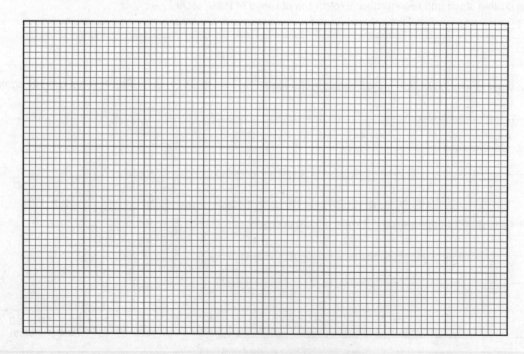

Related activities: Constructing Tables, Describing Data

Periodicals: Drawing graphs, It's a plot!

Drawing Histograms

Guidelines for Histograms

Histograms are plots of **continuous** data and are often used to represent frequency distributions, where the y-axis shows the number of times a particular measurement or value was obtained. For this reason, they are often called frequency histograms. Important features of this type of graph include:

- The data are numerical and continuous (e.g. height or weight), so the bars touch.

- The x-axis usually records the class interval. The y-axis usually records the number of individuals in each class interval (frequency).

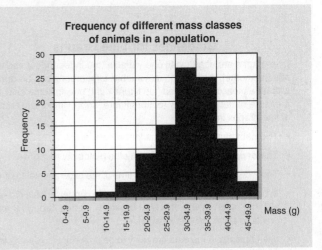

Frequency of different mass classes of animals in a population.

1. The weight data provided below were recorded from 95 individuals (male and female), older than 17 years.

(a) Create a tally chart (frequency table) in the frame provided, organising the weight data into a form suitable for plotting. An example of the tally for the weight grouping 55-59.9 kg has been completed for you as an example. Note that the raw data values, once they are recorded as counts on the tally chart, are crossed off the data set in the notebook. It is important to do this in order to prevent data entry errors.

(b) Plot a **frequency histogram** of the tallied data on the grid provided below.

Weight (kg)	Tally	Total
45-49.9		
50-54.9		
55-59.9	LHT //	7
60-64.9		
65-69.9		
70-74.9		
75-79.9		
80-84.9		
85-89.9		
90-94.9		
95-99.9		
100-104.9		
105-109.9		

Lab notebook

Weight (in kg) of 95 individuals

63.4	81.2	65
56.5	83.3	75.6
84	95	76.8
81.5	105.5	67.8
73.4	82	68.3
56	73.5	63.5
60.4	75.2	58
83.5	63	58.5
82	70.4	50
61	82.2	92
55.2	87.8	91.5
48	86.5	88.3
53.5	85.5	81
63.8	87	72
69	98	66.5
82.8	71	61.5
68.5	76	66
67.2	72.5	65.5
82.5	61	67.4
83	60.5	73
78.4	67	67
76.5	86	71
83.4	85	70.5
77.5	93.5	65.5
77	62	68
87	62.5	90
89	63	83.5
93.4	60	73
83	71.5	66
80	73.8	57.5
76	77.5	76
56	74	

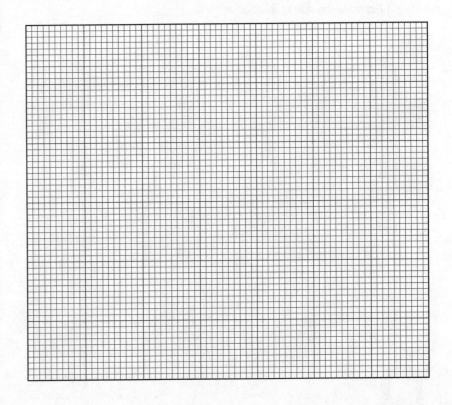

Periodicals:
Drawing graphs, It's a plot!

Related activities: *Constructing Tables*

DA 2

Drawing Line Graphs

Guidelines for Line Graphs

Line graphs are used when one variable (the independent variable) affects another, the dependent variable. Line graphs can be drawn without a measure of spread (top figure, right) or with some calculated measure of data variability (bottom figure, right). Important features of line graphs include:

- The data must be continuous for both variables.

- The dependent variable is usually the biological response.

- The independent variable is often time or the experimental treatment.

- In cases where there is an implied trend (e.g. one variable increases with the other), a line of best fit is usually plotted through the data points to show the relationship.

- If fluctuations in the data are likely to be important (e.g. with climate and other environmental data) the data points are usually connected directly (point to point).

- Line graphs may be drawn with measure of error. The data are presented as points (the calculated means), with bars above and below, indicating a measure of variability or spread in the data (e.g. standard error, standard deviation, or 95% confidence intervals).

- Where no error value has been calculated, the scatter can be shown by plotting the individual data points vertically above and below the mean. By convention, bars are not used to indicate the range of raw values in a data set.

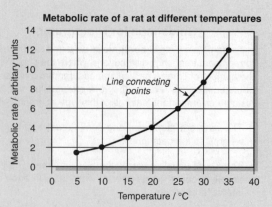

Metabolic rate of a rat at different temperatures

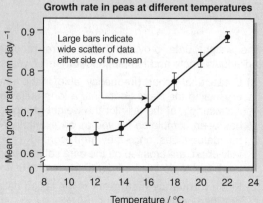

Growth rate in peas at different temperatures

1. The results (shown right) were collected in a study investigating the effect of temperature on the activity of an enzyme.

   (a) Using the results provided in the table (right), plot a line graph on the grid below:

   (b) Estimate the rate of reaction at 15°C: _____

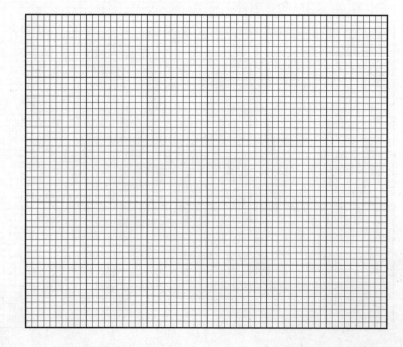

Lab Notebook

An enzyme's activity at different temperatures

Temperature /°C	Rate of reaction /mg of product formed per minute
10	1.0
20	2.1
30	3.2
35	3.7
40	4.1
45	3.7
50	2.7
60	0

Related activities: Manipulating Raw Data, Constructing Graphs, Interpreting Line Graphs

Periodicals: Dealing with data, It's a plot!

© Biozone International 2010

Plotting Multiple Data Sets

A single figure can be used to show two or more data sets, i.e. more than one curve can be plotted per set of axes. This type of presentation is useful when you want to visually compare the trends for two or more treatments, or the response of one species against the response of another. Important points regarding this format are:

- If the two data sets use the same measurement units and a similar range of values for the independent variable, one scale on the y axis is used.

- If the two data sets use different units and/or have a very different range of values for the independent variable, two scales for the y axis are used (see example provided). The scales can be adjusted if necessary to avoid overlapping plots

- The two curves must be distinguished with a key.

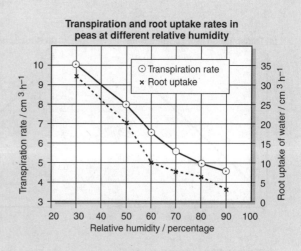

Transpiration and root uptake rates in peas at different relative humidity

2. A census of a deer population on an island indicated a population of 2000 animals in 1960. In 1961, ten wolves (natural predators of deer) were brought to the island in an attempt to control deer numbers. Over the next nine years, the numbers of deer and wolves were monitored. The results of these population surveys are presented in the table, right.

Plot a line graph (joining the data points) for the tabulated results. Use one scale (on the left) for numbers of deer and another scale (on the right) for the number of wolves. Use different symbols or colours to distinguish the lines and include a key.

Field data notebook
Results of a population survey on an island

Time/ year	Wolf numbers	Deer numbers
1961	10	2000
1962	12	2300
1963	16	2500
1964	22	2360
1965	28	2244
1966	24	2094
1967	21	1968
1968	18	1916
1969	19	1952

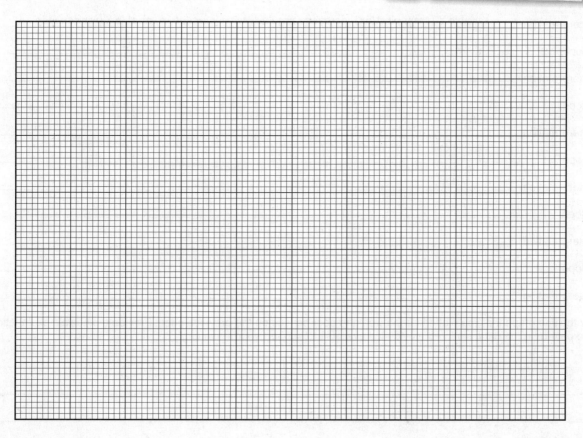

Practical Biology & Research Skills

Interpreting Line Graphs

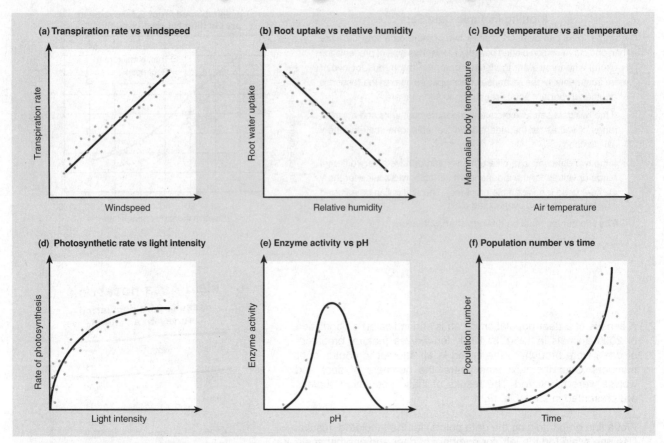

(a) Transpiration rate vs windspeed

(b) Root uptake vs relative humidity

(c) Body temperature vs air temperature

(d) Photosynthetic rate vs light intensity

(e) Enzyme activity vs pH

(f) Population number vs time

1. For each of the graphs (b-f) above, give a description of the slope and an interpretation of how one variable changes with respect to the other. For the purposes of your description, call the independent variable (horizontal or x-axis) in each example "variable X" and the dependent variable (vertical or y-axis) "variable Y". Be aware that the existence of a relationship between two variables does not necessarily mean that the relationship is causative (although it may be).

(a) Slope: _____Positive linear relationship, with constantly rising slope_____

Interpretation: _____Variable Y (transpiration) increases regularly with increase in variable X (windspeed)_____

(b) Slope: _____

Interpretation: _____

(c) Slope: _____

Interpretation: _____

(d) Slope: _____

Interpretation: _____

(e) Slope: _____

Interpretation: _____

(f) Slope: _____

Interpretation: _____

2. Study the line graph that you plotted for the wolf and deer census on the previous page. Provide a plausible explanation for the pattern in the data, stating the evidence available to support your reasoning:

_____

_____

_____

_____

Related activities: *Drawing Line Graphs*

Periodicals:
Dealing with data,
It's a plot!

Drawing Kite Graphs

Guidelines for Kite Graphs

Kite graphs are ideal for representing distributional data, e.g. abundance along an environmental gradient. They are elongated figures drawn along a baseline. Important features of kite graphs include:

- Each kite represents changes in species abundance across a landscape. The abundance can be calculated from the kite width.

- They often involve plots for more than one species; this makes them good for highlighting probable differences in habitat preferences between species.

- A thin line on a kite graph represents species absence.

- The axes can be reversed depending on preference.

- Kite graphs may also be used to show changes in distribution with time, for example, with daily or seasonal cycles of movement.

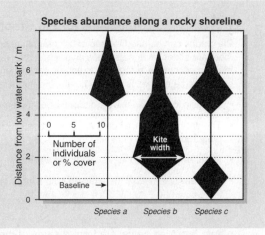

Species abundance along a rocky shoreline

1. The following data were collected from three streams of different lengths and flow rates. Invertebrates were collected at 0.5 km intervals from the headwaters (0 km) to the stream mouth. Their wet weight was measured and recorded (per m²).

   (a) Tabulate the data below for plotting.

   (b) Plot a **kite graph** of the data from all three streams on the grid provided below. Do not forget to include a scale so that the weight at each point on the kite can be calculated.

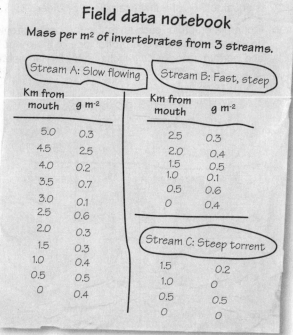

Field data notebook

Mass per m² of invertebrates from 3 streams.

Stream A: Slow flowing

Km from mouth	g m⁻²
5.0	0.3
4.5	2.5
4.0	0.2
3.5	0.7
3.0	0.1
2.5	0.6
2.0	0.3
1.5	0.3
1.0	0.4
0.5	0.5
0	0.4

Stream B: Fast, steep

Km from mouth	g m⁻²
2.5	0.3
2.0	0.4
1.5	0.5
1.0	0.1
0.5	0.6
0	0.4

Stream C: Steep torrent

Km from mouth	g m⁻²
1.5	0.2
1.0	0
0.5	0.5
0	0

Wet mass of invertebrates along three different streams

Distance from mouth/ km	Wet weight/g m⁻²		
	Stream A	Stream B	Stream C

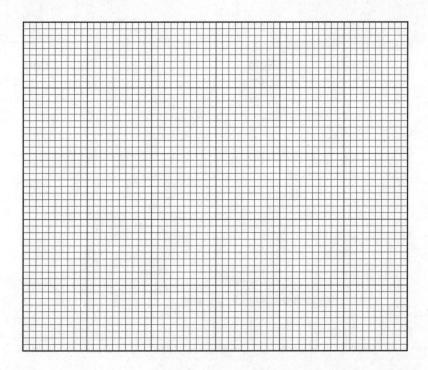

Periodicals:
Drawing graphs,
It's a plot!

Related activities: Constructing Tables

DA 2

Practical Biology & Research Skills

Drawing Pie Graphs

Guidelines for Pie Graphs

Pie graphs can be used instead of bar graphs, generally in cases where there are six or fewer categories involved. A pie graph provides strong visual impact of the relative proportions in each category, particularly where one of the categories is very dominant. Features of pie graphs include:

- The data for one variable are discontinuous (non-numerical or categories).

- The data for the dependent variable are usually in the form of counts, proportions, or percentages.

- Pie graphs are good for visual impact and showing relative proportions.

- They are not suitable for data sets with a large number of categories.

Average residential water use

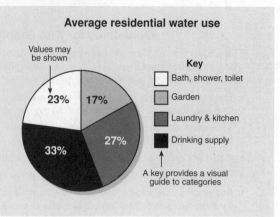

Values may be shown

Key
- Bath, shower, toilet
- Garden
- Laundry & kitchen
- Drinking supply

A key provides a visual guide to categories

1. The data provided below are from a study of the diets of three vertebrates.

   (a) Tabulate the data from the notebook shown. Calculate the angle for each percentage, given that each percentage point is equal to 3.6° (the first example is provided: 23.6 x 3.6 = 85).

   (b) Plot a pie graph for each animal in the circles provided. The circles have been marked at 5° intervals to enable you to do this exercise without a protractor. For the purposes of this exercise, begin your pie graphs at the 0° (= 360°) mark and work in a clockwise direction from the largest to the smallest percentage. Use one key for all three pie graphs.

Field data notebook

% of different food items in the diet

Food item	Stoats	Rats	Cats
Birds	23.6	1.4	6.9
Crickets	15.3	23.6	0
Other insects (not crickets)	15.3	20.8	1.9
Voles	9.2	0	19.4
Rabbits	8.3	0	18.1
Rats	6.1	0	43.1
Mice	13.9	0	10.6
Fruits and seeds	0	40.3	0
Green leaves	0	13.9	0
Unidentified	8.3	0	0

Percentage occurrence of different foods in the diet of stoats, rats, and cats. Graph angle representing the % is shown to assist plotting.

Food item in diet	Stoats		Rats		Cats	
	% in diet	Angle / °	% in diet	Angle / °	% in diet	Angle / °
Birds	23.6	85				

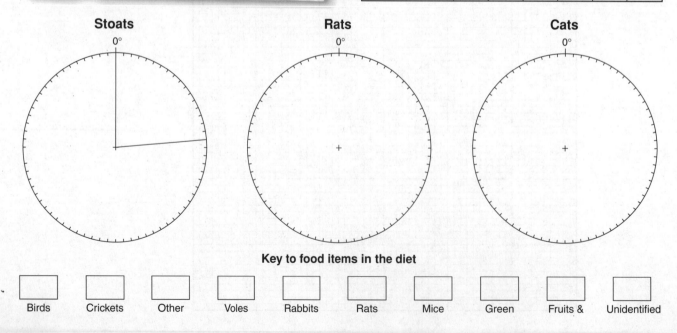

Stoats
0°

Rats
0°

Cats
0°

Key to food items in the diet

| Birds | Crickets | Other | Voles | Rabbits | Rats | Mice | Green | Fruits & | Unidentified |

Related activities: Manipulating Raw Data

Periodicals: Drawing graphs, It's a plot!

Describing Data

For most investigations, measures of the biological response are made from more than one sampling unit. The sample size (the number of sampling units) will vary depending on the resources available. In lab based investigations, the sample size may be as small as two or three (e.g. two test-tubes in each treatment). In field studies, each individual may be a sampling unit, and the sample size can be very large (e.g. 100 individuals). It is useful to summarise the data collected using **descriptive statistics**. Descriptive statistics, such as mean, median, and mode, can help to highlight trends or patterns in the data. Each of these statistics is appropriate to certain types of data or distributions, e.g. a mean is not appropriate for data with a skewed distribution (see below). Frequency graphs are useful for indicating the distribution of data.

Variation in Data

Whether they are obtained from observation or experiments, most biological data show variability. In a set of data values, it is useful to know the value about which most of the data are grouped; the centre value. This value can be the mean, median, or mode depending on the type of variable involved (see schematic below). The main purpose of these statistics is to summarise important trends in your data and to provide the basis for statistical analyses.

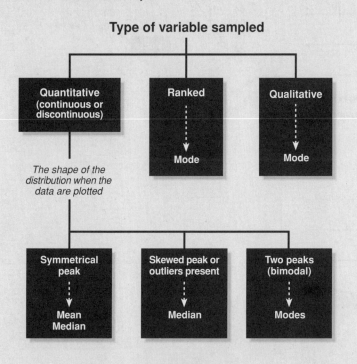

Type of variable sampled

- Quantitative (continuous or discontinuous)
- Ranked → Mode
- Qualitative → Mode

The shape of the distribution when the data are plotted

- Symmetrical peak → Mean / Median
- Skewed peak or outliers present → Median
- Two peaks (bimodal) → Modes

Variability in continuous data is often displayed as a **frequency distribution**. A frequency plot will indicate whether the data have a normal distribution (A), with a symmetrical spread of data about the mean, or whether the distribution is skewed (B), or bimodal (C). The shape of the distribution will determine which statistic (mean, median, or mode) best describes the central tendency of the sample data.

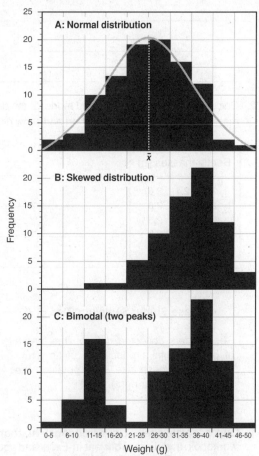

Statistic	Definition and use	Method of calculation
Mean	• The average of all data entries. • Measure of central tendency for normally distributed data.	• Add up all the data entries. • Divide by the total number of data entries.
Median	• The middle value when data entries are placed in rank order. • A good measure of central tendency for skewed distributions.	• Arrange the data in increasing rank order. • Identify the middle value. • For an even number of entries, find the mid point of the two middle values.
Mode	• The most common data value. • Suitable for bimodal distributions and qualitative data.	• Identify the category with the highest number of data entries using a tally chart or a bar graph.
Range	• The difference between the smallest and largest data values. • Provides a crude indication of data spread.	• Identify the smallest and largest values and find the difference between them.

When NOT to calculate a mean:

In certain situations, calculation of a simple arithmetic mean is inappropriate.

Remember:

- *DO NOT* calculate a mean from values that are already means (averages) themselves.

- *DO NOT* calculate a mean of ratios (e.g. percentages) for several groups of different sizes; go back to the raw values and recalculate.

- *DO NOT* calculate a mean when the measurement scale is not linear, e.g. pH units are not measured on a linear scale.

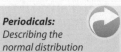

Periodicals: Describing the normal distribution

Related activities: Variables and Data

DA 2

Case Study: Fern Reproduction

Raw data (below) and descriptive statistics (right) from a survey of the number of spores found on the fronds of a fern plant.

$$\frac{\text{Total of data entries}}{\text{Number of entries}} = \frac{1641}{25} = 66 \text{ spores}$$

Mean

Raw data: Number of spores per frond

64	60	64	62	68	66	63
69	70	63	70	70	63	62
71	69	59	70	66	61	70
67	64	63	64

Fern spores

Number of spores per frond (in rank order)	
59	66
60	66
61	67
62	68
62	69
63	69
63	70 — Median
63	70
63	70
64	70
64	70
64	71
64 — Mode	

Spores per frond	Tally	Total
59	✔	1
60	✔	1
61	✔	1
62	✔✔	2
63	✔✔✔✔	4
64	✔✔✔✔	4
65		0
66	✔✔	2
67	✔	1
68	✔	1
69	✔✔	2
70	✔✔✔✔✔	5
71	✔	1

1. Give a reason for the difference between the mean, median, and mode for the fern spore data:

_____

_____

_____

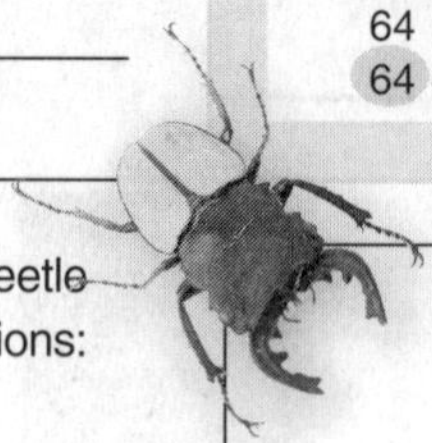

2. Calculate the mean, median, and mode for the data on beetle masses below. Draw up a tally chart and show all calculations:

Beetle masses (g)

2.2	2.1	2.6
2.5	2.4	2.8
2.5	2.7	2.5
2.6	2.6	2.5
2.2	2.8	2.4

Case Study: Sampling Bias

Researchers collected data on the length of perch within a population using a net with a large mesh size. They obtained the following data (far right).

1. On a separate sheet, plot a **frequency histogram** for the data set. Staple it into your workbook. If you are proficient in *Excel* and you have the *Data Analysis* plug in loaded, you can use *Excel* to plot the histogram once you have entered the data.

2. (a) Describe the distribution of your plotted histogram: _____

_____

(b) Explain why the researchers sampling method has resulted in this distribution:

_____

_____

(c) Explain how this sampling bias could be corrected: _____

_____

3. Calculate the following for the data set:

(a) Mean: _____ (b) Median: _____ (c) Mode: _____

Length in mm	Frequency
46	1
47	0
48	0
49	1
50	0
51	0
52	1
53	1
54	1
55	1
56	0
57	2
58	2
59	4
60	1
61	0
62	8
63	10
64	13
65	2
66	0
67	2
Total	**50**

Preparing A Research Report

The practical component of Edexcel AS Biology requires you to prepare a written report. Your report should be between 1500 and 2000 words long about a biological site visit arranged by your teacher, or a biological topic that you have investigated. You will not have to carry out any practical work yourself, but you will need to understand and describe the research methodologies used and identify any relevant biological applications. The information below will help you to identify and address the key components required to meet the assessment criteria. For clarity, the term research refers to both the research carried out at the site visit, and to the research topic undertaken by the student.

Carry out a site visit

- Collect any additional information handed out during the visit.
- Ask relevant biological questions to gain further information.
- Keep accurate notes of what you have seen and heard.
- Be mindful of health and safety considerations during the visit.

Selection of a biological topic

- Check with your teacher that your topic meets the assessment criteria, and that there is sufficient information available to complete the project.

Description of the biological methods and processes

- Identify a specific question or problem related to the research.
- Describe the methods and processes used to produce the data.
- Explain how the method and processes used are appropriate to the research. Explain how valid and reliable data were generated.
- Describe any problems or questions that were identified during the research, and explain how the researchers came up with solutions to address them.

Identify any applications or implications

- Identify two implications of the applied biology associated with the research. They may be social, economic, ethical, or environmental.
- Identify and evaluate any risks associated with the research. Consider risks to humans, other organisms, and the environment.
- Discuss any alternative views or solutions to the identified risks or implications.

Carry out additional research

- Find three or more sources of information related to your biological topic. You must include at least one web-based source and one non web-based source.
- List your additional sources of information accurately (i.e produce a bibliography). Ensure that your text is appropriately referenced.
- Evaluate at least two references cited in your report.

Investigation into
Light Avoidance Behaviour
in the
Pond Springtail
Podura aquatica

By Jane Smith

Write your report

- The report must be generated using a word processor and be between 1500 and 2000 words in length.
- Identify the target audience your report is written for. Use appropriate technical terminology where required.
- Present your report in a logical and concise manner.
- Use correct spelling, punctuation, and grammar.
- Use appropriate graphs, tables, and images (e.g diagrams and photographs) to enhance your report.

Related activities: Constructing Tables, Constructing Graphs, Report Checklist, Citing and Listing References, Ethical Considerations

Practical Biology & Research Skills

A 1

Report Checklist

Title:

☐   Gives a clear indication of what the study is about.

☐   Includes the species name and a common name of all organisms (if applicable).

Report body:

☐   The report is presented in a logical manner.

☐   The target audience is identified.

☐   The aspect of biology you have chosen to write about is clearly identified.

☐   The purpose, significance and implications of the biological aspect chosen are identified.

☐   The methods used to produce the data are clearly presented.

☐   The validity and reliability of the data are explained.

☐   Problems and/or questions raised by the research are identified.

☐   Solutions to the problems and questions raised by the research are discussed.

☐   Relevant graphs, tables, and other artwork are included, and have appropriate titles.

☐   Appropriate technical terminology has been used.

References:

☐   Three additional sources of information are utilised. At least one is from a web source, and at least one is from a non-web source.

☐   Additional source information is clearly identified in the report body.

☐   A bibliography of all additional information and assistance obtained is included.

☐   The report includes analysis of at least two additional references.

General:

☐   The report is generated on a word processor.

☐   The report contains between 1500 and 2000 words (excluding the reference list).

☐   The report has been checked for grammar, punctuation, and spelling.

Citing and Listing References

Referencing sources of information shows that you have explored the topic and recognise the work of others. There are two aspects to consider: **citing sources** within the text (making reference to other work to support a statement or compare results) and **compiling a reference list** at the end of the report. A **bibliography** lists all sources of information, but these may not necessarily appear as citations in the report. In contrast, a reference list should contain only those texts cited in the report. Citations in the main body of the report should include only the authors' surnames, publication date, and page numbers (or internet site) and should be relevant to the statement it claims to support. Accepted methods for referencing vary, but your reference list should provide all the information necessary to locate the source material, it should be consistently presented, and it should contain only the references that you have *yourself* read. A suggested format using the **APA** referencing system is described below.

Preparing a Reference List

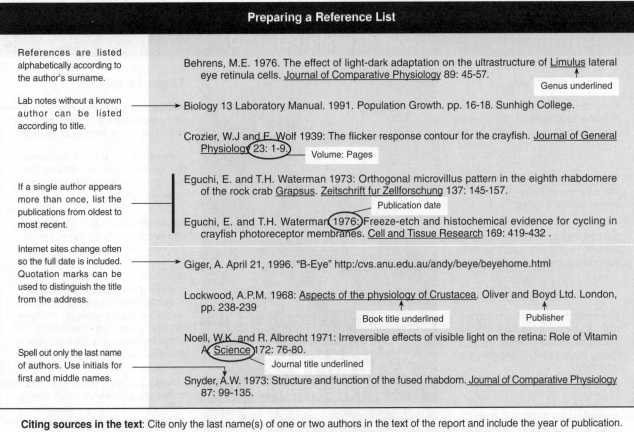

References are listed alphabetically according to the author's surname.

Behrens, M.E. 1976. The effect of light-dark adaptation on the ultrastructure of Limulus lateral eye retinula cells. Journal of Comparative Physiology 89: 45-57.

Genus underlined

Lab notes without a known author can be listed according to title.

Biology 13 Laboratory Manual. 1991. Population Growth. pp. 16-18. Sunhigh College.

Crozier, W.J and E. Wolf 1939: The flicker response contour for the crayfish. Journal of General Physiology 23: 1-9.

Volume: Pages

If a single author appears more than once, list the publications from oldest to most recent.

Eguchi, E. and T.H. Waterman 1973: Orthogonal microvillus pattern in the eighth rhabdomere of the rock crab Grapsus. Zeitschrift fur Zellforschung 137: 145-157.

Publication date

Eguchi, E. and T.H. Waterman 1976: Freeze-etch and histochemical evidence for cycling in crayfish photoreceptor membranes. Cell and Tissue Research 169: 419-432 .

Internet sites change often so the full date is included. Quotation marks can be used to distinguish the title from the address.

Giger, A. April 21, 1996. "B-Eye" http://cvs.anu.edu.au/andy/beye/beyehome.html

Lockwood, A.P.M. 1968: Aspects of the physiology of Crustacea. Oliver and Boyd Ltd. London, pp. 238-239

Book title underlined   *Publisher*

Noell, W.K. and R. Albrecht 1971: Irreversible effects of visible light on the retina: Role of Vitamin A. Science 172: 76-80.

Journal title underlined

Spell out only the last name of authors. Use initials for first and middle names.

Snyder, A.W. 1973: Structure and function of the fused rhabdom. Journal of Comparative Physiology 87: 99-135.

Citing sources in the text: Cite only the last name(s) of one or two authors in the text of the report and include the year of publication. *Example: In many invertebrates oxygen is carried in the blood in simple physical solution (Schmidt-Nielsen, 1979).* For more than two authors use the notation *et al.* (and others). You can cite personal conversations using *personal communication* after the source of the information. *Example: (Mr Jury, personal communication).* All references cited in the text should appear in the reference list.

1. Following are the details of references and source material used by a student in preparing a report on enzymes and their uses in biotechnology. He provided his reference list in prose. From it, compile a correctly formatted reference list:

REFERENCE LIST

Pages 18-23 in the sixth edition of the textbook "Biology" by Neil Campbell. Published by Benjamin/Cummings in California (2002). New Scientist article by Peter Moore called "Fuelled for life" (January 1996, volume 2012, supplement). "Food biotechnology" published in the journal Biological Sciences Review, page 25, volume 8 (number 3) 1996, by Liam and Katherine O'Hare. An article called "Living factories" by Philip Ball in New Scientist, volume 2015 1996, pages 28-31. Pages 75-85 in the book "The cell: a molecular approach" by Geoffrey Cooper, published in 1997 by ASM Press, Washington D.C. An article called "Development of a procedure for purification of a recombinant therapeutic protein" in the journal "Australasian Biotechnology", by I Roberts and S. Taylor, pages 93-99 in volume 6, number 2, 1996.

_____

_____

_____

_____

_____

_____

_____

Related activities: Writing Your Discussion

A 3

Practical Biology & Research Skills

Ethical Considerations

Animal testing is carried out by many institutions for reasons that include genetic research, developmental studies, medical research, toxicology studies, and education. Approximately 100,000 vertebrates (mainly mice) are used in scientific experiments each year. Because researchers are dealing with living organisms, ethical guidelines have been established to maintain **animal welfare** standards. In England, the Animal Procedures Committee (APC) ensures that the welfare of animals used in scientific procedures is maintained. Special ethical standards have also been drafted for the use of humans in scientific testing (e.g. for clinical trials in drug development). Specifically, participants must give **informed consent**. This means that they are participating voluntarily, are fully informed, and that their privacy and confidentiality are maintained. **Environmental ethics** places similar guidelines for and restrictions on human use of the environment (e.g. forestry, mining) to minimise the impact on the ecology of the region involved.

Animal Testing: YES or NO?

Those against say...

Many thousands of animals are used in animal testing each year. Most are killed at the end of the testing regime.

The animals experience stress, pain, and suffering while being kept in animal testing facilities.

Animals maintained in testing facilities are often highly stressed. The stress could affect the results of the testing.

Alternative testing methods that do not use animals are available. These include experiments on human tissue, *in vitro* assays, and computer modelling.

Animals do not respond to new drugs in the same way as humans. Therefore, animal testing is not only a waste of time, it kills animals needlessly.

Researchers are unwilling to change their research methods from the outdated animal models, to modern non-animal research methods.

Janet Stephens

HUNTINGDON LIFE SCIEN
ANIMAL KILLERS
CLOSE THEM DOWN !

SHAC

RDC/Welcome Trust

Those for say...

Animal testing has the potential to benefit many thousands of humans who suffer from a particular disease or condition.

Regulations are in place to ensure that animal welfare is maintained to the highest level possible to minimise stress and suffering.

Researchers keep animals as healthy and stress free as possible to obtain the most realistic results.

Alternative tests are used in initial screening, they reduce the numbers of animals used in testing. However, animal testing is necessary because the alternative tests do not mimic the whole organism response required to determine safety.

The reaction of particular animal species to new drugs and treatments closely mimics that of a human response. Researchers are able to test and refine a new drug on an animal model before safely testing it on human subjects.

Researchers are continually trying to minimise the use of animals in scientific research by following the 3R principle; reducing the number of animals used, refining the testing to minimise distress, and replacing animal tests with assays where possible.

1. Identify any environmental, social, economic, or ethical considerations associated with your chosen research topic. Discuss the actions being taken to address these issues, and adapt this discussion for your report:

_____

_____

_____

_____

_____

_____

_____

_____

Related activities: Preparing A Research Report, Testing New Drugs

Periodicals:
Experimental animal research

Terms and Notation

The definitions for some commonly encountered terms related to making biological investigations are provided below. Use these as you would use a biology dictionary when planning your investigation and writing up your report. It is important to be consistent with the use of terms i.e. use the same term for the same procedure or unit throughout your study. Be sure, when using a term with a specific statistical meaning, such as sample, that you are using the term correctly.

General Terms

Data: Facts collected for analysis.

Qualitative: Not quantitative. Described in words or terms rather than by numbers. Includes subjective descriptions in terms of variables such as colour or shape.

Quantitative: Able to be expressed in numbers. Numerical values derived from counts or measurements.

The Design of Investigations

Hypothesis: A tentative explanation of an observation, capable of being tested by experimentation. Hypotheses are written as clear statements, not as questions.

Control treatment (control): A standard (reference) treatment that helps to ensure that responses to other treatments can be reliably interpreted. There may be more than one control in an investigation.

Dependent variable: A variable whose values are determined by another variable (the independent variable). In practice, the dependent variable is the variable representing the biological response.

Independent variable: A variable whose values are set, or systematically altered, by the investigator.

Controlled variables: Variables that may take on different values in different situations, but are controlled (fixed) as part of the design of the investigation.

Experiment: A contrived situation designed to test (one or more) hypotheses and their predictions. It is good practice to use sample sizes that are as large as possible for experiments.

Investigation: A very broad term applied to scientific studies; investigations may be controlled experiments or field based studies involving population sampling.

Parameter: A numerical value that describes a characteristic of a population (e.g. the mean height of all 17 year-old males).

Prediction: The prediction of the response (Y) variable on the basis of changes in the independent (X) variable.

Random sample: A method of choosing a sample from a population that avoids any subjective element. It is the equivalent to drawing numbers out of a hat, but using random number tables. For field based studies involving quadrats or transects, random numbers can be used to determine the positioning of the sampling unit.

Repeat / Trial: The entire investigation is carried out again at a different time. This ensures that the results are reproducible. Note that repeats or trials are not **replicates** in the true sense unless they are run at the same time.

Replicate: A duplication of the entire experimental design run at the same time.

Sample: A sub-set of a whole used to estimate the values that might have been obtained if every individual or response was measured. A sample is made up of **sampling units**, In lab based investigations, the sampling unit might be a test-tube, while in field based studies, the sampling unit might be an individual organism or a quadrat.

Sample size (*n*): The number of samples taken. In a field study, a typical sample size may involve 20-50 individuals or 20 quadrats. In a lab based investigation, a typical sample size may be two to three sampling units, e.g. two test-tubes held at 10°C.

Sampling unit: Sampling units make up the sample size. Examples of sampling units in different investigations are an individual organism, a test tube undergoing a particular treatment, an area (e.g. quadrat size), or a volume. The size of the sampling unit is an important consideration in studies where the area or volume of a habitat is being sampled.

Statistic: An estimate of a parameter obtained from a sample (e.g. the mean height of all 17 year-old males in your class). A precise (reliable) statistic will be close to the value of the parameter being estimated.

Treatments: Well defined conditions applied to the sample units. The response of sample units to a treatment is intended to shed light on the hypothesis under investigation. What is often of most interest is the comparison of the responses to different treatments.

Variable: A factor in an experiment that is subject to change. Variables may be controlled (fixed), manipulated (systematically altered), or represent a biological response.

Precision and Significance

Accuracy: The correctness of the measurement (the closeness of the measured value to the true value). Accuracy is often a function of the calibration of the instrument used for measuring.

Measurement errors: When measuring or setting the value of a variable, there may be some difference between your answer and the 'right' answer. These errors are often as a result of poor technique or poorly set up equipment.

Objective measurement: Measurement not significantly involving subjective (or personal) judgment. If a second person repeats the measurement they should get the same answer.

Precision (of a measurement): The repeatability of the measurement. As there is usually no reason to suspect that a piece of equipment is giving inaccurate measures, making precise measurements is usually the most important consideration. You can assess or quantify the precision of any measurement system by taking repeated measurements from individual samples.

The Expression of Units

The value of a variable must be written with its units where possible. Common ways of recording measurements in biology are: volume in litres, mass in grams, length in metres, time in seconds. The following example shows different ways to express the same term. Note that ml and cm^3 are equivalent.

Oxygen consumption (millilitres per gram per hour)

Oxygen consumption ($ml g^{-1} h^{-1}$) or ($mL g^{-1} h^{-1}$)

Oxygen consumption ($ml/g/h$) or ($mL/g/h$)

Oxygen consumption/$cm^3 g^{-1} h^{-1}$

Validity: Whether or not you are truly measuring the right thing.

Practical Biology & Research Skills

KEY TERMS: Mix and Match

INSTRUCTIONS: Test your vocab by matching each term to its correct definition, as identified by its preceding letter code.

ACCURACY	A	Note normally appearing directly after a new fact or data that states the author of the information and the date it was published.
BIBLIOGRAPHY	B	A variable whose values are set, or systematically altered, by the investigator.
BIOLOGICAL DRAWING	C	A diagram drawn to accurately show what has been seen by the observer.
CITATION	D	Facts collected for analysis.
CONTROL	E	The value that occurs most often in a data set.
CONTROLLED VARIABLE	F	A pattern observed in processed data showing that data values may be linked.
DATA	G	A standard (reference) treatment that helps to ensure that the responses to the other treatments can be reliably interpreted.
DEPENDENT VARIABLE	H	The sum of the data divided by the number of data entries (n).
GRAPH	I	A variable whose values are determined by another variable.
HISTOGRAM	J	Data able to be expressed in numbers. Numerical values derived from counts or measurements.
HYPOTHESIS	K	Variable that is fixed at a specific amount as part of the design of experiment.
INDEPENDENT VARIABLE	L	A type of column graph used to display frequency distributions.
MEAN	M	A tentative explanation of an observation, capable of being tested by experimentation.
MEASUREMENT	N	The number that occurs in the middle of a set of sorted numbers. It divides the upper half of the number data set from the lower half.
MEDIAN	O	How close a statistic is to the value of the parameter being estimated.
MODE	P	The sampling of an object or substance to record numerical data that describes some aspect of the it, e.g. length or temperature.
OBSERVATION	Q	Data that have not been processed or manipulated in any way.
PRECISION	R	A diagram which often displays numerical information in a way that can be used to identify trends in the data.
QUALITATIVE DATA	S	The act of seeing and noting an occurrence in the object or substance being studied.
QUANTITATIVE DATA	T	The use of an ordered, repeatable method to investigate, manipulate, gather, and record data.
RAW DATA	U	Data described in descriptors or terms rather than by numbers.
REPORT	V	A set of data arranged in rows and columns.
SCIENTIFIC METHOD	W	A factor in an experiment that is subject to change.
TABLE	X	The completed study including methods, results and discussion of the data obtained.
TREND (OF DATA)	Y	A list displaying the titles and publication information of resources used in the gathering of information.
VARIABLE	Z	The degree of closeness of a measured value to its true amount.

Healthy Lifestyle, Healthy Heart

| The circulatory system | • The need for mass transport systems
• The human cardiovascular system |

| CVD | • What is cardiovascular disease?
• Risk factors for CVD
• Treating CVD
• Risk evaluation studies |

| Diet and weight | • Energy budgets
• A balanced diet
• Maintaining a healthy weight |

Biological Molecules

| Water and biological molecules | • The structure and properties of water
• The structure and properties of carbohydrates and lipids
• Condensation and hydrolysis reactions |

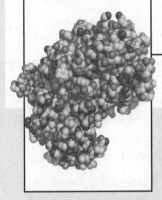

Understanding the biochemistry of molecules is crucial to understanding biological processes.

Lifestyle factors can affect human health and energy balances.

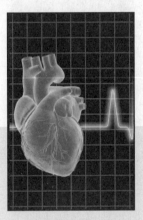

Unit 1

Lifestyle, Transport, Genes and Health

Biological molecules have roles in biological processes, and cell structure and function.
Human health can be affected by changes to the genetic code (mutations), or by lifestyle choices.

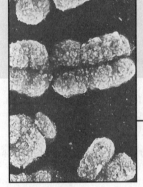

The genetic code is the blueprint for life. Changes to it may affect human health.

Membranes regulate cell transport. Gas exchange membranes are adapted for rapid gas transfer.

| Amino acids and proteins | • The structure of amino acids
• Protein structure and function
• Enzymes
• Protein synthesis |

| Nucleic acids | • The structure of DNA and RNA
• The base pairing rule
• DNA replication
• Meselson and Stahl's experiment
• The genetic code |

| Inheritance and mutations | • Principles of Mendelian genetics
• Mutations and metabolic disorders
• Testing for genetic disorders
• Gene therapy |

| Cellular membranes | • Evidence for membrane structure
• The structure of cell membranes
• The fluid mosaic model |

| Cell transport processes | • Surface area: volume ratios
• Fick's law
• Active transport processes
• Passive transport processes |

| Gas exchange system | • Properties of exchange surfaces
• Structure of the human respiratory system
• Adaptations for rapid gas exchange |

Proteins, Genes & Health

Membranes & Exchange Surfaces

Biological Molecules

KEY CONCEPTS

▶ Organic molecules are carbon-containing molecules and are central to living systems.

▶ Water's properties make it essential to life.

▶ Biological molecules have specific roles in cells, including providing energy to the cell.

KEY TERMS

biological molecule
carbohydrate
condensation reaction
dipole
disaccharide
fatty acid
glycogen
hydrolysis reaction
inorganic ion
isomer
lipid
macromolecule
molecular formula
monomer
monosaccharide
organic molecule
polymer
polysaccharide
saturated fatty acid
solvent
starch
structural formula
triglyceride
unsaturated fatty acids
water

OBJECTIVES

☐ 1. Use the **KEY TERMS** to help you understand and complete these objectives.

Water and inorganic Ions   pages 39-42

☐ 2. Identify the common elements found in organisms and give examples of where these elements occur in cells.

☐ 3. Describe the importance of **organic molecules** in biological systems.

☐ 4. Describe the importance of **inorganic ions** in biological systems.

☐ 5. Describe the structure of water, including reference to its **polar** nature and the physical properties that are important in biological systems, including the role it plays in transporting substances around the body.

Biological Molecules   pages 39, 43-46

☐ 6. Distinguish between **monomers** and **polymers**. Describe the range of **macromolecules** produced by cells.

☐ 7. Describe the basic composition and general formula of **carbohydrates**. Appreciate the difference between mono- and disaccharides, and polysaccharides).

☐ 8. Describe the roles of **glycogen** and **starch** in providing and storing energy for animal and plant cells respectively. Understand how their structure relates to their functional role.

☐ 9. Describe examples of **disaccharides** and their functions.

☐ 10. Describe how **condensation reactions** result in formation of polymers, and water is produced as a result. Describe how the opposite reaction (**hydrolysis**) splits polymers into smaller units, and requires water to proceed.

☐ 11. Recognise that an **isomer** has the same molecular formula, but a different structural formula. Explain how this affects the properties of the molecule.

☐ 12. Describe the structure of a **triglyceride** and explain how they are formed by the condensation reaction.

☐ 13. Understand the differences between a **saturated** and **unsaturated** fatty acid. Describe the consequence to the triglyceride that results.

☐ 14. Appreciate the diverse roles of **lipids** in biological systems.

Periodicals:
listings for this
chapter are on page 228

Weblinks:
www.biozone.co.uk/
weblink/Edx-AS-2542.html

*Teacher Resource
CD-ROM:*
Biochemical Tests

The Biochemical Nature of the Cell

The molecules that make up living things can be grouped into five broad classes: carbohydrates, lipids, proteins, nucleic acids, and water. Water is the main component of organisms and provides an environment in which metabolic reactions can occur. An important feature of water is its dipole nature. Water molecules attract each other, forming large numbers of hydrogen bonds. It is this feature that gives water many of its unique properties, including its low viscosity and its chemical behaviour as a **universal solvent**. Apart from water, most other substances in cells are compounds of carbon, hydrogen, oxygen, and nitrogen. These elements form strong, stable covalent bonds by sharing electrons. The combination of carbon atoms with the atoms of other elements provides a huge variety of molecular structures. Many of these **biological molecules**, e.g. DNA, are very large and contain millions of atoms. The role of these molecules in cells is outlined below.

Carbohydrates form the structural components of cells, e.g. cellulose cell walls (arrowed), they are important in energy storage, and they are involved in cellular recognition.

Nucleotides and nucleic acids
Nucleic acids encode information for the construction and functioning of an organism. The nucleotide, ATP, is the energy currency of the cell.

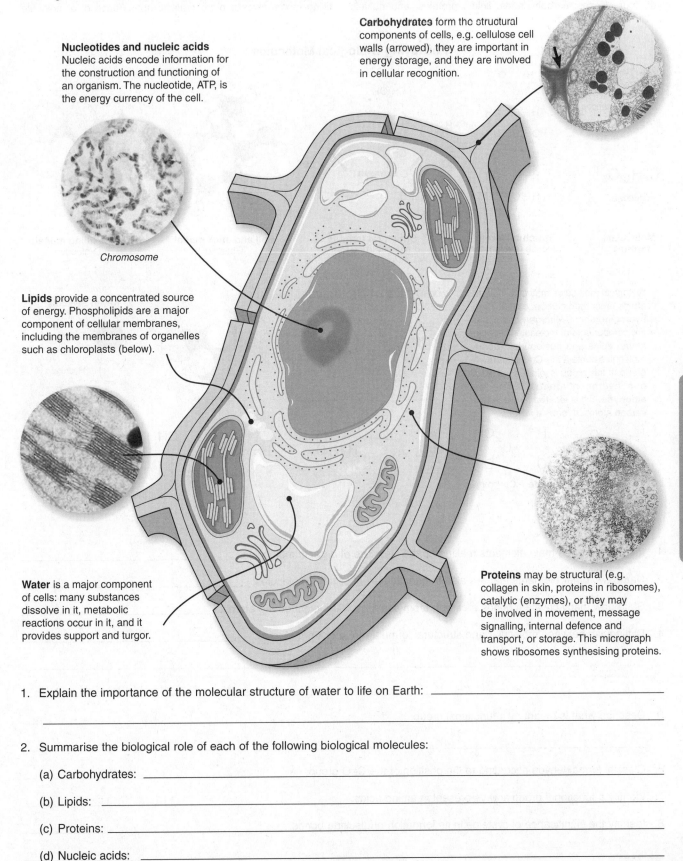

Chromosome

Lipids provide a concentrated source of energy. Phospholipids are a major component of cellular membranes, including the membranes of organelles such as chloroplasts (below).

Water is a major component of cells: many substances dissolve in it, metabolic reactions occur in it, and it provides support and turgor.

Proteins may be structural (e.g. collagen in skin, proteins in ribosomes), catalytic (enzymes), or they may be involved in movement, message signalling, internal defence and transport, or storage. This micrograph shows ribosomes synthesising proteins.

Biological Molecules

1. Explain the importance of the molecular structure of water to life on Earth: _____

_____

2. Summarise the biological role of each of the following biological molecules:

   (a) Carbohydrates: _____

   (b) Lipids: _____

   (c) Proteins: _____

   (d) Nucleic acids: _____

Periodicals:
Water, life and H bonding

Related activities: Water and Inorganic Ions, Carbohydrates , Lipids,
Proteins, Nucleic Acids   ***Web links**: a Closer Look at Water*

A 1

Organic Molecules

Organic molecules are those chemical compounds containing carbon that are found in living things. Specific groups of atoms, called **functional groups**, attach to a carbon-hydrogen core and confer specific chemical properties on the molecule. Some organic molecules in organisms are small and simple, containing only one or a few functional groups, while others are large complex assemblies called **macromolecules**. The macromolecules that make up living things can be grouped into four classes: carbohydrates, lipids, proteins, and nucleic acids. An understanding of the structure and function of these molecules is necessary to many branches of biology, especially biochemistry, physiology, and molecular genetics. The diagram below illustrates some of the common ways in which biological molecules are portrayed. Note that the **molecular formula** expresses the number of atoms in a molecule, but does not convey its structure; this is indicated by the **structural formula**. Molecules can also be represented as **models**. A ball and stick model shows the arrangement and type of bonds while a space filling model gives a more realistic appearance of a molecule, showing how close the atoms really are.

Portraying Biological Molecules

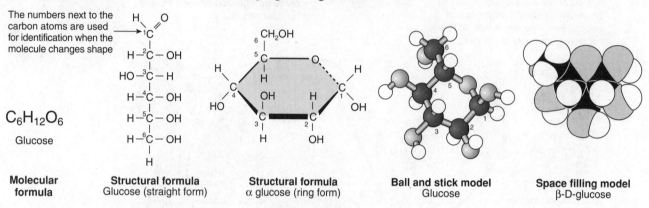

The numbers next to the carbon atoms are used for identification when the molecule changes shape

$C_6H_{12}O_6$
Glucose

Molecular formula

Structural formula
Glucose (straight form)

Structural formula
α glucose (ring form)

Ball and stick model
Glucose

Space filling model
β-D-glucose

Examples of Biological Molecules

Biological molecules may also include atoms other than carbon, oxygen, and hydrogen atoms. Nitrogen and sulfur are components of molecules such as amino acids and nucleotides. Some molecules contain the **C=O** (carbonyl) group. If this group is joined to at least one hydrogen atom it forms an **aldehyde**. If it is located between two carbon atoms, it forms a **ketone**.

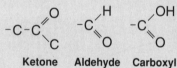

Ketone **Aldehyde** **Carboxyl**

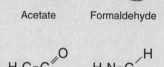

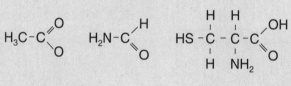

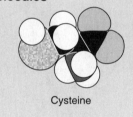

Acetate Formaldehyde Cysteine

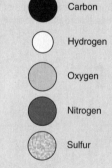

Key to Symbols

● Carbon
○ Hydrogen
● Oxygen
● Nitrogen
● Sulfur

---

1. Identify the three main elements making up the structure of organic molecules: _____

2. Name two other elements that are also frequently part of organic molecules: _____

3. State how many covalent bonds a carbon atom can form with neighbouring atoms: _____

4. Distinguish between molecular and structural formulae for a given molecule: _____
_____
_____

5. Describe what is meant by a functional group: _____
_____

6. Classify formaldehyde according to the position of the C=O group: _____

7. Identify a functional group always present in amino acids: _____

8. Identify the significance of cysteine in its formation of disulfide bonds: _____
_____

© Biozone International 2010

Related activities: Biochemical Nature of the Cell, Amino Acids, Proteins

Water and Inorganic Ions

Water is essential for life. It provides an environment in which metabolic reactions take place in, and is itself involved in many biochemical reactions. Water is the most abundant of the smaller molecules making up living things. The adult human body is typically about 60% water, but some cells and organs contain significantly more water than this (e.g. the human brain is 85% water). Many of the physical and chemical properties of water are significant for life (see below). Water is an excellent **solvent**. Molecules such as **inorganic ions**, proteins, sugars, and gases readily dissolve in water and can be transported throughout the body.

Important Properties of Water

A lot of energy is required before water will change state so aquatic environments are thermally stable and sweating and transpiration cause rapid cooling.

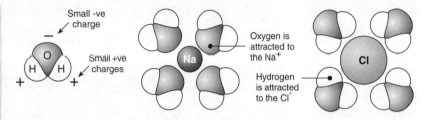

Water molecule
Formula: H_2O

Water surrounding a positive ion (Na⁺)

Water surrounding a negative ion (Cl⁻)

The most important feature of the chemical behaviour of water is its **dipole** nature. It has a small positive charge on each of the two hydrogens and a small negative charge on the oxygen.

Water is colourless, with a high transmission of visible light, so light penetrates tissue and aquatic environments.

Ice is less dense than water. Consequently ice floats, insulating the underlying water and providing valuable habitat.

Water has low viscosity, strong cohesive properties, and high surface tension. It can flow freely through small spaces.

Biological Molecules

Water Properties	Significance for Life
Low viscosity	Water will flow through very small spaces and capillaries.
Liquid at room temperature	Provides a liquid medium for aquatic life and inside cells.
Polar nature	Many substances can dissolve in water. It provides a medium for the chemical reactions of life (metabolism), and a transport medium.
High latent heat of fusion	A significant amount of energy is required before water will change state. This means that cell contents are unlikely to freeze, environments are thermally stable.
High latent heat of evaporation	Water absorbs energy to evaporate. When an animal sweats, heat is lost by evaporation of the water to cause cooling. In plants, transpiration causes cooling.
High specific heat capacity	Can absorb a lot of energy for only a small rise in temperature. This allows organisms to maintain stable internal temperatures despite fluctuations in external temperatures.

1. On the diagram above, showing a positive and a negative ion surrounded by water molecules, draw the positive and negative charges on the water molecules (as shown in the example provided).

2. Explain the importance of the **dipole nature** of water molecules to the chemistry of life: _____

_____

_____

3. Water is often referred to as a universal solvent. Explain why this is not quite true: _____ : _____

_____

_____

Related activities: Biochemical Nature of the Cell, Organic Molecules
Web links: Hydrogen Bonds and Water, Water and pH

RA 2

Inorganic ions are important for the structure and metabolism of all living organisms. An ion is simply an atom (or group of atoms) that has gained or lost one or more electrons. Many of these ions are water soluble.

Some of the inorganic ions required by organisms and examples of their biological roles are described in this table (right). A deficiency in any of these ions can result in specific deficiency disorders. Some examples are described below.

Ion	Name	Example of Biological Roles
Ca^{2+}	Calcium	Component of bones and teeth, required for muscle contraction
Mg^{2+}	Magnesium	Component of chlorophyll, role in energy metabolism
Fe^{2+}	Iron (II)	Component of haemoglobin and cytochromes
NO_3^-	Nitrate	Component of amino acids
PO_4^{3-}	Phosphate	Component of phospholipids, and nucleotides, including ATP
Na^+	Sodium	Component of extracellular fluid and the need for nerve function
K^+	Potassium	Important intracellular ion, needed for heart and nerve function
Cl^-	Chloride	Component of extracellular fluid in multicellular organisms

Common Mineral Ion Deficiencies

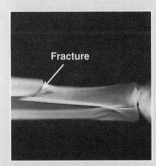

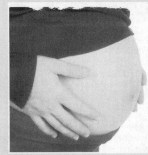

Calcium Deficiency

Calcium deficiency causes poor bone growth and structure, increasing the tendency for bone damage. Muscular spasms and poor blood clotting also occur.

Iron Deficiency

Dietary iron is required to produce haemoglobin. Lower than normal levels result in anaemia. Women, specially when pregnant, are most at risk.

Potassium Deficiency

Plants require potassium for protein synthesis and for opening and closing the stomata. Plants develop brownish-yellow spots on the leaf tips (see arrows).

Magnesium Deficiency

Magnesium (Mg) is required for chlorophyll production, and for the activation of some enzymes. Leaves appear yellow (arrow above) if magnesium-deficient.

4. Distinguish between inorganic and organic compounds: _____

_____

5. Describe <u>one</u> biological role for each of the following elements. For each, also describe the consequence of a deficiency:

(a) Calcium: _____

_____

(b) Iron: _____

_____

(c) Phosphorus: _____

_____

(d) Sodium: _____

_____

(e) Nitrogen: _____

_____

Carbohydrates for Energy

Carbohydrates are a family of organic molecules made up of carbon, hydrogen, and oxygen atoms with the general formula $(CH_2O)_x$. The most common arrangements found in sugars are hexose (6 sided) or pentose (5 sided) rings. Simple sugars, or **monosaccharides**, may join together to form compound sugars (**disaccharides** and **polysaccharides**), and compound sugars can be broken down into their constituent monosaccharides.

Sugars have numerous roles in cells, one of which is providing energy. Plants and animals store carbohydrates as polysaccharides, which can be broken down to provide glucose for energy. In plants, the main storage polysaccharide is **starch**; in animals it is **glycogen**. The structure of a carbohydrate is closely related to its functional properties.

Monosaccharides

Monosaccharides are used as a primary energy source for fuelling cell metabolism. They are **single-sugar** molecules and include glucose (grape sugar and blood sugar) and fructose (honey and fruit juices). The commonly occurring monosaccharides contain between three and seven carbon atoms in their carbon chains and, of these, the 6C hexose sugars occur most frequently. All monosaccharides are classified as **reducing** sugars (i.e. they can participate in reduction reactions).

Single sugars (monosaccharides)

Triose

e.g. glyceraldehyde

Pentose

e.g. ribose, deoxyribose

Hexose

e.g. glucose, fructose, galactose

Disaccharides

Disaccharides are **double-sugar** molecules and are used as energy sources and as building blocks for larger molecules. The type of disaccharide formed depends on the monomers involved and whether they are in their α- or β- form. Only a few disaccharides (e.g. lactose) are classified as reducing sugars.

Sucrose = α-glucose + β-fructose (simple sugar found in plant sap)
Maltose = α-glucose + α-glucose (a product of starch hydrolysis)
Lactose = β-glucose + β-galactose (milk sugar)
Cellobiose = β-glucose + β-glucose (from cellulose hydrolysis)

Double sugars (disaccharides)

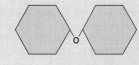

Examples
sucrose,
lactose,
maltose,
cellobiose

Polysaccharides

Polysaccharides are **macromolecules**: polymers made from many repeating monosaccharide units linked together.

Starch: Starch is a polymer of glucose, it is made up of long chains of α-**glucose** molecules linked together. It contains a mixture of 25-30% **amylose** (unbranched chains linked by α-1, 4 glycosidic bonds) and 70-75% **amylopectin** (branched chains with α-1, 6 glycosidic bonds every 24-30 glucose units). Starch is an energy storage molecule in plants and is found concentrated in insoluble **starch granules** within plant cells (photo, right). Starch can be easily hydrolysed by enzymes to soluble sugars when required.

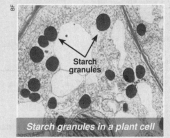

Starch granules

Starch granules in a plant cell

Glycogen: Glycogen, like starch, is a branched polysaccharide. It is chemically similar to amylopectin, being composed of α-**glucose** molecules, but there are more α-1,6 glycosidic links mixed with α-1,4 links. This makes it more highly branched and water-soluble than starch. Glycogen is a storage compound in animal tissues and is found mainly in **liver** and **muscle** cells (photo, right). It is readily hydrolysed by enzymes to form glucose.

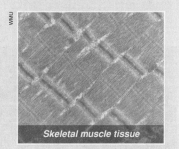

Skeletal muscle tissue

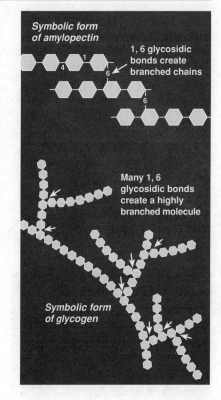

Symbolic form of amylopectin

1, 6 glycosidic bonds create branched chains

Many 1, 6 glycosidic bonds create a highly branched molecule

Symbolic form of glycogen

1. Explain why polysaccharides are such a good source of energy: _____

_____

2. Discuss the structural differences between the polysaccharides starch and glycogen, explaining how the differences in structure contribute to the functional properties of the molecule:

_____

_____

_____

_____

_____

Periodicals:
Glucose,
Designer starches

Related activities: Carbohydrate Chemistry

A 2

Carbohydrate Chemistry

Carbohydrates are versatile macromolecules. Carbohydrate monomers are linked together in different ways to provide a variety of structurally and functionally different molecules. Monomers are linked together by **condensation reactions**, so called because the reaction produces a water molecule. The reverse **hydrolysis reaction** splits polymers into smaller units by breaking the bond between two monomers. Hydrolysis literally means breaking with water, and so requires the addition of a water molecule.

Disaccharides are two monomers joined together. Several disaccharides (lactose, sucrose and maltose) have important roles in human nutrition. Carbohydrates also exist as **isomers**. Isomers are compounds with the same molecular formula, but different structural formulae. Because of this they have different properties. For example, when α–glucose polymers are linked together they form starch, but linked β–glucose polymers form cellulose.

Isomerism

Compounds with the same chemical formula (same types and numbers of atoms) may differ in the arrangement of their atoms. Such variations in the arrangement of atoms in molecules are called **isomers**. In **structural isomers** (such as fructose and glucose, and the α and β glucose, right), the atoms are linked in different sequences. **Optical isomers** are identical in every way but are mirror images of each other.

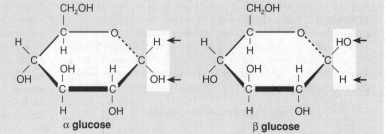

α glucose   β glucose

Condensation and Hydrolysis Reactions

Monosaccharides can combine to form compound sugars in what is called a **condensation** reaction. Compound sugars can be broken down by **hydrolysis** to simple monosaccharides.

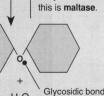

2 mono-saccharides

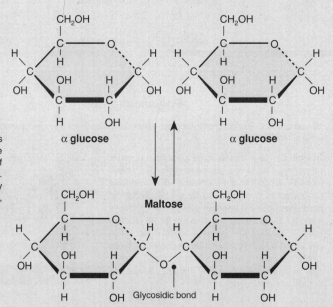

α glucose   α glucose

Maltose

Condensation reaction

Two monosaccharides are joined together to form a disaccharide with the release of a water molecule (hence its name). Energy is supplied by a nucleotide sugar (e.g. ADP-glucose).

Hydrolysis reaction

When a disaccharide is split, as in digestion, a water molecule is used as a source of hydrogen and a hydroxyl group. The reaction is catalysed by enzymes. For maltose (right), this is **maltase**.

+
H_2O — Glycosidic bond

Disaccharide + water

Glycosidic bond

Disaccharide + water

Lactose, or milk sugar, is made up of β-glucose + β-galactose. Milk contains 2-8% lactose by weight. It is the primary carbohydrate source for suckling mammals

Maltose is composed of two α-glucose molecules. These germinating wheat seeds (above) contain maltose because the plant breaks down their starch stores to use it for food.

Sucrose (table sugar) is a simple sugar derived from plants such as sugar cane (above), sugar beet, or maple sap. It is composed of an α-glucose molecule and a β-fructose molecule.

1. Explain how the isomeric structure of a carbohydrate may affect its chemical behaviour: _____

_____

_____

2. Explain briefly how compound sugars are formed and broken down: _____

_____

_____

© Biozone International 2010
Photocopying Prohibited

Related activities: *Carbohydrates for Energy, Cellulose and Starch*
Web links: *Condensation and Hydrolysis*

Lipids

Lipids are a group of organic compounds with an oily, greasy, or waxy consistency. Like carbohydrates they contain carbon, hydrogen, and oxygen, but in lipids the proportion of oxygen is much lower. They are relatively insoluble in water and tend to be hydrophobic and act to repel water (e.g. cuticle on leaf surfaces). This hydrophobic (water hating) property causes the lipids to aggregate into globules. Lipids are important biological fuels, some are hormones, and some serve as structural components in plasma membranes. Proteins and carbohydrates may be converted into fats by enzymes and stored within cells of adipose tissue. During times of plenty, this store is increased, to be used during times of food shortage.

Neutral Fats and Oils

The most abundant lipids in living things are **neutral fats**. They make up the fats and oils found in plants and animals. Fats are an economical way to store fuel reserves, since they yield more than twice as much energy as the same quantity of carbohydrate. Neutral fats are composed of a glycerol molecule attached to one (monoglyceride), two (diglyceride) or three (triglyceride) fatty acids. The fatty acid chains may be saturated or unsaturated (see below). **Waxes** are similar in structure to fats and oils, but they are formed with a complex alcohol instead of glycerol.

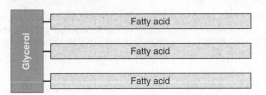

Triglyceride: an example of a neutral fat

Condensation

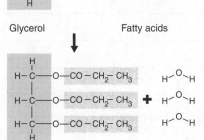

Glycerol   Fatty acids

Triglycerides form when glycerol bonds with three fatty acids. Glycerol is an alcohol containing three carbons. Each of these carbons is bonded to a hydroxyl (-OH) group.

When glycerol bonds with the fatty acid, an **ester bond** is formed and water is released. Three separate condensation reactions are involved in producing a triglyceride.

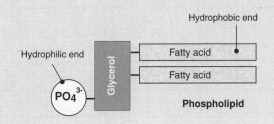

Triglyceride   Water

Saturated and Unsaturated Fatty Acids

Fatty acids are a major component of neutral fats and phospholipids. About 30 different kinds are found in animal lipids. **Saturated fatty acids** contain the maximum number of hydrogen atoms. **Unsaturated fatty acids** contain some carbon atoms that are double-bonded with each other and are not fully saturated with hydrogens. Lipids containing a high proportion of saturated fatty acids tend to be solids at room temperature (e.g. butter). Lipids with a high proportion of unsaturated fatty acids are oils and tend to be liquid at room temperature. This is because the unsaturation causes kinks in the straight chains so that the fatty acids do not pack closely together. Regardless of their degree of saturation, fatty acids yield a large amount of energy when oxidised.

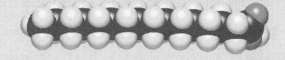

Formula (above) and molecular model (below) for **palmitic acid** (a saturated fatty acid)

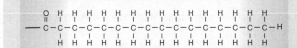

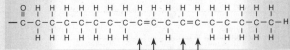

Formula (above) and molecular model (below) for **linoleic acid** (an unsaturated fatty acid)

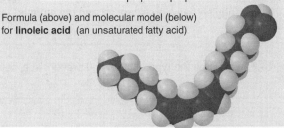

Phospholipids

Phospholipids are the main component of cellular membranes. They consist of a glycerol attached to two fatty acid chains and a phosphate (PO_4^{3-}) group. The phosphate end of the molecule is attracted to water (it is hydrophilic) while the fatty acid end is repelled (hydrophobic). The hydrophobic ends turn inwards in the membrane to form a **phospholipid bilayer**.

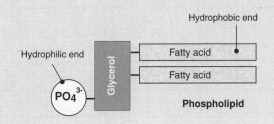

Steroids

Although steroids are classified as lipids, their structure is quite different from that of other lipids. Steroids have a basic structure of three rings made of 6 carbon atoms each and a fourth ring containing 5 carbon atoms. Examples of steroids include the male and female sex hormones (testosterone and oestrogen), and the hormones cortisol and aldosterone. Cholesterol, while not a steroid itself, is a sterol lipid and is a precursor to several steroid hormones.

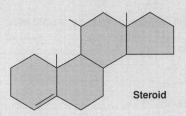

Steroid

Biological Molecules

Related activities: The Structure of Membranes
Web links: Biomolecules: Lipids, Formation of Triglycerides

A 2

Important Biological Functions of Lipids

Lipids are concentrated sources of energy and provide fuel for aerobic respiration.

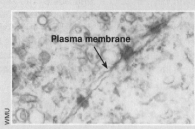

Plasma membrane

Phospholipids form the structural framework of cellular membranes.

Waxes and oils secreted on to surfaces provide waterproofing in plants and animals.

Fat absorbs shocks. Organs that are prone to bumps and shocks (e.g. kidneys) are cushioned with a relatively thick layer of fat.

Lipids are a source of metabolic water. During respiration, stored lipids are metabolised for energy, producing water and carbon dioxide.

Stored lipids provide insulation. Increased body fat reduces the amount of heat lost to the environment (e.g. in winter or in water).

1. Outline the key **chemical** difference between a phospholipid and a triglyceride: _____

_____

_____

2. Name the type of fatty acids found in lipids that form the following at room temperature:

   (a) Solid fats: _____ (b) Oils: _____

3. Relate the structure of phospholipids to their chemical properties and their functional role in cellular membranes:

_____

_____

4. (a) Distinguish between saturated and unsaturated fatty acids: _____

_____

   (b) Explain how the type of fatty acid present in a neutral fat or phospholipid is related to that molecule's properties:

_____

_____

   (c) Suggest how the cell membrane structure of an Arctic fish might differ from that of tropical fish species:

_____

_____

5. Identify two examples of steroids. For each example, describe its physiological function:

   (a) _____

   (b) _____

6. Explain how fats can provide an animal with:

   (a) Energy: _____

   (b) Water: _____

   (c) Insulation: _____

KEY TERMS: Word Find

Use the clues below to find the relevant key terms in the WORD FIND grid

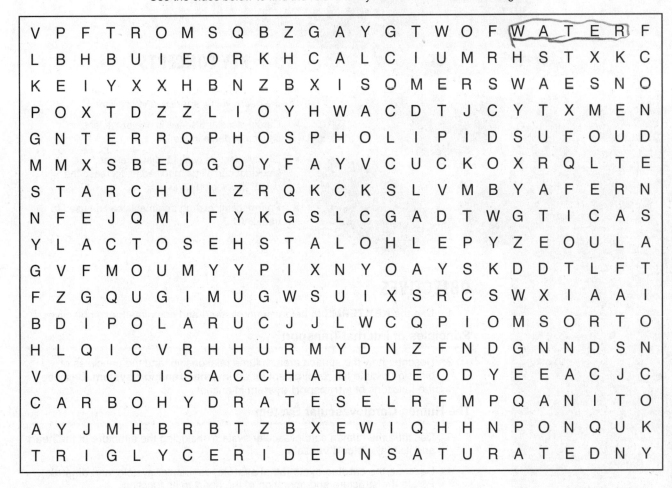

Known as the universal solvent.

Carbon-based compounds are known as this.

The formula that describes the number of atoms in a molecule.

This inorganic ion is a component of haemoglobin.

A molecule, like water, in which the opposite ends are oppositely charged.

A carbohydrate storage molecule found in muscle and liver tissue.

A storage polymer in plants made up of long chains of alpha-glucose.

A general term for a reaction in which water is released.

These lipid molecules naturally form bilayers.

A form a molecule can take are called this.

The most abundant lipids in living things (2 words).

A fatty acid containing the maximum number of hydrogen atoms is called this.

General name for a double sugar molecule.

A disaccharide sugar found in milk.

Lipids which are liquid at room temperature are called this.

General name for a single sugar molecule.

A general term for a reaction which splits molecules into smaller components.

This inorganic ion is an important component of teeth and bones.

Name given to a glycerol molecule attached to three fatty acids.

At atom which has lost or gained one or more electrons.

A family of organic molecules that includes simple sugars and complex molecules such as starch.

Healthy Lifestyle, Healthy Heart

KEY CONCEPTS

▶ The human cardiovascular system comprises the heart, veins, arteries, and capillaries.

▶ Cardiovascular disease is one of the leading causes of death in the UK.

▶ Lifestyle factors (weight, diet, exercise, smoking) can all be modified to decrease the risk of developing CVD.

▶ Pharmaceutical drug treatments can be used to treat CVD.

KEY TERMS

anticoagulant
antihypertensive
artery (pl. arteries)
atherosclerosis
atrium (pl. atria)
basal metabolic rate (BMR)
blood
blood clot
blood pressure
body mass index (BMI)
bulk flow
capillary (pl. capillaries)
cardiovascular disease (CVD)
cholesterol
coronary heart disease (CHD)
diastole
diffusion
energy budget
fibrin
fibrinogen
food pyramid
heart
high-density lipoprotein (HDL)
internal transport
LDL-HDL ratio
low-density lipoprotein (LDL)
obesity
plaque
platelet inhibitor
prothrombin
statins
systole
thrombin
tissue fluid
vein
ventricle

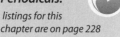

Periodicals:
listings for this
chapter are on page 228

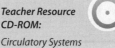

Weblinks:
www.biozone.co.uk/
weblink/Edx-AS-2542.html

Teacher Resource CD-ROM:
Circulatory Systems

OBJECTIVES

☐ 1. Use the **KEY TERMS** to help you understand and complete these objectives.

Principles of Internal Transport   page 49

☐ 2. Describe how the surface area: volume relationship, and the principles of diffusion dictate the requirement for an **internal transport system**. Describe the functions of a **transport system** in animals.

The Human Cardiovascular System   pages 50-57

☐ 3. Describe the human cardiovascular system, including the structure of the **heart** and its major **blood vessels**.

☐ 4. Describe the **cardiac cycle** (atrial **systole**, ventricular systole, and **diastole**). Relate the structure and operation of the heart to its function.

☐ 5. Describe the structure of blood vessels (**arteries**, **capillaries**, and **veins**) and explain how the structure relates to their function.

Cardiovascular Disease   pages 58-65

☐ 6. Distinguish between **cardiovascular disease (CVD)** and **coronary heart disease (CHD)**. Recognise these as a major causes of illness and death in the UK.

☐ 7. Describe blood clotting. Explain the role of it in CVD, in particular **thrombosis** and **embolism**, including the role of prothrombin/thrombin and fibrinogen/fibrin.

☐ 8. Explain the series of events which lead to the CVD **atherosclerosis**.

☐ 9. Describe the major **risk factors** for developing CVD. Analyse data to explain how **controllable** lifestyle factors (diet, exercise, smoking, overweight/obesity) can be altered to decrease risk of developing CVD.

☐ 10. Distinguish between **LDL** and **HDL** cholesterol. Explain why the **LDL-HDL ratio** is a more significant indicator for CVD than the total cholesterol level. Discuss the effect of blood **cholesterol** levels on CVD.

☐ 11. Describe the mode of action of commonly used drugs (**statins**, **antihypertensives**, **anticoagulants**, **platelet inhibitors**). Include reference to how they are used to treat patients with CVD.

Diet and Weight   pages 66-68

☐ 12. Explain the term **basal metabolic rate (BMR)**. Understand that an unbalanced **energy budget** results in either weight gain or weight loss.

Designing a Study   page 65

☐ 13. Evaluate the way a health risk study is designed and presented. Explain why people's perception of risk is often distorted even when presented with scientific data to the contrary.

Internal Transport in Animals

Animal cells require a constant supply of nutrients and oxygen, and continuous removal of wastes. Simple, small organisms (e.g. sponges, cnidarians, flatworms, nematodes) can achieve this through simple diffusion across moist body surfaces without requiring a specialised system (below). Larger, more complex organisms require a circulatory system to transport materials because diffusion is too inefficient and slow to supply all the cells

of the body adequately. The principal components of a circulatory system are blood, a heart, and blood vessels. Circulatory systems transport nutrients, oxygen, carbon dioxide, wastes, and hormones. They also help to maintain fluid balance, regulate body temperature, and may assist in the defence of the body against invading microorganisms. In the diagram below, simple diffusion is compared with transport by a circulatory system.

Transport via Diffusion

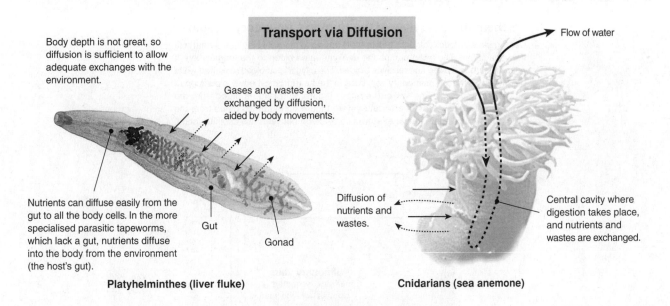

Body depth is not great, so diffusion is sufficient to allow adequate exchanges with the environment.

Gases and wastes are exchanged by diffusion, aided by body movements.

Flow of water

Nutrients can diffuse easily from the gut to all the body cells. In the more specialised parasitic tapeworms, which lack a gut, nutrients diffuse into the body from the environment (the host's gut).

Gut

Gonad

Diffusion of nutrients and wastes.

Central cavity where digestion takes place, and nutrients and wastes are exchanged.

Platyhelminthes (liver fluke)

Cnidarians (sea anemone)

Transport via a Circulatory System

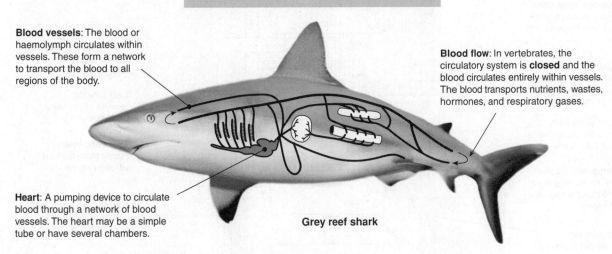

Blood vessels: The blood or haemolymph circulates within vessels. These form a network to transport the blood to all regions of the body.

Blood flow: In vertebrates, the circulatory system is **closed** and the blood circulates entirely within vessels. The blood transports nutrients, wastes, hormones, and respiratory gases.

Heart: A pumping device to circulate blood through a network of blood vessels. The heart may be a simple tube or have several chambers.

Grey reef shark

1. Explain why animals above a certain size require an internal transport system of some kind: _____

_____

2. Briefly describe the function of each of the three major components of a circulatory system in an animal:

(a) Blood vessels: _____

(b) Heart: _____

(c) Blood or haemolymph: _____

_____

3. For simple aquatic organisms, diffusion presents no problem because they are surrounded in a fluid medium. Explain how similar organisms living on land are able to use diffusion to obtain nutrients and dispose of wastes:

_____

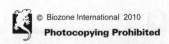
Healthy Lifestyle, Healthy Heart

The Human Cardiovascular System

The blood vessels of the circulatory system form a vast network of tubes that carry blood away from the heart, transport it to the tissues of the body, and then return it to the heart. The arteries, arterioles, capillaries, venules, and veins are organised into specific routes to circulate the blood throughout the body. The figure below shows a number of the basic **circulatory routes** through which the blood travels. Humans, like all mammals, have a **double circulatory system**: a **pulmonary system** (or circulation), which carries blood between the heart and lungs, and a **systemic system** (circulation), which carries blood between the heart and the rest of the body. The systemic circulation has many subdivisions. Two important subdivisions are the coronary (cardiac) circulation, which supplies the heart muscle, and the **hepatic portal circulation**, which runs from the gut to the liver.

Schematic Overview of the Human Circulatory System

Deoxygenated blood (coloured grey below) travels to the right side of the heart via the vena cavae. The heart pumps the deoxygenated blood to the lungs where it releases carbon dioxide and receives oxygen. The oxygenated blood (coloured white below) travels via the pulmonary vein back to the heart from where it is pumped to all parts of the body. The **venous system** (figure, left) returns blood from the capillaries to the heart. The **arterial system** (figure right) carries blood from the heart to the capillaries. **Portal systems** carry blood between two capillary beds.

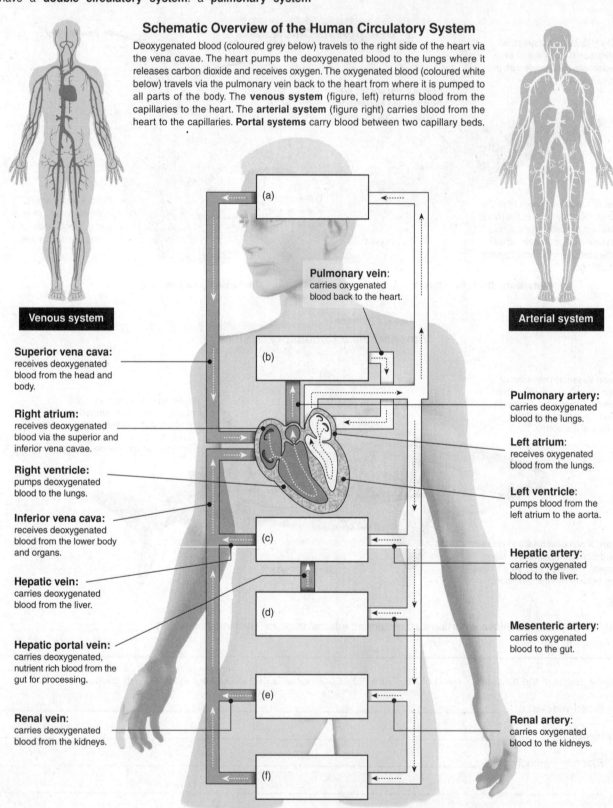

Venous system

Superior vena cava:
receives deoxygenated blood from the head and body.

Right atrium:
receives deoxygenated blood via the superior and inferior vena cavae.

Right ventricle:
pumps deoxygenated blood to the lungs.

Inferior vena cava:
receives deoxygenated blood from the lower body and organs.

Hepatic vein:
carries deoxygenated blood from the liver.

Hepatic portal vein:
carries deoxygenated, nutrient rich blood from the gut for processing.

Renal vein:
carries deoxygenated blood from the kidneys.

Pulmonary vein:
carries oxygenated blood back to the heart.

Arterial system

Pulmonary artery:
carries deoxygenated blood to the lungs.

Left atrium:
receives oxygenated blood from the lungs.

Left ventricle:
pumps blood from the left atrium to the aorta.

Hepatic artery:
carries oxygenated blood to the liver.

Mesenteric artery:
carries oxygenated blood to the gut.

Renal artery:
carries oxygenated blood to the kidneys.

1. Complete the diagram above by labelling the boxes with the organs or structures they represent.

Related activities: The Human Heart
Web links: Circulatory Systems

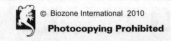

The Human Heart

The heart is the centre of the human cardiovascular system. It is a hollow, muscular organ, weighing on average 342 grams. Each day it beats over 100 000 times to pump 3780 litres of blood through 100 000 kilometres of blood vessels. It comprises a system of four muscular chambers (two **atria** and two **ventricles**) that alternately fill and empty of blood, acting as a double pump.

The left side pumps blood to the body tissues and the right side pumps blood to the lungs. The heart lies between the lungs, to the left of the body's midline, and it is surrounded by a double layered **pericardium** of tough fibrous connective tissue. The pericardium prevents overdistension of the heart and anchors the heart within the **mediastinum**.

Human Heart Structure

(sectioned, anterior view)

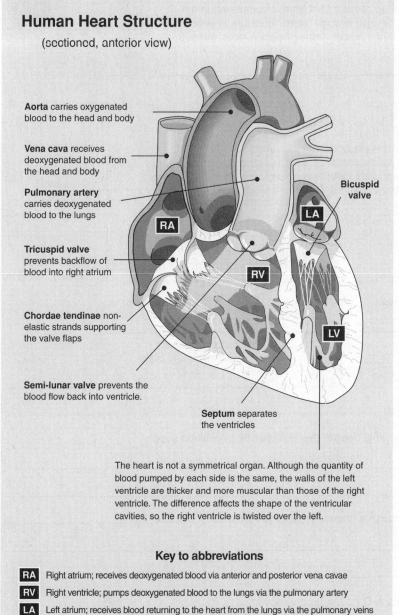

Aorta carries oxygenated blood to the head and body

Vena cava receives deoxygenated blood from the head and body

Pulmonary artery carries deoxygenated blood to the lungs

Tricuspid valve prevents backflow of blood into right atrium

Chordae tendinae non-elastic strands supporting the valve flaps

Semi-lunar valve prevents the blood flow back into ventricle.

Bicuspid valve

Septum separates the ventricles

The heart is not a symmetrical organ. Although the quantity of blood pumped by each side is the same, the walls of the left ventricle are thicker and more muscular than those of the right ventricle. The difference affects the shape of the ventricular cavities, so the right ventricle is twisted over the left.

Key to abbreviations

RA Right atrium; receives deoxygenated blood via anterior and posterior vena cavae

RV Right ventricle; pumps deoxygenated blood to the lungs via the pulmonary artery

LA Left atrium; receives blood returning to the heart from the lungs via the pulmonary veins

LV Left ventricle; pumps oxygenated blood to the head and body via the aorta

Top view of a heart in section, showing valves

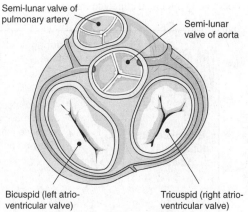

Semi-lunar valve of pulmonary artery

Semi-lunar valve of aorta

Bicuspid (left atrio-ventricular valve)

Tricuspid (right atrio-ventricular valve)

Posterior view of heart

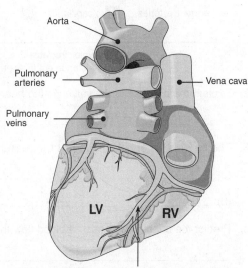

Aorta

Pulmonary arteries

Pulmonary veins

Vena cava

LV

RV

Coronary arteries: The high oxygen demands of the heart muscle are met by a dense capillary network. Coronary arteries arise from the aorta and spread over the surface of the heart supplying the cardiac muscle with oxygenated blood. Deoxygenated blood is collected by cardiac veins and returned to the right atrium via a large coronary sinus.

1. In the schematic diagram of the heart, below, label the four chambers and the main vessels entering and leaving them. The arrows indicate the direction of blood flow. Use large coloured circles to mark the position of each of the four valves.

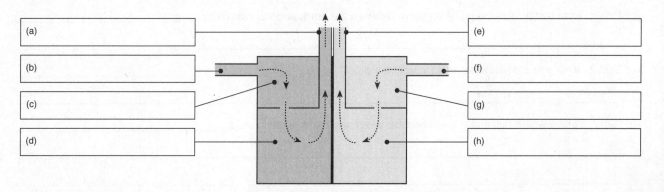

(a)
(b)
(c)
(d)
(e)
(f)
(g)
(h)

Periodicals:
The heart

Related activities: The Cardiac Cycle
Web links: Anatomy of the Heart

RA 2

Healthy Lifestyle, Healthy Heart

Pressure Changes and the Asymmetry of the Heart

The heart is not a symmetrical organ. The left ventricle and its associated arteries are thicker and more muscular than the corresponding structures on the right side. This asymmetry is related to the necessary pressure differences between the pulmonary (lung) and systemic (body) circulations (not to the distance over which the blood is pumped per se). The graph below shows changes blood pressure in each of the major blood vessel types in the systemic and pulmonary circuits (the horizontal distance not to scale). The pulmonary circuit must operate at a much lower pressure than the systemic circuit to prevent fluid from accumulating in the alveoli of the lungs. The left side of the heart must develop enough "spare" pressure to enable increased blood flow to the muscles of the body and maintain kidney filtration rates without decreasing the blood supply to the brain.

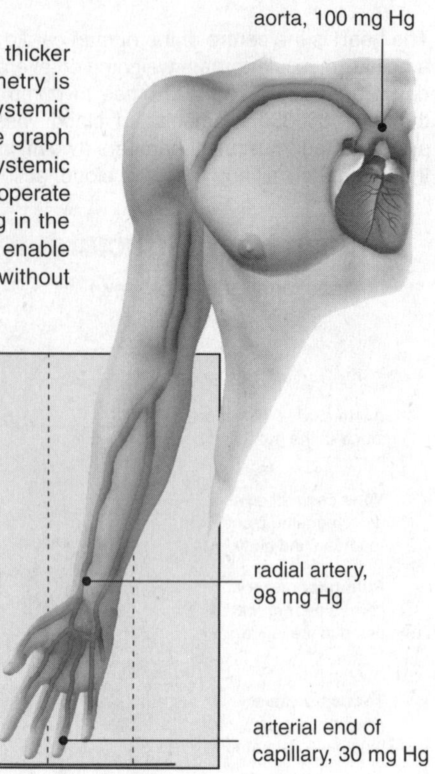

aorta, 100 mg Hg

Blood pressure during contraction (systole)

Blood pressure during relaxation (diastole)

The greatest fall in pressure occurs when the blood moves into the capillaries, even though the distance through the capillaries represents only a tiny proportion of the total distance travelled.

radial artery, 98 mg Hg

arterial end of capillary, 30 mg Hg

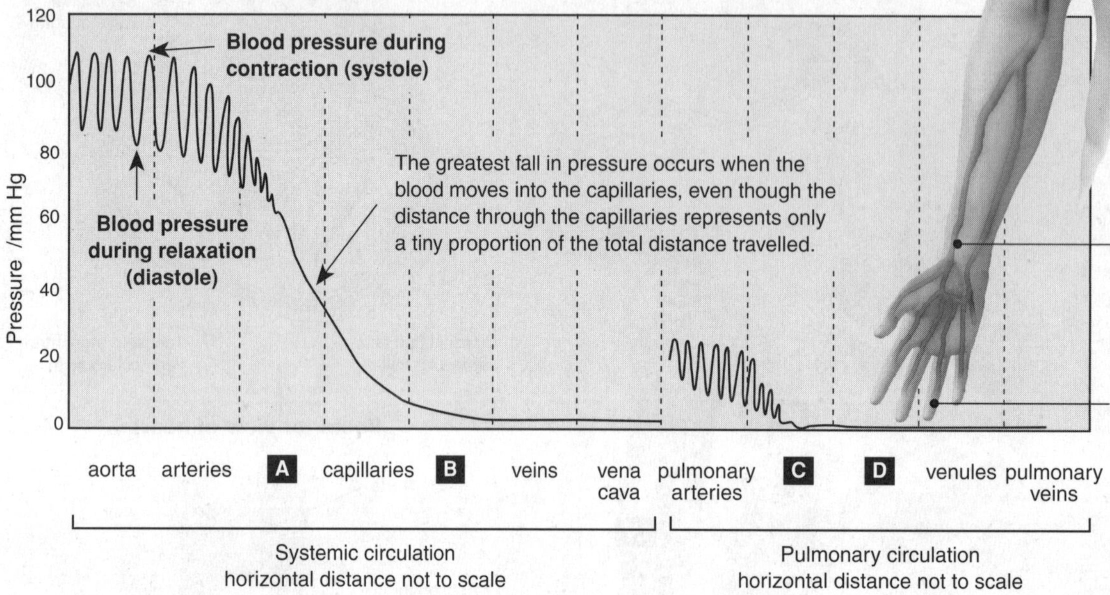

Pressure /mm Hg

aorta   arteries   **A**   capillaries   **B**   veins   vena cava   pulmonary arteries   **C**   **D**   venules   pulmonary veins

Systemic circulation horizontal distance not to scale

Pulmonary circulation horizontal distance not to scale

2. Explain the purpose of the valves in the heart: _____

3. The heart is full of blood. Suggest two reasons why, despite this, it needs its own blood supply:

(a) _____

(b) _____

4. Predict the effect on the heart if blood flow through a coronary artery is restricted or blocked: _____

5. Identify the vessels corresponding to the letters **A-D** on the graph above:

A: _____ B: _____ C: _____ D: _____

6. (a) Find out what is meant by the pulse pressure and explain how it is calculated: _____

(b) Predict what happens to the pulse pressure between the aorta and the capillaries: _____

7. (a) Explain what you are recording when you take a pulse: _____

(b) Name a place where pulse rate could best be taken and briefly explain why: _____

The Cardiac Cycle

The **cardiac cycle** refers to the sequence of events of a heartbeat The pumping of the heart consists of alternate contractions (**systole**) and relaxations (**diastole**). During a complete cycle, each chamber undergoes a systole and a diastole. For a heart beating at 75 beats per minute, one cardiac cycle lasts about 0.8 seconds. Pressure changes within the heart's chambers generated by the cycle of contraction and relaxation are responsible for blood movement and cause the heart valves to open and close, preventing the backflow of blood. The noise of the blood when the valves open and close produces the heartbeat sound (**lubb-dupp**).

The Cardiac Cycle

The **pulse** results from the rhythmic expansion of the arteries as the blood spurts from the left ventricle. Pulse rate therefore corresponds to heart rate.

Stage 1: **Atrial systole and ventricular filling** The ventricles relax and blood flows into them from the atria. Note that 70% of the blood from the atria flows passively into the ventricles. It is during the last third of ventricular filling that the atria contract.

Stage 2: **Ventricular systole** The atria relax, the ventricles contract, and blood is pumped from the ventricles into the aorta and the pulmonary artery. The start of ventricular contraction coincides with the first heart sound.

Stage 3: (not shown) There is a short period of atrial and ventricular relaxation (diastole). Semilunar valves (**SLV**) close to prevent backflow into the ventricles (see diagram, left). The cycle begins again. For a heart beating at 75 beats per minute, one cardiac cycle lasts about 0.8 seconds.

Atrio-ventricular valves closed

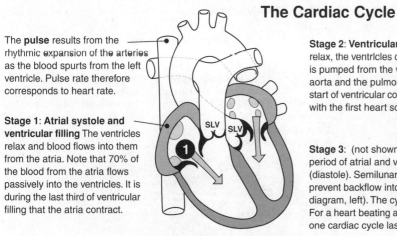
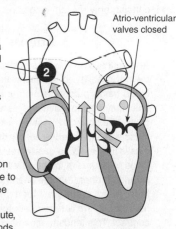

Heart during ventricular filling

Heart during ventricular contraction

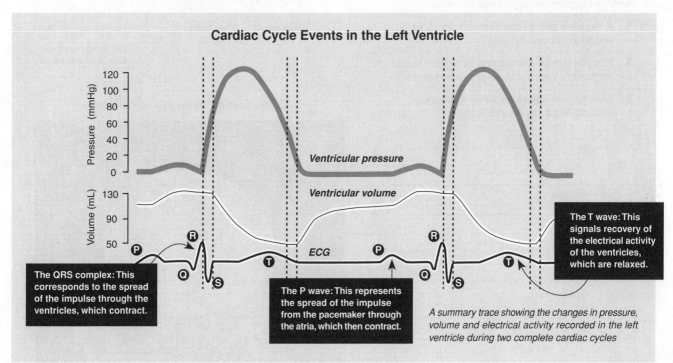

Cardiac Cycle Events in the Left Ventricle

Ventricular pressure

Ventricular volume

ECG

The T wave: This signals recovery of the electrical activity of the ventricles, which are relaxed.

The QRS complex: This corresponds to the spread of the impulse through the ventricles, which contract.

The P wave: This represents the spread of the impulse from the pacemaker through the atria, which then contract.

A summary trace showing the changes in pressure, volume and electrical activity recorded in the left ventricle during two complete cardiac cycles

1. Identify each of the following phases of an ECG by its international code:

    (a) Excitation of the ventricles and ventricular systole: _____

    (b) Electrical recovery of the ventricles and ventricular diastole: _____

    (c) Excitation of the atria and atrial systole: _____

2. Suggest the physiological reason for the period of electrical recovery experienced each cycle (the T wave):

    _____

    _____

3. Using the letters indicated, mark the points on the trace above corresponding to each of the following:

    (a) **E:** Ejection of blood from the ventricle

    (b) **AVC:** Closing of the atrioventricular valve

    (c) **FV:** Filling of the ventricle

    (d) **AVO:** Opening of the atrioventricular valve

Periodicals: The heart

Related activities: The Human Heart
Web links: Electrocardiogram, Cardiac Cycle Animation

RA 2

Healthy Lifestyle, Healthy Heart

Arteries

In vertebrates, arteries are the blood vessels that carry blood away from the heart to the capillaries within the tissues. The large arteries that leave the heart divide into medium-sized (distributing) arteries. Within the tissues and organs, these distribution arteries branch to form very small vessels called **arterioles**, which deliver blood to capillaries. Arterioles lack the thick layers of arteries and consist only of an endothelial layer wrapped by a few smooth muscle fibres at intervals along their length. Resistance to blood flow is altered by contraction (**vasoconstriction**) or relaxation (**vasodilation**) of the blood vessel walls, especially in the arterioles. Vasoconstriction increases resistance and leads to an increase in blood pressure whereas vasodilation has the opposite effect. This mechanism is important in regulating the blood flow into tissues.

Arteries

Arteries have an elastic, stretchy structure that gives them the ability to withstand the high pressure of blood being pumped from the heart. At the same time, they help to maintain pressure by having some contractile ability themselves (a feature of the central muscle layer). Arteries nearer the heart have more elastic tissue, giving greater resistance to the higher blood pressures of the blood leaving the left ventricle. Arteries further from the heart have more muscle to help them maintain blood pressure. Between heartbeats, the arteries undergo elastic recoil and contract. This tends to smooth out the flow of blood through the vessel.

Arteries comprise three main regions (right):

1. A thin inner layer of epithelial cells called the **endothelium** lines the artery.

2. A central layer (the **tunica media**) of elastic tissue and smooth muscle that can stretch and contract.

3. An outer connective tissue layer (the **tunica externa**) has a lot of elastic tissue.

(a)

(b)

(c)

(d)

Artery Structure

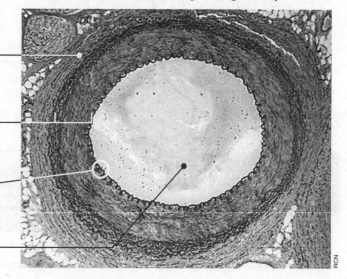

Thin inner layer is in contact with the blood

Layers of elastic tissue and smooth muscle give stretch and contraction

Thick layer of elastic and connective tissue allows for expansion of the artery

Blood flow

Endothelium

Thick tunica media

Thick tunica externa (elastic and collagen fibres)

Cross section through a large artery

RCN

1. Using the diagram to help you, label the photograph (a)-(d) of the cross section through an artery (above).

2. (a) Explain why the walls of arteries need to be thick with a lot of elastic tissue: _____

   _____

   (b) Explain why arterioles lack this elastic tissue layer: _____

   _____

3. Explain the purpose of the smooth muscle in the artery walls: _____

   _____

4. (a) Describe the effect of vasodilation on the diameter of an arteriole: _____

   (b) Describe the effect of vasodilation on blood pressure: _____

Related activities: *Veins, Capillaries and Tissue Fluid*
Web links: *Arteries*

Periodicals:
Cunning plumbing

Capillaries and Tissue Fluid

In vertebrates, capillaries are very small vessels that connect arterial and venous circulation and allow efficient exchange of nutrients and wastes between the blood and tissues. Fluid that leaks out of the capillaries also has an essential role in bathing the tissues. Capillaries form networks and are abundant where metabolic rates are high. The flow of blood through a capillary bed is called **microcirculation**. In most parts of the body, there are two types of vessels in a capillary bed: the **true capillaries**, where exchanges take place, and a vessel called a **vascular shunt**, which connects the arteriole and venule at either end of the bed. The shunt diverts blood past the true capillaries when the metabolic demands of the tissue are low (e.g. vasoconstriction in the skin when conserving body heat). When tissue activity increases, the entire network fills with blood.

Exchanges in Capillaries

Blood passes from the arterioles into the capillaries. Capillaries are small blood vessels with a diameter of just 4-10 µm. The only tissue present is an **endothelium** of squamous epithelial cells. Capillaries are so numerous that no cell is more than 25 µm from any capillary. It is in the capillaries that the exchange of materials between the body cells and the blood takes place.

Blood pressure causes fluid to leak from capillaries through small gaps where the endothelial cells join. This fluid bathes the tissues, supplying nutrients and oxygen, and removing wastes (right). The density of capillaries in a tissue is an indication of that tissue's metabolic activity. For example, cardiac muscle relies heavily on oxidative metabolism. It has a high demand for blood flow and is well supplied with capillaries. Smooth muscle is far less active than cardiac muscle, relies more on anaerobic metabolism, and does not require such an extensive blood supply.

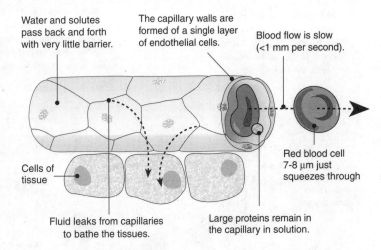

Water and solutes pass back and forth with very little barrier.

The capillary walls are formed of a single layer of endothelial cells.

Blood flow is slow (<1 mm per second).

Red blood cell 7-8 µm just squeezes through

Cells of tissue

Fluid leaks from capillaries to bathe the tissues.

Large proteins remain in the capillary in solution.

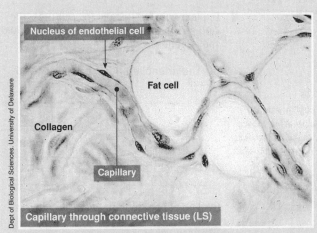

Nucleus of endothelial cell

Fat cell

Collagen

Capillary

Capillary through connective tissue (LS)

Dept of Biological Sciences. University of Delaware

Capillaries are found near almost every cell in the body. In many places, the capillaries form extensive branching networks. In most tissues, blood normally flows through only a small portion of a capillary network when the metabolic demands of the tissue are low. When the tissue becomes active, the entire capillary network fills with blood.

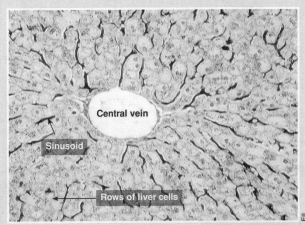

Central vein

Sinusoid

Rows of liver cells

Microscopic blood vessels in some dense organs, such as the liver (above), are called **sinusoids**. They are wider than capillaries and follow a more convoluted path through the tissue. Instead of the usual endothelial lining, they are lined with phagocytic cells. Like capillaries, sinusoids transport blood from arterioles to venules.

1. Describe the structure of a capillary, contrasting it with the structure of a vein and an artery:

_____

_____

_____

_____

2. Sinusoids provide a functional replacement for capillaries in some organs:

    (a) Describe how sinusoids differ structurally from capillaries: _____

    _____

    (b) Describe in what way capillaries and sinusoids are similar: _____

    _____

Related activities: Arteries, Veins
Web links: Microcirculation

RA 2

Healthy Lifestyle, Healthy Heart

The Formation of Tissue Fluid

The network of capillaries supplying the body's tissues ensures that no substance has to diffuse far to enter or leave a cell. Substances exchanged first diffuse through the interstitial fluid (or tissue fluid), which surrounds and bathes the cells. As with all cells, substances can move into and out of the endothelial cells of the capillary walls in several ways; by diffusion, by cytosis, and through gaps where the membranes are not joined by tight junctions. Some fenestrated capillaries are also more permeable than others. These specialised capillaries are important where absorption or filtration occurs (e.g. in the intestine or the kidney). Because capillaries are leaky, fluid flows across their plasma membranes. Whether fluid moves into or out of a capillary depends on the balance between the blood (hydrostatic) pressure and the concentration of solutes at each end of a capillary bed.

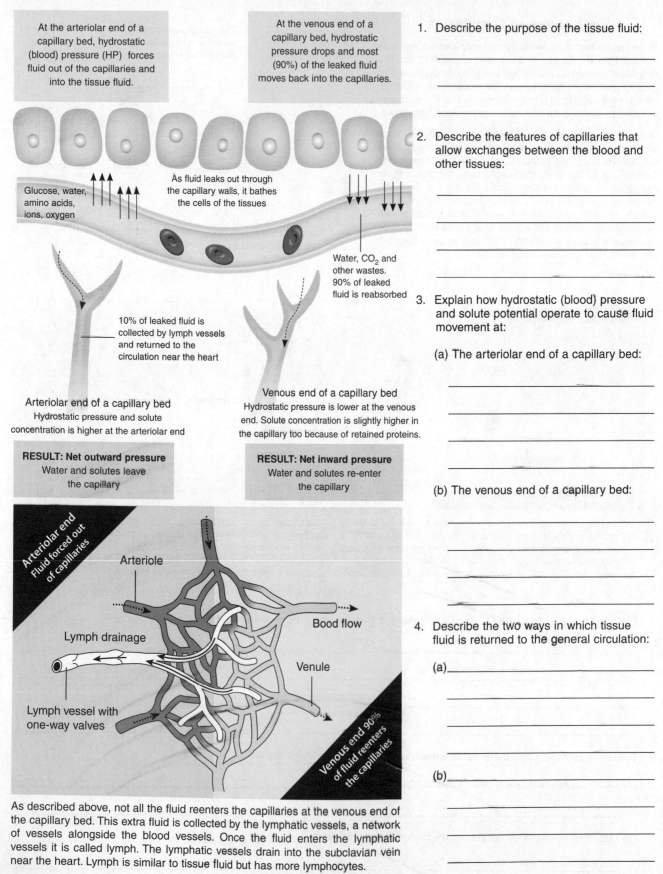

At the arteriolar end of a capillary bed, hydrostatic (blood) pressure (HP) forces fluid out of the capillaries and into the tissue fluid.

At the venous end of a capillary bed, hydrostatic pressure drops and most (90%) of the leaked fluid moves back into the capillaries.

Glucose, water, amino acids, ions, oxygen

As fluid leaks out through the capillary walls, it bathes the cells of the tissues

Water, CO$_2$ and other wastes. 90% of leaked fluid is reabsorbed

10% of leaked fluid is collected by lymph vessels and returned to the circulation near the heart

Arteriolar end of a capillary bed
Hydrostatic pressure and solute concentration is higher at the arteriolar end

Venous end of a capillary bed
Hydrostatic pressure is lower at the venous end. Solute concentration is slightly higher in the capillary too because of retained proteins.

RESULT: Net outward pressure
Water and solutes leave the capillary

RESULT: Net inward pressure
Water and solutes re-enter the capillary

Arteriolar end Fluid forced out of capillaries

Arteriole

Lymph drainage

Bood flow

Venule

Lymph vessel with one-way valves

Venous end 90% of fluid reenters the capillaries

As described above, not all the fluid reenters the capillaries at the venous end of the capillary bed. This extra fluid is collected by the lymphatic vessels, a network of vessels alongside the blood vessels. Once the fluid enters the lymphatic vessels it is called lymph. The lymphatic vessels drain into the subclavian vein near the heart. Lymph is similar to tissue fluid but has more lymphocytes.

1. Describe the purpose of the tissue fluid:

_____

_____

_____

2. Describe the features of capillaries that allow exchanges between the blood and other tissues:

_____

_____

_____

_____

3. Explain how hydrostatic (blood) pressure and solute potential operate to cause fluid movement at:

(a) The arteriolar end of a capillary bed:

_____

_____

_____

_____

(b) The venous end of a capillary bed:

_____

_____

_____

_____

4. Describe the two ways in which tissue fluid is returned to the general circulation:

(a)_____

_____

_____

_____

(b)_____

_____

_____

_____

Related activities: Arteries, Veins
Web links: Microcirculation

Periodicals:
A fair exchange

© Biozone International 2010
Photocopying Prohibited

Veins

Veins are the blood vessels that return blood to the heart from the tissues. The smallest veins (**vénules**) return blood from the capillary beds to the larger veins. Veins and their branches contain about 59% of the blood in the body. The structural differences between veins and arteries are mainly associated with differences in the relative thickness of the vessel layers and the diameter of the lumen. These, in turn, are related to the vessel's functional role.

Veins

When several capillaries unite, they form small veins called **venules**. The venules collect the blood from capillaries and drain it into **veins**. Veins are made up of essentially the same three layers as arteries but they have less elastic and muscle tissue and a larger **lumen**. The venules closest to the capillaries consist of an **endothelium** and a tunica externa of connective tissue. As the venules approach the veins, they also contain the tunica media characteristic of veins (right). Although veins are less elastic than arteries, they can still expand enough to adapt to changes in the pressure and volume of the blood passing through them. Blood flowing in the veins has lost a lot of pressure because it has passed through the narrow capillary vessels. The low pressure in veins means that many veins, especially those in the limbs, need to have valves to prevent backflow of the blood as it returns to the heart.

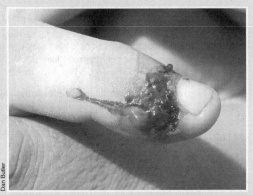

If a vein is cut, as is shown in this severe finger wound, the blood oozes out slowly in an even flow, and usually clots quickly as it leaves. In contrast, arterial blood spurts rapidly and requires pressure to staunch the flow.

Vein Structure

Inner thin layer of simple squamous epithelium lines the vein (**endothelium** or **tunica intima**).

Central thin layer of elastic and muscle tissue (**tunica media**). The smaller venules lack this inner layer.

Thin layer of elastic connective tissue (**tunica externa**)

One-way valves are located along the length of veins to prevent the blood from flowing backwards.

Blood flow

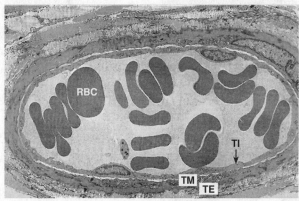

Above: TEM of a vein showing red blood cells (RBC) in the lumen, and the tunica intima (TI), tunica media (TM), and tunica externa (TE).

1. Contrast the structure of veins and arteries for each of the following properties:

   (a) Thickness of muscle and elastic tissue: _____

   _____

   (b) Size of the lumen (inside of the vessel): _____

   _____

2. With respect to their functional roles, give a reason for the differences you have described above: _____

   _____

   _____

3. Explain the role of the valves in assisting the veins to return blood back to the heart: _____

   _____

4. Blood oozes from a venous wound, rather than spurting as it does from an arterial wound. Account for this difference:

   _____

   _____

   _____

Related activities: Arteries, Capillaries and Tissue Fluid
Web links: Veins

RA 2

Healthy Lifestyle, Healthy Heart

Cardiovascular Disease

Cardiovascular disease (CVD) is a term describing all diseases involving the heart and blood vessels. It includes coronary heart disease (CHD), atherosclerosis, hypertension (high blood pressure), peripheral vascular disease, stroke, and congenital heart disorders. Most CVD develops as a result of lifestyle or environmental factors, but a small proportion of the population (< 1%), are born with the CVD. This is termed a **congenital disorder**. CVD is the single most common cause of death in the UK. The British Heart Foundation estimate CVD is responsible for 34% of deaths in the UK. The high mortality rate (just over 193,000 people in 2007) means that the economic cost of CVD is significant. In 2006, it was estimated that £14.4 billion was spent on treating people with CVD, and lost productivity (related to CVD) cost an additional £16 billion. Despite high mortality, CVD related deaths have been in decline since the late 1970s, mainly due to improvements in education about the disease and its risk factors, and advances in screening and treatment.

Types of Cardiovascular Disease

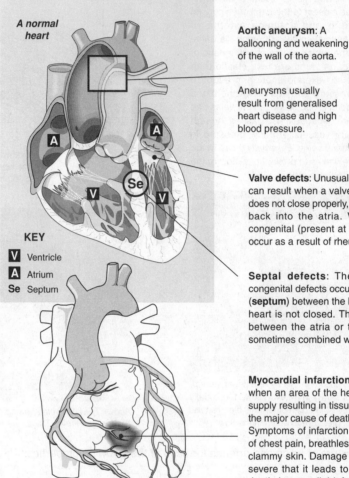

A normal heart

KEY

V Ventricle
A Atrium
Se Septum

Atherosclerosis (hardening of the arteries) is caused by deposits of fats and cholesterol on the inner walls of the arteries. Blood flow becomes restricted and increases the risk of blood clots (**thrombosis**). Complications arising as a result of atherosclerosis include heart attack (**infarction**), gangrene, and **stroke**. A stroke is the rapid loss of brain function due to a disturbance in the blood supply to the brain, and may result in death if the damage is severe. Speech, or vision and movement on one side of the body is often affected.

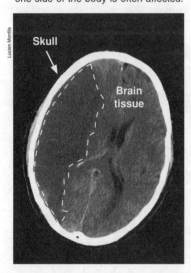

Skull

Brain tissue

The CT scan (above) shows a brain affected by a severe cerebral infarction or ischaemic stroke. The loss of blood supply results in tissue death (outlined area). Blood clots resulting from atherosclerosis are a common cause of ischaemic stroke.

Aortic aneurysm: A ballooning and weakening of the wall of the aorta.

Aneurysms usually result from generalised heart disease and high blood pressure.

Valve defects: Unusual heart sounds (murmurs) can result when a valve (often the mitral valve) does not close properly, allowing blood to bubble back into the atria. Valve defects may be congenital (present at birth) but they can also occur as a result of rheumatic fever.

Septal defects: These hole-in-the-heart congenital defects occur where the dividing wall (**septum**) between the left and right sides of the heart is not closed. These defects may occur between the atria or the ventricles, and are sometimes combined with valve problems.

Myocardial infarction (*heart attack*): Occurs when an area of the heart is deprived of blood supply resulting in tissue damage or death. It is the major cause of death in developed countries. Symptoms of infarction include a sudden onset of chest pain, breathlessness, nausea, and cold clammy skin. Damage to the heart may be so severe that it leads to heart failure and even death (myocardial infarction is fatal within 20 days in 40 to 50% of all cases).

Restricted supply of blood to heart muscle resulting in myocardial infarction

1. Define the term cardiovascular disease (CVD): _____

_____

2. Suggest why CVD kills more people in the UK than any other disease: _____

_____

_____

_____

_____

3. Explain the difference between a congenital cardiovascular defect and a defect that develops later in life:

_____

_____

_____

***Related activities**: Blood clotting and CVD, Atherosclerosis, CVD Risk Factors*

Periodicals: Coronary heart disease

 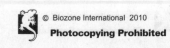

Blood Clotting and CVD

Blood clotting is an essential homeostatic process, preventing bleeding after injury and thereby helping to maintain blood volume. Clotting is initiated when a blood vessel is torn or punctured. It is normally a rapid process that seals off the tear, preventing blood loss and the invasion of bacteria into the site. Clot formation is triggered by the release **of clotting factors** from the damaged cells at the site of the tear or puncture. A hardened clot forms a scab, which acts to prevent further blood loss and acts as a mechanical barrier to the entry of pathogens. Internal blood clots also form in blood vessels, often at sites of injury. This is called **thrombosis**. If the clots become lodged in a blood vessel they can obstruct blood flow and result in tissue injury or death (necrosis). More than 25,000 Britons a year die from blood clots.

When tissue is wounded, the blood quickly coagulates to prevent further blood loss and maintain the integrity of the circulatory system. For external wounds, clotting also prevents the entry of pathogens. Blood clotting involves a cascade of reactions involving at least twelve clotting factors in the blood. The end result is the formation of an insoluble network of fibres, which traps red blood cells and seals the wound.

1 Injury to the lining of a blood vessels exposes collagen fibres to the blood. Platelets stick to the collagen fibres.

3 Platelets clump together. The platelet plug forms an emergency protection against blood loss.

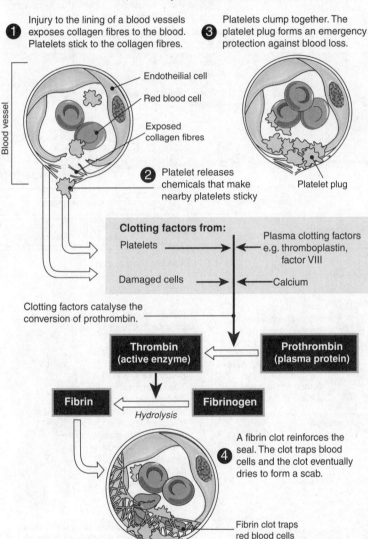

Endotheilial cell

Red blood cell

Exposed collagen fibres

Blood vessel

2 Platelet releases chemicals that make nearby platelets sticky

Platelet plug

Clotting factors from:

Platelets → Plasma clotting factors e.g. thromboplastin, factor VIII

Damaged cells → Calcium

Clotting factors catalyse the conversion of prothrombin.

Thrombin (active enzyme) ⇐ **Prothrombin (plasma protein)**

Fibrin ⇐ **Fibrinogen**
Hydrolysis

4 A fibrin clot reinforces the seal. The clot traps blood cells and the clot eventually dries to form a scab.

Fibrin clot traps red blood cells

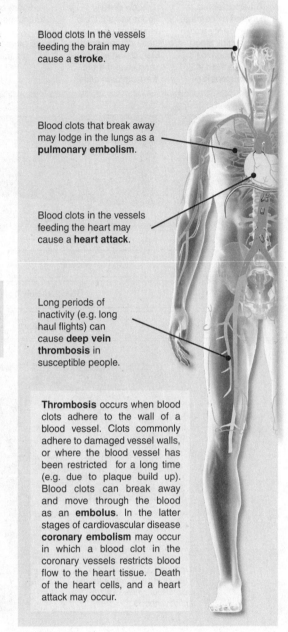

Blood clots in the vessels feeding the brain may cause a **stroke**.

Blood clots that break away may lodge in the lungs as a **pulmonary embolism**.

Blood clots in the vessels feeding the heart may cause a **heart attack**.

Long periods of inactivity (e.g. long haul flights) can cause **deep vein thrombosis** in susceptible people.

Thrombosis occurs when blood clots adhere to the wall of a blood vessel. Clots commonly adhere to damaged vessel walls, or where the blood vessel has been restricted for a long time (e.g. due to plaque build up). Blood clots can break away and move through the blood as an **embolus**. In the latter stages of cardiovascular disease **coronary embolism** may occur in which a blood clot in the coronary vessels restricts blood flow to the heart tissue. Death of the heart cells, and a heart attack may occur.

1. Describe the process the leads to the formation of a blood clot: _____

_____

_____

_____

2. Explain why clotting factors are not normally present in the blood: _____

_____

3. Explain why adhesion of a blood clot within an artery or vein can have serious consequences: ____

_____

_____

Healthy Lifestyle, Healthy Heart

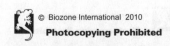

Periodicals:
Heart stopping,
Skin, scabs, and scars

Related activities: Atherosclerosis
Web links: Clot Formation Animation

A 2

Atherosclerosis

Atherosclerosis is a disease of the arteries caused by **atheromas** (fatty deposits) on the inner arterial walls. An atheroma is made up of cells (mostly macrophages) or cell debris, with associated fatty acids, cholesterol, calcium, and varying amounts of fibrous connective tissue. The accumulation of fat and plaques causes the lining of the arteries to degenerate. Atheromas weaken the arterial walls and eventually restrict blood flow through the arteries, increasing the risk of **aneurysm** (swelling of the artery wall) and **thrombosis** (blood clots). Complications arising as a result of atherosclerosis include heart attacks, strokes, and gangrene. A typical progression for the formation of an atheroma is illustrated below.

Initial lesion	Fatty streak	Intermediate lesion	Atheroma	Fibroatheroma	Complicated plaque
Atherosclerosis is triggered by damage to an artery wall caused by blood borne chemicals or persistent **hypertension**.	Low density lipoproteins (LDLs) accumulate beneath the endothelial cells. Macrophages follow and absorb them, forming foam cells.	Foam cells accumulate forming greasy yellow lesions called atherosclerotic plaques.	A core of extracellular lipids under a cap of fibrous tissue forms.	Lipid core and fibrous layers. Accumulated smooth muscle cells die. Fibres deteriorate and are replaced with scar tissue.	Calcification of plaque. Arterial wall may ulcerate. Hypertension may worsen. Plaque may break away causing a clot.

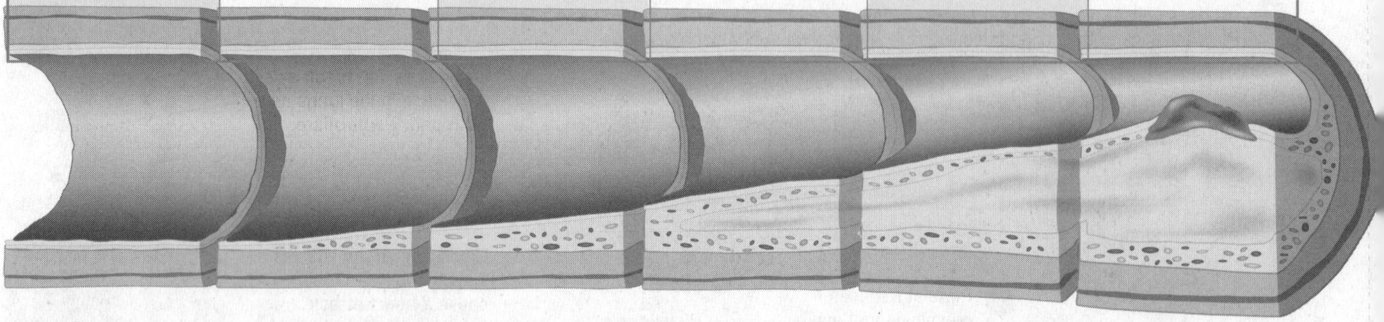

Earliest onset	From first decade	From third decade	From fourth decade
Growth mechanism	Growth mainly by lipid accumulation		Smooth muscle/ collagen increase	Thrombosis, haematoma
Clinical correlation	Clinically silent		Clinically silent or overt

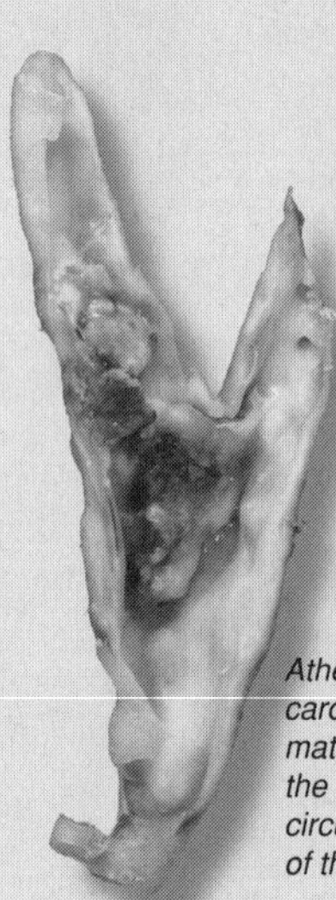

PEIR Digital Library

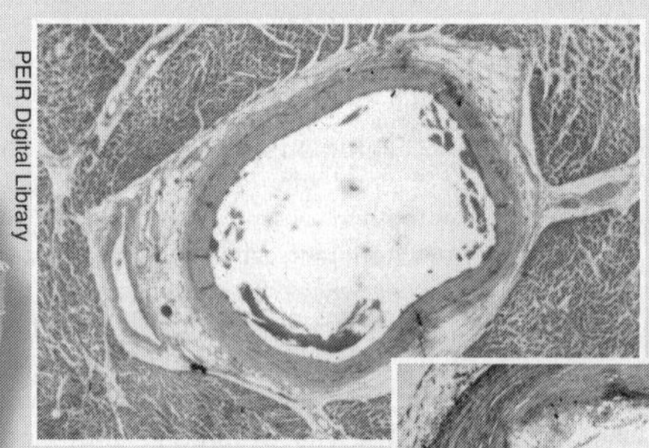

Normal unobstructed coronary artery (left), and a coronary artery with moderately severe atheroma (below). Note the formation of the plaque on the inside surface of the artery.

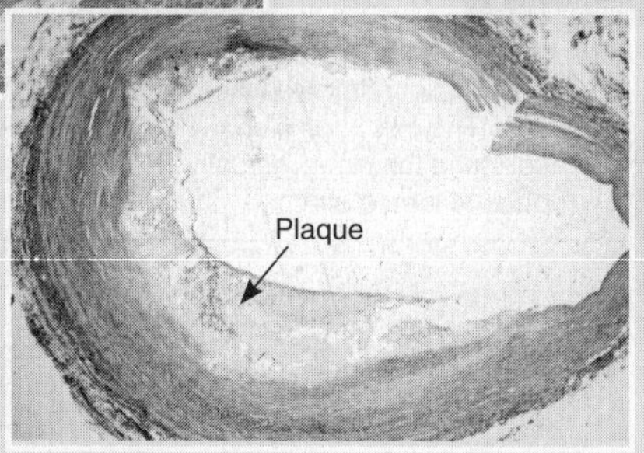

Recent studies indicate that most heart attacks are caused by the body's **inflammatory response** to a plaque. The inflammatory process causes young, soft, cholesterol-rich plaques to rupture and break into pieces. If these block blood vessels they can cause lethal heart attacks, even in previously healthy people.

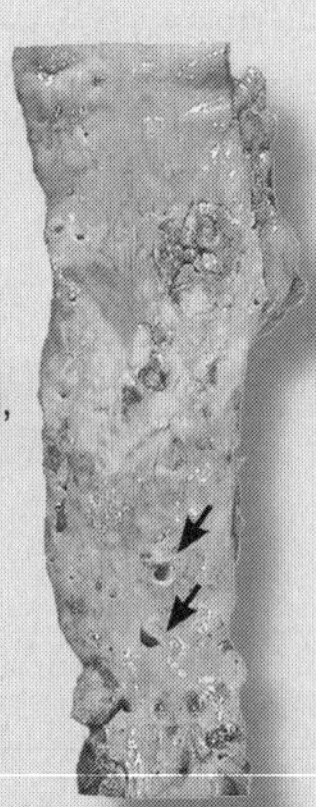

Atherosclerotic plaque in the carotid artery (left). Plaque material can detach from the artery wall and enter the circulation, increasing the risk of thrombosis.

Plaque

Aorta opened lengthwise (above), with extensive atherosclerotic lesions (arrowed).

1. Explain why most people are unlikely to realise they developing atherosclerosis until are serious complications arise:

_____

_____

_____

2. Explain how an atherosclerostic plaque changes over time: _____

_____

_____

3. Describe some of the consequences of developing atherosclerosis: _____

_____

Related activities: Blood Clotting and CVD, CVD Risk Factors, Cholesterol as a Risk Factor

Periodicals: Atherosclerosis: the new view

CVD Risk Factors

Cardiovascular disease (CVD) is the leading cause of death in the UK. Studies have identified several **risk factors** that increase the likelihood of a person developing CVD. Some risk factors are **controllable** in that they can be modified by lifestyle changes. Controllable risk factors include cigarette smoking, obesity, high blood cholesterol, high blood pressure, diabetes, and physical inactivity. **Uncontrollable risk factors** (advancing age, gender, and heredity) cannot be modified, but overall risk can minimised by reducing the number of controllable risk factors. The more risk factors a person has, the greater the likelihood they will develop CVD (below). Increased levels of education and awareness about CVD and its risk factors have helped to reduce levels of the disease. Public education programs in the UK, such as the **National School Fruit Scheme** and the **Be Active, Be Healthy** scheme have helped raise public awareness about the importance of a healthy lifestyle in reducing the risk of developing CVD.

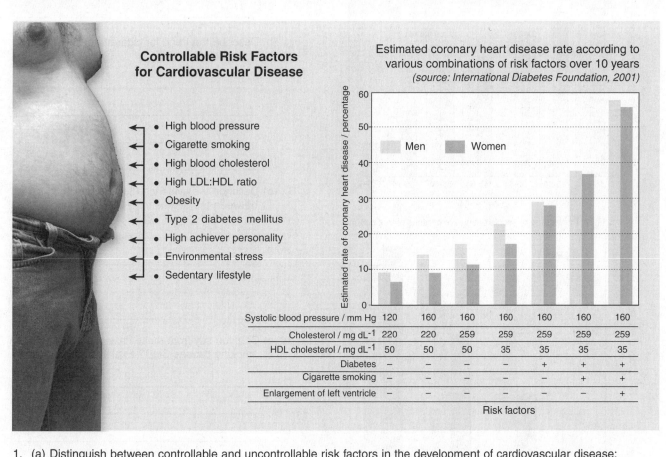

Controllable Risk Factors for Cardiovascular Disease

- High blood pressure
- Cigarette smoking
- High blood cholesterol
- High LDL:HDL ratio
- Obesity
- Type 2 diabetes mellitus
- High achiever personality
- Environmental stress
- Sedentary lifestyle

Estimated coronary heart disease rate according to various combinations of risk factors over 10 years
(source: International Diabetes Foundation, 2001)

Risk factors						
Systolic blood pressure / mm Hg	120	160	160	160	160	160	160
Cholesterol / mg dL^{-1}	220	220	259	259	259	259	259
HDL cholesterol / mg dL^{-1}	50	50	50	35	35	35	35
Diabetes	–	–	–	–	+	+	+
Cigarette smoking	–	–	–	–	–	+	+
Enlargement of left ventricle	–	–	–	–	–	–	+

1. (a) Distinguish between controllable and uncontrollable risk factors in the development of cardiovascular disease:

_____

_____

_____

_____

(b) Suggest why some controllable risk factors often occur together: _____

_____

_____

_____

(c) Evaluate the evidence supporting the observation that patients with several risk factors are at a higher risk of CVD:

_____

_____

_____

_____

_____

_____

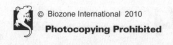
Periodicals: Coronary heart disease

Related activities: Cholesterol and CVD
Web links: Five a Day

RA 2

Healthy Lifestyle, Healthy Heart

Controlling Risk Factors

Fig 1: Physically active people are less likely to develop CHD. For adults, aerobic activity lasting at least thirty minutes five time a week is required for maximum benefit. Children should be physically active for at least one hour a day, five times a week. In developed countries, over 20% of all CHD and 10% of strokes are directly linked to physical inactivity.

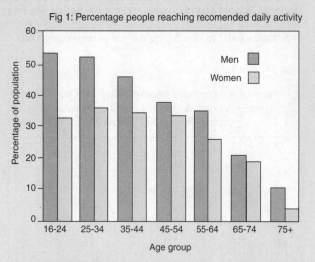

Fig 1: Percentage people reaching recomended daily activity

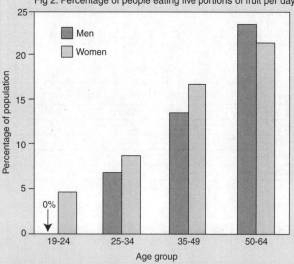

Fig 2: Percentage of people eating five portions of fruit per day

Fig. 2: Eating less than 600 g of fruit and vegetables a day is estimated to cause 30% of CHD and 20% of strokes in developed countries (WHO, 2002). The National School Fruit Scheme, which provides one free piece of fruit a day and encourages the eating of five portions of fruit or vegetables per day, resulted in the number of children achieving the five a day target increase from 13% to 44% in two years. However, despite these public education programmes, there has been little improvement in people's overall eating habits in the UK. Currently, only 13% of men and 15% of women consume five or more portions of fruits or vegetables a day.

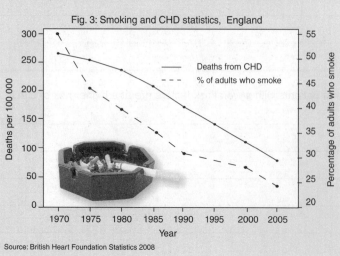

Fig. 3: Smoking and CHD statistics, England

Source: British Heart Foundation Statistics 2008

2. (a) Describe the general trend for deaths from CHD since 1970:

_____

_____

_____

_____

(b) Describe the trend for cigarette smoking during this time:

_____

_____

_____

_____

3. (a) Describe the evidence (fig. 3) suggesting a link between the deaths from CHD and smoking:

_____

_____

_____

_____

(b) Can you say from these data alone that smoking causes CHD? Explain your answer:

_____

_____

_____

_____

4. Discuss the need for, and the effectiveness of public health education programs:

_____

_____

_____

_____

_____

_____

_____

_____

_____

_____

_____

_____

Cholesterol as a Risk Factor

Cholesterol is a sterol lipid found in all animal tissues as part of cellular membranes. It regulates membrane fluidity and is an important precursor molecule in vitamin D synthesis, as well as many steroid hormones (e.g. testosterone and oestrogen). Cholesterol is not soluble in the blood and is transported within spherical particles called **lipoproteins**. There are various compositions of lipoproteins (e.g. HDL and LDL, below left) and the composition determines how cholesterol will be metabolised. Cholesterol has an essential role in body chemistry and is made by the body even when it is not taken in as part of the diet. However, high levels of cholesterol (particularly the ratio of LDL to HDL) are associated with cardiovascular disease.

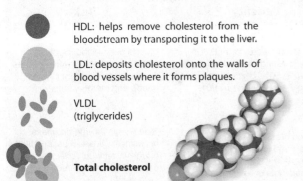

HDL: helps remove cholesterol from the bloodstream by transporting it to the liver.

LDL: deposits cholesterol onto the walls of blood vessels where it forms plaques.

VLDL (triglycerides)

Total cholesterol

Total blood cholesterol comprises low density lipoprotein (LDL), high density lipoprotein (HDL), and very low density lipoprotein (VLDL), which is the triglyceride carrying component in the blood.

Cholesterol and Risk of CVD

Abnormally high concentrations of LDL and lower concentrations of functional HDL are strongly associated with the development of atheroma. It is the **LDL:HDL ratio**, rather than total cholesterol itself, that provides the best indicator of risk for developing cardiovascular disease, and the risk profile is different for men and women (table below). The LDL:HDL ratio is mostly genetically determined but can be influenced by body composition, diet, and exercise.

Ratio of LDL to HDL		
Risk	**Men**	**Women**
Very low (half average)	1.0	1.5
Average risk	3.6	3.2
Moderate risk (2X average risk)	6.3	5.0
High (3X average)	8.0	6.1

Cholesterol Levels in Food

Food	Cholesterol / mg 100g⁻¹	Food	Cholesterol / mg 100g⁻¹
Bacon rasher	80	Fruit	0
Beans	0	Margarine	2
Boiled egg	450	Milk	10
Bread	1	Rice	0
Butter	260	Roast chicken	90
Cheese	70	Sausage	55
Cheese burger	34	Tuna (canned)	90
Cornflakes	0	Vegetables	0

The recommended cholesterol intake is between 200-400 mg per day. The amount required varies depending upon level of activity and general body type. Eating a balanced diet will keep most people within these levels, but too many animal based products quickly drives up a person's cholesterol intake. Cholesterol levels in some commonly eaten foods are presented in the table above.

Effects of HDL and LDL on CVD

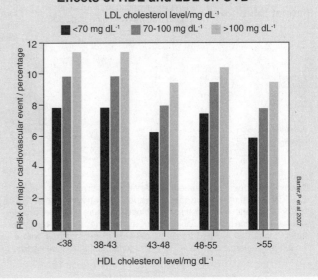

1. (a) Explain the link between high LDL:HDL ratio and the risk of cardiovascular disease: _____

_____

_____

_____

(b) Explain why this ratio is more important to medical practitioners than total blood cholesterol *per se*:

_____

(c) Suggest how this ratio could be lowered in at-risk individuals: _____

_____

2. From the table above showing cholesterol levels in food, recommend a day's menu that would keep a person within the recommended daily cholesterol intake level:

_____

_____

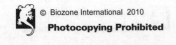

Periodicals: Heart disease and cholesterol

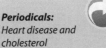 *Related activities:* CVD Risk Factors, A Balanced Diet

A 2

Healthy Lifestyle, Healthy Heart

Treating CVD with Drugs

Treating cardiovascular disease (CVD) costs the UK around £30 billion a year in treatment and lost productivity. Depending on the nature and severity of the disease, treatment can include changing lifestyle factors (e.g. diet, exercise, weight), drug treatments, surgery, or a combination of these methods. Drug treatments are used in cases where lifestyle changes have been ineffective. The use of CVD drugs in the UK increased four-fold between 1986 and 2006. CVD drugs are categorised according to their mode of action (e.g. **antihypertensives, anticoagulants, statins**, or **platelet inhibitors**). Many can be easily obtained on prescription or purchased over the counter. For example, aspirin, commonly used to treat headaches, is equally useful as a platelet inhibitor.

Drug Group	Mode of Action	Benefit	Risk
Statins	Decreases cholesterol production in liver cells by blocking cholesterol synthesis. Increases LDL receptors on liver cells so more LDL cholesterol is removed from the blood.	Reduces risk of plaque formation in arteries by lowering blood cholesterol levels by up to 50%. Decreases cardiac events (e.g. heart attacks) by 60%.	Muscle pain, rhabdomyolysis (rapid breakdown of muscle fibres and release of content into the bood), and memory loss.
Antihypertensives	Lowers blood pressure. For example, diuretics reduce blood volume by promoting elimination of salts and water in the kidney. Beta blockers and calcium channel blockers cause vasodilation.	Lower blood pressure reduces strain on the heart and lowers the risk of damaging artery walls (which could lead to blood clots or atheroma).	Side effects include dizziness, drowsiness, headaches, and fluid retention, although effects vary depending on the person and drug used.
Anticoagulants	Decreases the ability of blood to clot. Different stages of the clotting process are affected depending on the drug.	Work relatively quickly to prevent clots forming or to prevent existing clots from becoming larger.	An increased risk of bleeding complications, (e.g. internal bleeding) because of reduced ability to clot.
Platelet inhibitors	Prevents the aggregation of platelets, often by preventing the formation of fibrin that helps to bind platelets together.	Lowers risk of heart attack or stroke by inhibiting thrombus (blood clot) formation. Chewing an aspirin at the onset of a heart attack can significantly lower the risk of death.	As with anticoagulants, their use comes with an increased risk of bleeding complications.

Lovastatin molecule

gone, gone, gone

Trounce

Statins have been researched as a CVD treatment since the 1970s. Lovastatin (above) was the first commercially available statin. It was first isolated from the fungus *Aspergillus terreus*.

Antihypertensives act in a variety of ways to treat high blood pressure. A wide range of antihypertensive drugs allow patients to be matched with a drug that best suits their needs.

Warfarin (above) is a commonly prescribed anticoagulant. Many medicines interact with warfarin, so frequent monitoring by blood testing is required to obtain a safe, effective dosage.

Platelet inhibiting drugs include clopidogrel, which acts on the ADP receptor to inhibit platelet formation, and aspirin which acts on the COX enzyme to reduce platelet production.

1. Explain the need for drug therapies in treating CVD: _____

_____

_____

2. Explain why statins are so effective at lowering the risk of CVD: _____

_____

_____

3. Describe the similarities in benefits and risks of anticoagulants and platelet inhibitor drugs: _____

_____

_____

4. Explain why chewing an aspirin at the onset of a heart attack can lower the risk of the heart attack being fatal:

_____

_____

Related activities: CVD Risk Factors

Periodicals:
Mending broken hearts,
The statin story

Evaluating the Risk

Most health investigations of population death rates involve data collected after the event (often going back many years) and analysed for patterns. Some studies follow a cohort of patients and record aspects of their lifestyle and health at set intervals (e.g. yearly). These studies are useful but costly and they take many years to complete. The studies are more reliable when they involve a large group encompassing a range of people and lifestyles and controlled for those factors that are not of interest (nuisance factors). The results of the studies help health organisations alert the public to various lifestyle risks. However, how the public perceives and acts on these risks varies greatly. The actual versus the perceived risk of an event are often quite different. This is true not only for different events, but for different groups of people (e.g. sex, ethnicity, socio-economic group).

Passive Smoking and Coronary Heart Disease

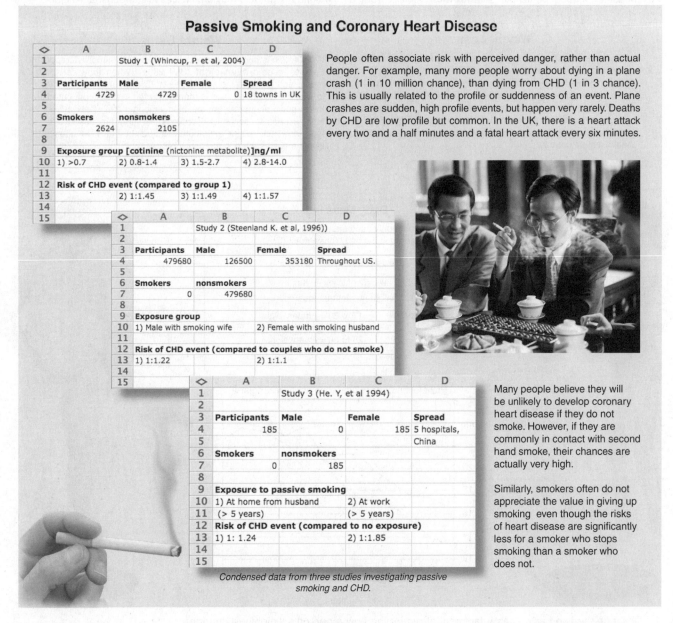

◇	A	B	C	D
1		Study 1 (Whincup, P. et al, 2004)		
2				
3	**Participants**	**Male**	**Female**	**Spread**
4	4729	4729	0	18 towns in UK
5				
6	**Smokers**	**nonsmokers**		
7	2624	2105		
8				
9	**Exposure group [cotinine (nictonine metabolite)]ng/ml**			
10	1) >0.7	2) 0.8-1.4	3) 1.5-2.7	4) 2.8-14.0
11				
12	**Risk of CHD event (compared to group 1)**			
13		2) 1:1.45	3) 1:1.49	4) 1:1.57
14				
15				

◇	A	B	C	D
1		Study 2 (Steenland K. et al, 1996))		
2				
3	**Participants**	**Male**	**Female**	**Spread**
4	479680	126500	353180	Throughout US.
5				
6	**Smokers**	**nonsmokers**		
7	0	479680		
8				
9	**Exposure group**			
10	1) Male with smoking wife		2) Female with smoking husband	
11				
12	**Risk of CHD event (compared to couples who do not smoke)**			
13	1) 1:1.22		2) 1:1.1	
14				
15				

◇	A	B	C	D
1		Study 3 (He. Y, et al 1994)		
2				
3	**Participants**	**Male**	**Female**	**Spread**
4	185	0	185	5 hospitals, China
5				
6	**Smokers**	**nonsmokers**		
7	0	185		
8				
9	**Exposure to passive smoking**			
10	1) At home from husband		2) At work	
11	(> 5 years)		(> 5 years)	
12	**Risk of CHD event (compared to no exposure)**			
13	1) 1: 1.24		2) 1:1.85	
14				
15				

People often associate risk with perceived danger, rather than actual danger. For example, many more people worry about dying in a plane crash (1 in 10 million chance), than dying from CHD (1 in 3 chance). This is usually related to the profile or suddenness of an event. Plane crashes are sudden, high profile events, but happen very rarely. Deaths by CHD are low profile but common. In the UK, there is a heart attack every two and a half minutes and a fatal heart attack every six minutes.

Many people believe they will be unlikely to develop coronary heart disease if they do not smoke. However, if they are commonly in contact with second hand smoke, their chances are actually very high.

Similarly, smokers often do not appreciate the value in giving up smoking even though the risks of heart disease are significantly less for a smoker who stops smoking than a smoker who does not.

Condensed data from three studies investigating passive smoking and CHD.

1. Explain why many people ignore the risks of various lifestyles increasing their chances of coronary heart disease:

_____

_____

2. Based on the data above, identify the study that makes the most reliable link between passive smoking and CHD and explain why:

_____

_____

3. Describe how studies of this kind produce reliable results: _____

_____

_____

Related activities: CVD Risk Factors

DA 2

Healthy Lifestyle, Healthy Heart

Maintaining a Healthy Weight

The **basal metabolic rate** (**BMR**) is the amount of energy your body requires to meet its daily metabolic needs. BMR varies depending on the age and lifestyle of an individual. To maintain a stable weight, energy intake and energy output must be balanced. Weight is gained or lost if the balance is not maintained. Physiological factors (blood nutrient levels, hormones, body temperature) and psychological factors have an impact on eating behaviour. **Malnutrition** results from a nutritional imbalance of any sort. In economically developed countries, most forms of malnutrition are the result of poorly balanced nutrient intakes rather than a lack of food *per se*. The **body mass index** (**BMI**) is used to determine if a person is a healthy weight. It is a commonly used statistical measure which compares the relationship between a person's weight and height (below). The waist-to-hip ratio is also used to assess healthy weight; a high ratio indicates excessive abdominal weight.

Weight Loss

Weight loss occurs when energy output exceeds energy intake. A person is **underweight** (BMI <20) if too much weight is lost. One billion people (mostly in developing nations) do not have an adequate energy intake, and their BMR requirements are not met. Many will die from malnutrition or because their immune systems can not combat disease. An adult BMI <17.5 is considered severely underweight and may result in a number of health issues (below).

Weight Gain

Weight gain occurs when energy intake exceeds energy output. People become **overweight** (BMI 25-30) when they eat too much, and exercise too little. **Obesity** (BMI >30) is widespread in western countries, and is more common in poorly educated, lower socio-economic groups than amongst the wealthy, who have access to better food choices. A highly processed diet, high in fat and sugar contributes to obesity. Adult UK obesity levels are presented in the graph below.

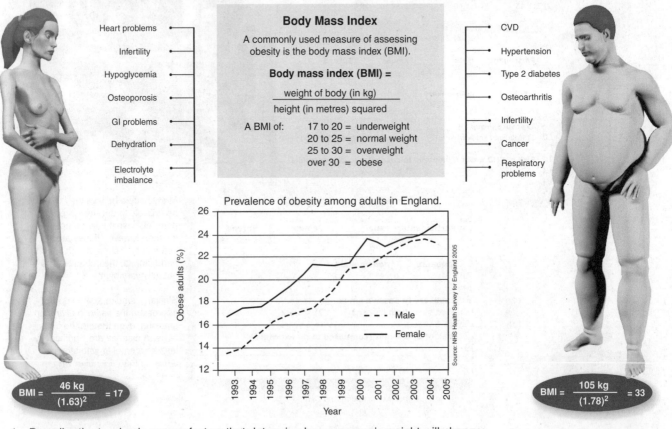

Body Mass Index

A commonly used measure of assessing obesity is the body mass index (BMI).

Body mass index (BMI) =

$$\frac{\text{weight of body (in kg)}}{\text{height (in metres) squared}}$$

A BMI of: 17 to 20 = underweight
20 to 25 = normal weight
25 to 30 = overweight
over 30 = obese

Heart problems
Infertility
Hypoglycemia
Osteoporosis
GI problems
Dehydration
Electrolyte imbalance

CVD
Hypertension
Type 2 diabetes
Osteoarthritis
Infertility
Cancer
Respiratory problems

Prevalence of obesity among adults in England.

Obese adults (%)

- - - Male
—— Female

Source: NHS Health Survey for England 2005

Year

$$\text{BMI} = \frac{46 \text{ kg}}{(1.63)^2} = 17$$

$$\text{BMI} = \frac{105 \text{ kg}}{(1.78)^2} = 33$$

1. Describe the two basic energy factors that determine how a person's weight will change: _____

_____

_____

2. Explain why a teenager requires a greater daily energy intake than an adult: _____

_____

3. Using the BMI, calculate the minimum weight at which a 1.85 m tall man would be considered:

(a) Overweight: _____   (c) Obese: _____

(b) Normal weight: _____   (d) Underweight: _____

4. Suggest a reason for the prevalence of obesity in the developed world: _____

_____

Periodicals:
Why are we so fat?,
Obesity: size matters

A Balanced Diet

Humans require a wide variety of foods to obtain all of their nutritional requirements. Nutrients are required for metabolism, tissue growth and repair, and as an energy source. Good nutrition (provided by a **balanced diet**) is recognised as a key factor in good health. Conversely poor nutrition (**malnutrition**) contributes to ill-health. The term diet refers to the quantity and nature of the food eaten. In a recent overhaul of previous dietary recommendations, the health benefits of monounsaturated fats (such as olive and canola oils), fish oils, and whole grains in the diet have been recognised, and people are being urged to reduce their consumption of highly processed foods and saturated (rather than total) fat. Those on diets that restrict certain food groups (e.g. vegans) must take care to balance their intake of foods to ensure an adequate supply of protein and other nutrients (e.g. iron and B vitamins). **Reference Nutrient Intakes** (RNIs) provide nutritional guidelines for different sectors of the population in the UK. RNIs help to define the upper and lower limits of adequate nutrient intake for most people, but they are not recommendations for intakes by individuals.

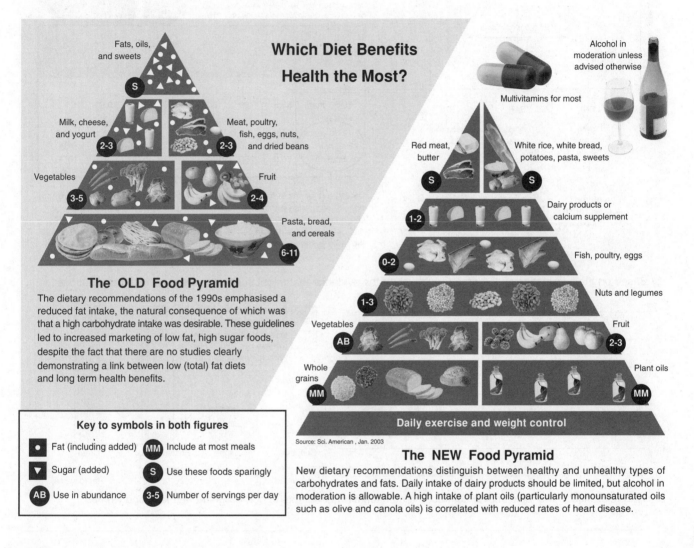

Which Diet Benefits Health the Most?

The OLD Food Pyramid

The dietary recommendations of the 1990s emphasised a reduced fat intake, the natural consequence of which was that a high carbohydrate intake was desirable. These guidelines led to increased marketing of low fat, high sugar foods, despite the fact that there are no studies clearly demonstrating a link between low (total) fat diets and long term health benefits.

Source: Sci. American , Jan. 2003

The NEW Food Pyramid

New dietary recommendations distinguish between healthy and unhealthy types of carbohydrates and fats. Daily intake of dairy products should be limited, but alcohol in moderation is allowable. A high intake of plant oils (particularly monounsaturated oils such as olive and canola oils) is correlated with reduced rates of heart disease.

Key to symbols in both figures

- **·** Fat (including added)
- **▼** Sugar (added)
- **AB** Use in abundance
- **MM** Include at most meals
- **S** Use these foods sparingly
- **3-5** Number of servings per day

---

1. Identify two major roles of **nutrients** in the diet:

    (a) _____

    (b) _____

2. (a) Compare the two food pyramids (above) and discuss how they differ in their recommendations for good nutrition:

    _____
    _____
    _____
    _____
    _____
    _____
    _____

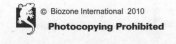

Periodicals:
Feast and famine,
The good, the fad, & the unhealthy

Related activities: Maintaining a Healthy Weight

DA 2

Healthy Lifestyle, Healthy Heart

Nutritional Guidelines in the UK

In the UK, Dietary Reference Values (DRVs) provide guidelines for nutrient and energy intake for particular groups of the population. In a population, it is assumed that the nutritional requirements of the population as a whole are represented by a normal, bell-shaped, curve (below). DRVs collectively encompass RNIs, LRNIs, and EARs, and replace the earlier Recommended Daily Amounts (RDAs), which *recommended* nutrient intakes for particular groups in the population, including those with very high needs.

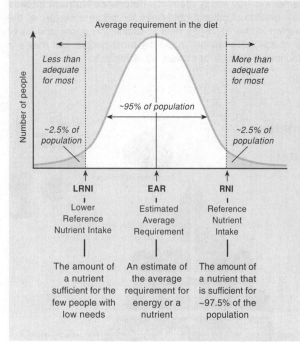

Average requirement in the diet

Less than adequate for most

More than adequate for most

~95% of population

~2.5% of population

~2.5% of population

Number of people

LRNI	EAR	RNI
Lower Reference Nutrient Intake	Estimated Average Requirement	Reference Nutrient Intake
The amount of a nutrient sufficient for the few people with low needs	An estimate of the average requirement for energy or a nutrient	The amount of a nutrient that is sufficient for ~97.5% of the population

Table 1 (below): Estimated Average Requirements (EAR) for energy, and Reference Nutrient Intakes (RNIs) for selected nutrients, for UK males and females aged 19-50 years (per day).

Source: Dept of Health. Dietary Reference Values for Food Energy and Nutrients for the UK, 1991.

Age range	Reference Nutrient Intakes (RNIs)					EARs	
	Protein (g)	Calcium (mg)	Iron (mg)	Folate (µg)	Vit.C (mg)	EAR (MJ) Males	Females
Males							
19-50 years	55.5	700	8.7	700	40	10.60	
Females							
19-50 years	45.0	700	14.8	600	40		8.10
Pregnant	51.0	1250	14.8	700	50		8.90
Lactating	56.0	1250	14.8	950	70		10.20

DRVs have been set for population groups within the UK, taking into account age and gender. Only a portion of the table is shown here.

RNIs are provided for each constituent of a balanced diet

EARs for energy are based on the present lifestyles and activity levels of the UK population.

(b) Based on the information on the graph (right), state the evidence that might support the revised recommendations:

_____

_____

_____

_____

_____

_____

_____

Percentage of calories from fat in the traditional diet

Incidence of coronary heart disease per 10 000 men (over a 10 year period)

	40%
38%	
	3000
10%	
500	200

Country	Japan	Eastern Finland	Crete
Type of fat consumed	Low fat	Saturated fat	Monounsaturated fat (olive oil)

3. With reference to the table above, contrast the nutritional requirements of non-pregnant and lactating women:

_____

_____

4. (a) Suggest in which way the older RDAs could have been misleading for many people: _____

_____

(b) Explain how the DRVs differ from the older RDAs: _____

_____

_____

5. Suggest how **DRVs** can be applied in each of the following situations:

(a) Dietary planning and assessment: _____

_____

(b) Food labelling and consumer information: _____

_____

KEY TERMS: Memory Card Game

The cards below have a keyword or term printed on one side and its definition printed on the opposite side. The aim is to win as many cards as possible from the table. To play the game.....

1) Cut out the cards and lay them definition side down on the desk. You will need one set of cards between two students.

2) Taking turns, choose a card and, BEFORE you pick it up, state

your own best definition of the keyword to your opponent.

3) Check the definition on the opposite side of the card. If both you and your opponent agree that your stated definition matches, then keep the card. If your definition does not match then return the card to the desk.

4) Once your turn is over, your opponent may choose a card.

Anticoagulant	Artery	Atherosclerosis
Atrium	Blood	Body mass index
Bulk flow	Cardiovascular disease	Cholesterol
Diastole	Fibrin	Obesity
Plaque	Statins	Systole
Tissue fluid	Vein	Ventricle

Healthy Lifestyle, Healthy Heart

R 2

KEY TERMS: Memory Card Game

When you've finished the game keep these cutouts and use them as flash cards!

Disease of the arteries caused by fatty deposits on the artery walls which cause the restriction of blood flow and may lead to blood clots or aneurysm.

A large blood vessel with a thick, muscled wall which carries blood away from the heart.

Drug that decreases the clotting ability of the blood by interfering with the clotting chain reaction. Generally used in the treatment of CVD and CHD.

Scale that takes into account the height and mass of a person and uses this to identify if their mass is in correct proportion to their height. A healthy BMI ranges between 20 and 25.

Circulatory fluid comprising numerous cell types, which moves respiratory gases and nutrients around the body.

Chamber of the heart that receives blood from the body or lungs.

A sterol lipid found in all animal tissues as part of cellular membranes, it regulates membrane fluidity. and is an important precursor in the synthesis of vitamin D and steroid hormones.

General term that includes diseases of the heart and blood vessels.

The process by which fluids move as a single mass around the body. The movement may be induced by pressure and/or gravity.

Condition of being overweight with an extreme excess of body fat. Has a BMI value of > 30.

A fibrous protein involved in blood clotting and made from fibrinogen. Forms a mesh over a wound during clotting.

The period of time in which the heart relaxes and fills with blood.

The period of time where the heart muscle is contracting in order to move blood through it.

Drug group developed for the treatment of high levels of cholesterol. Blocks the production of cholesterol in the liver and encourages the liver to increase LDL receptors.

A build up of material upon a surface. In CVD a plaque may refer to the build up of fatty deposits on the artery walls.

Chamber of the heart that pumps blood into the arteries.

Large blood vessel that returns blood to the heart.

A fluid derived from the blood plasma by leakage through capillaries. It bathes the tissues and is also called interstitial fluid.

Membranes and Exchange Surfaces

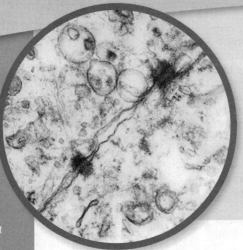

KEY CONCEPTS

▶ The plasma membrane is best described by the fluid mosaic model.

▶ The plasma membrane is partially permeable. It regulates the movement of molecules into a cell.

▶ The active transport of substances into a cell requires ATP. Passive transport mechanisms do not require any energy expenditure.

▶ The structure of the human lung is highly adapted to aid rapid rates of gas exchange.

OBJECTIVES

☐ 1. Use the **KEY TERMS** to help you understand and complete these objectives.

Membranes and Transport   page 72-74, 77-82

☐ 2. Describe the current view of a plasma membrane (the **fluid mosaic model**).

☐ 3. Explain how data collected from techniques such as **freeze fracture**, have helped scientists develop and refine the fluid mosaic model.

☐ 4. Explain the terms **active transport** and **passive transport**. Understand that passive transport requires no energy expenditure, but active transport uses energy in the form of **ATP**.

☐ 5. Explain what is meant by a **partially permeable membrane**, and describe its role in regulating the movement of molecules across a plasma membrane.

☐ 6. Explain how water moves across a partially permeable membrane by **osmosis**.

☐ 7. Explain the differences between **diffusion** and **facilitated diffusion**.

☐ 8. Describe the role of ion channels and carrier proteins in facilitated diffusion. Describe and explain a situation where facilitated diffusion is important.

☐ 9. Explain that **endocytosis** engulfs and draws molecules into a cell. Recognise **phagocytosis** and **pinocytosis** as forms of endocytosis. Explain that **exocytosis** is the release of material from the cell.

☐ 10. Describe the role of **ion pumps** in the active transport of molecules. Explain how ion pumps regulate movements across membranes.

The Human Gas Exchange System   page 75-76, 83-84

☐ 11. Explain how the **surface area to volume ratio** affects diffusion rates. Understand that diffusion rates are governed by the principles of **Fick's law**.

☐ 12. Describe the **mammalian respiratory system,** including the structure and location of the (lungs, trachea, bronchi, bronchiole, alveolus.

☐ 13. Explain how the structure of the lung tissue facilitates rapid gas exchange between air and blood. Include reference to; large surface area of the lungs, the thin membranes of the alveoli and neighbouring capillaries, and the differential concentration gradient between the lungs and blood.

☐ 14. Describe how the **lung alveoli** act to increase the surface area of the lung.

Periodicals:
listings for this
chapter are on page 228

Weblinks:
www.biozone.co.uk/
weblink/Edx-AS-2542.html

*Teacher Resource
CD-ROM:*
*Osmosis & Diffusion (Water
Potential Version)*

How Do We Know? Membrane Structure

Cellular membranes play many extremely important roles in cells, and understanding their structure is central to understanding cellular function. Moreover, understanding the structure and function of membrane proteins is essential to understanding cellular transport processes, and cell recognition and signalling. Cellular membranes are far too small to be seen clearly using light microscopy, and certainly any detail is impossible to resolve. Since early last century, scientists have known that membranes comprised a lipid bilayer with associated proteins. But how did they elucidate just how these molecules were organised?

The answers were provided with electron microscopy, and one technique in particular – **freeze fracture**. As the name implies, freeze fracture, at its very simplest level, is the freezing of a cell and then cleaving it so that it fractures in a certain way. Scientists can then use electron microscopy to see the indentations and outlines of the structures remaining after cleavage. Membranes are composed of two layers held together by weak intermolecular bonds, so they cleave into two halves when fractured. This provides views of the inner surfaces of the membrane.

The procedure involves several steps:

▶ The tissue is prefixed using a cross linking agent. This alters the strength of the internal and external parts of the membrane.
▶ The cell is fixed to immobilise any mobile macromolecules.
▶ The specimen is passed through a sequential series of glycerol solutions of increasing strength. This protects the cells from bursting when placed into the cryomaterial.
▶ The specimen is frozen using liquid propane cooled by liquid nitrogen. The specimens are mounted on gold supports and cooled briefly before transfer to the freeze-etch machine.
▶ Specimen is cleaved in a helium-vented vacuum at -150ºC. A razor blade cooled to -170ºC acts as both a cold trap for water and the cleaving instrument.
▶ At this stage the specimen may be evaporated a little to produce some relief in the surface of the fracture (known as etching) so that a 3-dimensional effect occurs.
▶ For viewing under EM, a replica of the specimen is made and coated in gold or platinum to ~3 nm thick. This produces a shadow effect that allows structures to be seen clearly. A layer of carbon around 30 nm thick is used to stabilise the specimen.
▶ The samples are then raised to room temperature and placed into distilled water or digestive enzymes, which allows the replica to separate from the sample. The replica is then rinsed several times in distilled water before it is ready for viewing.

The freeze fracture technique provided the necessary supporting evidence for the current fluid mosaic model of membrane structure. When cleaved, proteins in the membrane left impressions that showed they were embedded into the membrane and not a continuous layer on the outside as earlier models proposed.

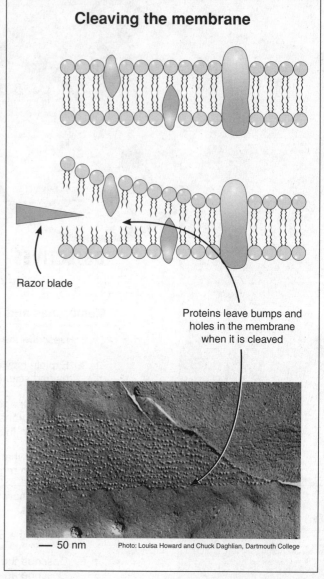

Cleaving the membrane

Razor blade

Proteins leave bumps and holes in the membrane when it is cleaved

— 50 nm   Photo: Louisa Howard and Chuck Daghlian, Dartmouth College

1. Describe the principle of freeze fracture and explain why it is such a useful technique for studying membrane structure:

_____

_____

_____

2. Explain how this freeze-fracture studies provided evidence for our current model of membrane structure: _____

_____

_____

3. An earlier model of membrane structure was the unit membrane; a phospholipid bilayer with a protein coat. Explain how the freeze-fracture studies showed this model to be flawed:

_____

_____

_____

The Structure of Membranes

All cells have a plasma membrane that forms the outer limit of the cell. Membranes are also found inside eukaryotic cells as part of membranous **organelles**. Our knowledge of membrane structure has been built up as a result of many observations and experiments. The now-accepted **fluid-mosaic model** of membrane structure (below) satisfies the observed properties of

membranes. The self-orientating properties of the phospholipids allows cellular membranes to reseal themselves when disrupted. The double layer of lipids is also quite fluid, and proteins move quite freely within it. The plasma membrane is more than just a passive envelope; it is a dynamic structure actively involved in cellular activities.

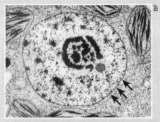

The **nuclear membrane** that surrounds the nucleus helps to control the passage of genetic information to the cytoplasm. It may also serve to protect the DNA.

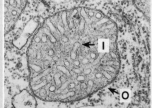

Mitochondria have an outer membrane (**O**) which controls the entry and exit of materials involved in aerobic respiration. Inner membranes (**I**) provide attachment sites for enzyme activity.

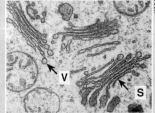

The **Golgi apparatus** comprises stacks of membrane-bound sacs (**S**). It is involved in packaging materials for transport or export from the cell as secretory vesicles (**V**).

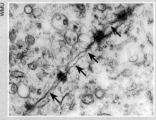

The cell is surrounded by a **plasma membrane** which controls the movement of most substances into and out of the cell. This photo shows two neighbouring cells (arrows).

The Fluid Mosaic Model of Membrane Structure

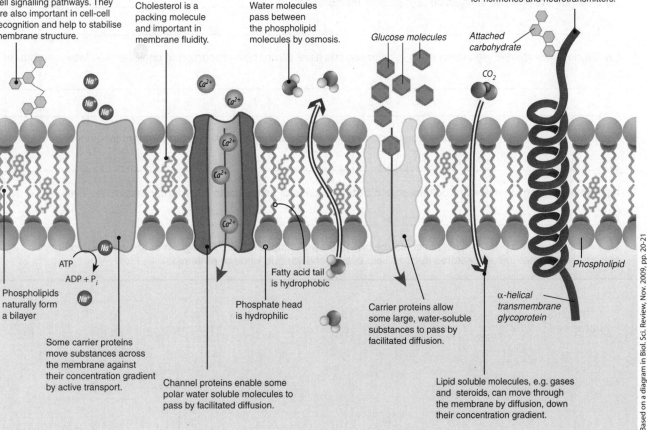

Glycolipids in membranes are phospholipids with attached carbohydrate. Like glycoproteins, they act as surface receptors in cell signalling pathways. They are also important in cell-cell recognition and help to stabilise membrane structure.

Cholesterol is a packing molecule and important in membrane fluidity.

Water molecules pass between the phospholipid molecules by osmosis.

Glucose molecules

Attached carbohydrate

CO_2

Glycoproteins are proteins with attached carbohydrate. They are important in membrane stability, in cell-cell recognition, and in cell signalling, acting as receptors for hormones and neurotransmitters.

Phospholipids naturally form a bilayer

ATP
ADP + P_i

Some carrier proteins move substances across the membrane against their concentration gradient by active transport.

Fatty acid tail is hydrophobic

Phosphate head is hydrophilic

Channel proteins enable some polar water soluble molecules to pass by facilitated diffusion.

Carrier proteins allow some large, water-soluble substances to pass by facilitated diffusion.

Lipid soluble molecules, e.g. gases and steroids, can move through the membrane by diffusion, down their concentration gradient.

α-helical transmembrane glycoprotein

Phospholipid

Based on a diagram in Biol. Sci. Review, Nov. 2009, pp. 20-21

1. Identify the component(s) of the plasma membrane involved in:

   (a) Facilitated diffusion: _____   (c) Cell signalling: _____

   (b) Active transport: _____   (d) Regulating membrane fluidity: _____

2. (a) Describe the modern fluid mosaic model of membrane structure: _____

   _____

   _____

Periodicals:
The fluid mosaic model,
Border control

Related activities: How do We Know? Membrane Structure
Web links: *Membrane Structure Tutorial*

RA 2

(b) Explain how the fluid mosaic model accounts for the observed properties of cellular membranes:

_____

_____

_____

_____

_____

_____

3. Discuss the various functional roles of membranes in cells: _____

_____

_____

_____

4. (a) Name a cellular organelle that possesses a membrane: _____

   (b) Describe the membrane's purpose in this organelle: _____

   _____

5. (a) Describe the purpose of cholesterol in plasma membranes: _____

   _____

   (b) Suggest why marine organisms living in polar regions have a very high proportion of cholesterol in their membranes:

   _____

   _____

6. List three substances that need to be transported **into** all kinds of animal cells, in order for them to survive:

   (a) _____ (b) _____ (c) _____

7. List two substances that need to be transported **out** of all kinds of animal cells, in order for them to survive:

   (a) _____ (b) _____

8. Use the symbol for a phospholipid molecule (below) to draw a **simple labelled diagram** to show the structure of a plasma membrane (include features such as lipid bilayer and various kinds of proteins):

Surface Area and Volume

When an object (e.g. a cell) is small it has a large surface area in comparison to its volume. In this case diffusion will be an effective way to transport materials (e.g. gases) into the cell. As an object becomes larger, its surface area compared to its volume is smaller. Diffusion is no longer an effective way to transport materials to the inside. For this reason, there is a physical limit for the size of a cell, with the effectiveness of diffusion being the controlling factor.

Diffusion in Organisms of Different Sizes

Respiratory gases and some other substances are exchanged with the surroundings by diffusion or active transport across the plasma membrane.

The **plasma membrane**, which surrounds every cell, functions as a selective barrier that regulates the cell's chemical composition. For each square micrometer of membrane, only so much of a particular substance can cross per second.

The surface area of an elephant is increased, for radiating body heat, by large flat ears.

The nucleus can control a smaller cell more efficiently.

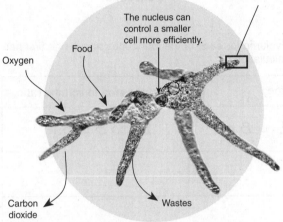

Oxygen

Food

Carbon dioxide

Wastes

A specialised gas exchange surface (lungs) and circulatory (blood) system are required to speed up the movement of substances through the body.

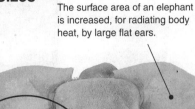

Respiratory gases cannot reach body tissues by diffusion alone.

Amoeba: The small size of single-celled protoctists, such as *Amoeba*, provides a large surface area relative to the cell's volume. This is adequate for many materials to be moved into and out of the cell by diffusion or active transport.

Multicellular organisms: To overcome the problems of small cell size, plants and animals became multicellular. They provide a small surface area compared to their volume but have evolved various adaptive features to improve their effective surface area.

Smaller is Better for Diffusion

One large cube

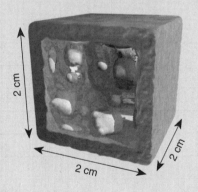

2 cm

2 cm

2 cm

Volume: = 8 cm³
Surface area: = 24 cm²

Eight small cubes

1 cm

1 cm

1 cm

Volume: = 8 cm³ for 8 cubes
Surface area: = 6 cm² for 1 cube
  = 48 cm² for 8 cubes

The eight small cells and the single large cell have the same total volume, but their surface areas are different. The small cells together have twice the total surface area of the large cell, because there are more exposed (inner) surfaces. Real organisms have complex shapes, but the same principles apply.

The surface-area volume relationship has important implications for processes involving transport into and out of cells across membranes. For activities such as gas exchange, the surface area available for diffusion is a major factor limiting the rate at which oxygen can be supplied to tissues.

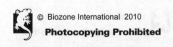
Periodicals:
Size does matter!

Related activities: Diffusion

DA 1

Membranes & Exchange Surfaces

The diagram below shows four hypothetical cells of different sizes (cells do not actually grow to this size, their large size is for the sake of the exercise). They range from a small 2 cm cube to a larger 5 cm cube. This exercise investigates the effect of cell size on the efficiency of diffusion.

2 cm cube   **3 cm cube**   **4 cm cube**   **5 cm cube**

1. Calculate the volume, surface area and the ratio of surface area to volume for each of the four cubes above (the first has been done for you). When completing the table below, show your calculations.

Cube size	Surface area	Volume	Surface area to volume ratio
2 cm cube	$2 \times 2 \times 6 = 24 \, cm^2$ (2 cm x 2 cm x 6 sides)	$2 \times 2 \times 2 = 8 \, cm^3$ (height x width x depth)	$24 \, to \, 8 = 3{:}1$
3 cm cube			
4 cm cube			
5 cm cube			

2. Create a graph, plotting the surface area against the volume of each cube, on the grid on the right. Draw a line connecting the points and label axes and units.

3. State which increases the fastest with increasing size, the **volume** or **surface area**.

_____

_____

4. Explain what happens to the ratio of surface area to volume with increasing size:

_____

_____

_____

_____

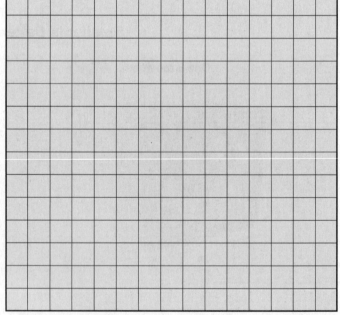

5. (a) Diffusion of substances into and out of a cell occurs across the cell surface. Describe how increasing the size of a cell affects the ability of diffusion to transport materials into and out of a cell:

_____

_____

(b) Describe how this places constraints on cell size and explain how multicellular organisms have overcome this:

_____

_____

_____

Active and Passive Transport

Cells have a need to move materials both into and out of the cell. Raw materials and other molecules necessary for metabolism must be accumulated from outside the cell. Some of these substances are scarce outside of the cell and some effort is required to accumulate them. Waste products and molecules for use in other parts of the body must be 'exported' out of the cell.

Some materials (e.g. gases and water) move into and out of the cell by **passive transport** processes, without the expenditure of energy on the part of the cell. Other molecules (e.g. sucrose) are moved into and out of the cell using **active transport**. Active transport processes involve the expenditure of energy in the form of ATP, and therefore use oxygen.

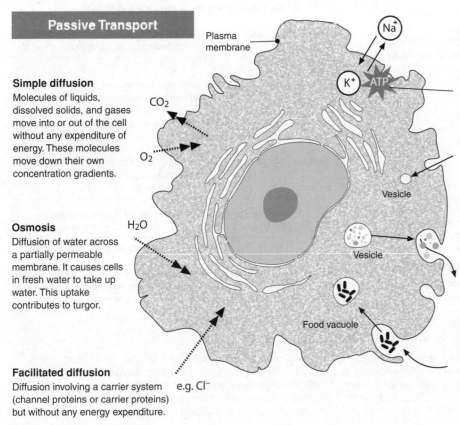

Passive Transport

Plasma membrane

Simple diffusion
Molecules of liquids, dissolved solids, and gases move into or out of the cell without any expenditure of energy. These molecules move down their own concentration gradients.

CO_2

O_2

Osmosis
Diffusion of water across a partially permeable membrane. It causes cells in fresh water to take up water. This uptake contributes to turgor.

H_2O

Vesicle

Vesicle

Food vacuole

Facilitated diffusion
Diffusion involving a carrier system (channel proteins or carrier proteins) but without any energy expenditure.

e.g. Cl^-

Na^+

K^+

ATP

Active Transport

Sodium-potassium ion pump
A specific protein in the plasma membrane that uses energy (ATP) to exchange sodium for potassium ions (3 Na+ out for every 2 K+ in). The concentration gradient can be used to drive other active transport processes.

Pinocytosis
Fluid or a suspension is taken into the cell. This process is non-specific to the substances taken in and is primarily used for absorbing extracellular fluid. The plasma membrane encloses some of the fluid to form a small vesicle, which is then fused with a lysosome and broken down.

Exocytosis
Vesicles budded off the Golgi or ER fuse with the plasma membrane, expelling their contents.

Phagocytosis
A form of endocytosis where solids are taken in to the cell. The plasma membrane encloses a particle and buds off to form a vacuole. Lysosomes will fuse with it to enable digestion of the contents.

1. In general terms, describe the energy requirements of **passive** and **active** transport: _____

_____

_____

_____

2. Name two gases that move into or out of living eukaryotic organisms by **diffusion**: _____

3. Name a gland (or glandular tissue) which has cells where **exocytosis** takes place for the purpose of secretion:

_____

4. **Phagocytosis** is a process where solid particles are enveloped by the cell membrane and drawn inside the cell.

   (a) Name a protozoan that would use this technique for feeding: _____

   (b) Describe how it uses the technique: _____

   _____

   _____

   (c) Name a type of cell found in human blood that is actively phagocytic: _____

   (d) Describe the functional role of the cell that is related to this feature: _____

Diffusion

The molecules that make up substances are constantly moving about in a random way. This random motion causes molecules to disperse from areas of high to low concentration; a process called **diffusion**. The molecules move down a **concentration gradient**. Diffusion and osmosis (diffusion of water molecules across a partially permeable membrane) are **passive** processes, and use no energy. Diffusion occurs freely across membranes, as long as the membrane is permeable to that molecule (partially permeable membranes allow the passage of some molecules but not others). Each type of molecule diffuses down its own concentration gradient. Diffusion of molecules in one direction does not hinder the movement of other molecules. Diffusion is important in allowing exchanges with the environment and in the regulation of cell water content.

Diffusion is the movement of particles from regions of high to low concentration (down a **concentration gradient**), with the end result being that the molecules become evenly distributed. In biological systems, diffusion often occurs across partially permeable membranes.

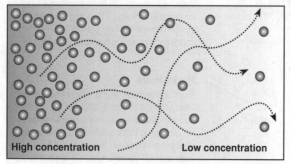

High concentration   **Low concentration**

Concentration gradient

If molecules are free to move, they move from high to low concentration until they are evenly dispersed.

Factors affecting rates of diffusion

Concentration gradient:	Diffusion rates will be higher when there is a greater difference in concentration between two regions.
The distance involved:	Diffusion over shorter distances occurs at a greater rate than diffusion over larger distances.
The area involved:	The larger the area across which diffusion occurs, the greater the rate of diffusion.
Barriers to diffusion:	Thicker barriers slow diffusion rate. Pores in a barrier enhance diffusion.

FICK'S LAW	$\dfrac{\text{Surface area of membrane} \quad X \quad \text{Difference in concentration across the membrane}}{\text{Length of the diffusion path (thickness of the membrane)}}$

These factors are expressed in **Fick's law**, which governs the rate of diffusion of substances within a system. Temperature also affects diffusion rates; at higher temperatures molecules have more energy and move more rapidly.

Diffusion through Membranes

Each type of diffusing molecule (gas, solvent, solute) moves **down its own concentration gradient**. Two-way diffusion (below) is common in biological systems, e.g. at the lung surface, carbon dioxide diffuses out and oxygen diffuses into the blood. Facilitated diffusion (below, right) increases the diffusion rate selectively and is important for larger molecules (e.g. glucose, amino acids) where a higher diffusion rate is desirable (e.g. transport of glucose into skeletal muscle fibres, transport of ADP into mitochondria). Neither type of diffusion requires energy expenditure because the molecules are not moving against their concentration gradient.

Unaided diffusion   **Facilitated diffusion**

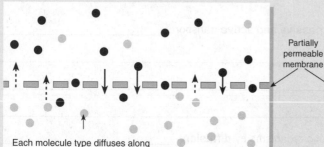

Partially permeable membrane

Each molecule type diffuses along its own concentration gradient.

Ionophore

Ionophore preferentially allows passage of certain molecules.

Diffusion rates depend on the concentration gradient. Diffusion can occur in either direction but **net** movement is down the concentration gradient. An equilibrium is reached when concentrations are equal.

Facilitated diffusion occurs when a substance is aided across a membrane by a special protein called an **ionophore**. Ionophores allow some molecules to diffuse but not others, effectively speeding up the rate of diffusion of that molecule.

1. Describe two properties of an exchange surface that would facilitate rapid diffusion rates:

    (a) _____ (b) _____

2. Identify one way in which organisms maintain concentration gradients across membranes: _____

    _____

3. State how facilitated diffusion is achieved: _____

    _____

Related activities: Active and Passive Transport, Osmosis in Cells
Web links: Cellular transport, Osmosis and Diffusion

Periodicals:
Getting in and out

© Biozone International 2010
Photocopying Prohibited

Osmosis in Cells

Osmosis is the term describing the diffusion of water down its concentration gradient across a partially permeable membrane. It is the principal mechanism by which water enters and leaves cells in living organisms. As it is a type of diffusion, the rate at which osmosis occurs is affected by the same factors that affect all diffusion rates. In animal biology and medicine, the terms osmotic potential and osmotic pressure are often used to express the water relations of animal cells (which, unlike plant cells, lack a rigid cell wall). The **osmotic potential** of a solution is a measure of the tendency of the solution to gain water by osmosis. The **osmotic pressure** is a measure of the tendency for water to move into a solution by osmosis. Because water movements in plant cells are also affected by the pressure exerted by the rigid cell wall, they are often described in terms of the **water potential** (ψ) of the solutions involved. Water potential takes account of the influence of the water concentration and the wall pressure, and is particularly appropriate for explaining water movements in plant cells. We have not used this terminology here, but coverage of water potential is provided for those who want it as a web link and on the *Teacher Resource CD-ROM*.

Osmotic Gradients and Water Movement

Osmosis is the diffusion of water molecules, across a partially permeable membrane, from higher to lower concentration of water molecules (sometimes described as from lower to higher solute concentration). The direction of net movement can be predicted on the basis of the relative concentrations of water and solute molecules in the solutions involved. Water always diffuses from regions of higher concentration to lower concentration of water molecules (from lower to higher solute concentration).

The cytoplasm contains dissolved substances (**solutes**). When cells are placed in a solution of different concentration, there is an osmotic gradient between the external environment and the inside of the cell. In plant cells, the rigid cell wall is also important. When a plant cell takes up water, it swells until the cell contents exert a pressure on the cell wall. The cell wall is rigid and the pressure exerted on it by the cytoplasm is sometimes called the wall or **turgor pressure**. Turgor is important in plant support.

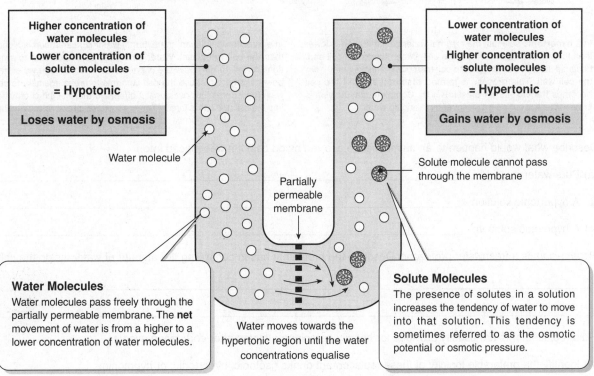

Higher concentration of water molecules

Lower concentration of solute molecules

= Hypotonic

Loses water by osmosis

Water molecule

Partially permeable membrane

Lower concentration of water molecules

Higher concentration of solute molecules

= Hypertonic

Gains water by osmosis

Solute molecule cannot pass through the membrane

Water Molecules

Water molecules pass freely through the partially permeable membrane. The **net** movement of water is from a higher to a lower concentration of water molecules.

Water moves towards the hypertonic region until the water concentrations equalise

Solute Molecules

The presence of solutes in a solution increases the tendency of water to move into that solution. This tendency is sometimes referred to as the osmotic potential or osmotic pressure.

1. Explain what is meant by partially permeable membrane: _____

_____

_____

2. Identify the factors influencing the net direction of water movement in:

    (a) Animal cells: _____

    (b) Plant cells: _____

3. Explain how animal cells differ from plant cells with respect to the effects of net water movements: _____

_____

_____

Related activities: Active and Passive Transport, Diffusion
Web links: Cellular Transport, Osmosis and Water Potential

RA 2

Water Relations in Plant Cells

The plasma membrane of cells is a partially permeable membrane and osmosis is the main way by which water enters and leaves the cell. When the external water concentration is the same as that of the cell there is no net movement of water. Two systems (cell and environment) with the same water concentration are termed **isotonic**. The diagram below illustrates two different situations: when the external water concentration is higher than the cell (**hypotonic**) and when it is lower than the cell (**hypertonic**).

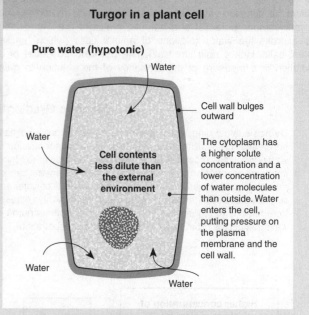

Plasmolysis in a plant cell

Hypertonic salt solution

Water

Water

Cell contents more dilute than the external environment

Cell wall is freely permeable to water molecules.

Water concentration in the cell is higher than outside.

Cytoplasm

Plasma membrane

Water

Water

Turgor in a plant cell

Pure water (hypotonic)

Water

Water

Cell contents less dilute than the external environment

Cell wall bulges outward

The cytoplasm has a higher solute concentration and a lower concentration of water molecules than outside. Water enters the cell, putting pressure on the plasma membrane and the cell wall.

Water

Water

In a **hypertonic** solution, the external water concentration is lower than the water concentration of the cell. Water leaves the cell and, because the cell wall is rigid, the cell membrane shrinks away from the cell wall. This process is termed **plasmolysis** and the cell becomes **flaccid** (turgor pressure = 0). Complete plasmolysis is irreversible; the cell cannot recover by taking up water.

In a **hypotonic** solution, the external water concentration is higher than the cell cytoplasm. Water enters the cell, causing it to swell tight. A wall (turgor) pressure is generated when enough water has been taken up to cause the cell contents to press against the cell wall. Turgor pressure rises until it offsets further net influx of water into the cell (the cell is turgid). The rigid cell wall prevents cell rupture.

4. Describe what would happen to an animal cell (e.g. a red blood cell) if it was placed into:

   (a) Pure water: _____

   (b) A hypertonic solution: _____

   (c) A hypotonic solution: _____

5. *Paramecium* is a freshwater protozoan. Describe the problem it has in controlling the amount of water inside the cell:

   _____

   _____

6. Fluid replacements are usually provided for heavily perspiring athletes after endurance events.

   (a) Identify the preferable tonicity of these replacement drinks (isotonic, hypertonic, or hypotonic): _____

   (b) Give a reason for your answer: _____

   _____

7. The malarial parasite lives in human blood. Relative to the tonicity of the blood, the parasite's cell contents would be hypertonic / isotonic / hypotonic (circle the correct answer).

8. (a) Explain the role of cell wall pressure in generating cell turgor in plants: _____

   _____

   (b) Discuss the role of cell turgor to plants: _____

   _____

   _____

   _____

Exocytosis and Endocytosis

Most cells carry out **cytosis**: a form of **active transport** involving the in- or outfolding of the plasma membrane. The ability of cells to do this is a function of the flexibility of the plasma membrane. Cytosis results in the bulk transport into or out of the cell and is achieved through the localised activity of microfilaments and microtubules in the cell cytoskeleton. Engulfment of material is

termed **endocytosis.** Endocytosis typically occurs in protozoans and certain white blood cells of the mammalian defence system (e.g. neutrophils, macrophages). **Exocytosis** is the reverse of endocytosis and involves the release of material from vesicles or vacuoles that have fused with the plasma membrane. Exocytosis is typical of cells that export material (secretory cells).

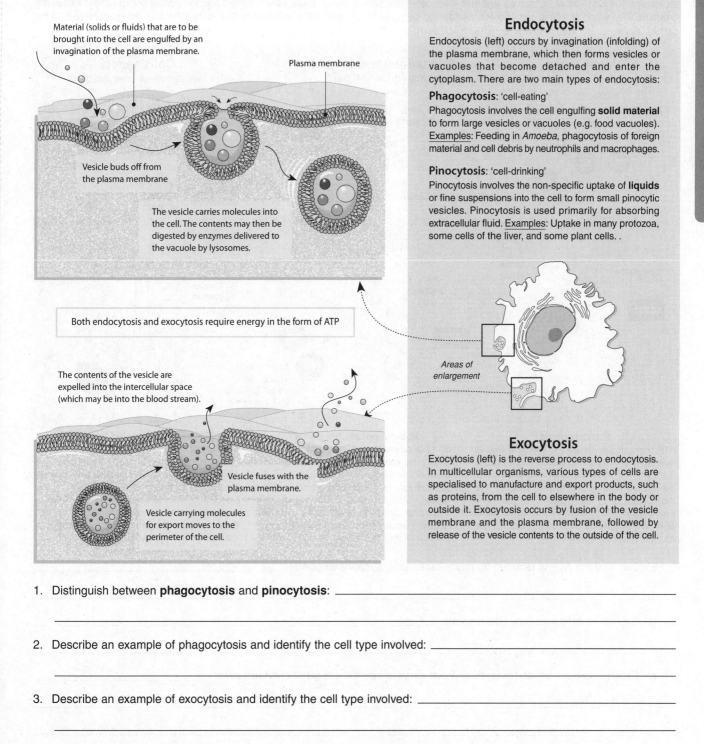

Material (solids or fluids) that are to be brought into the cell are engulfed by an invagination of the plasma membrane.

Plasma membrane

Vesicle buds off from the plasma membrane

The vesicle carries molecules into the cell. The contents may then be digested by enzymes delivered to the vacuole by lysosomes.

Both endocytosis and exocytosis require energy in the form of ATP

The contents of the vesicle are expelled into the intercellular space (which may be into the blood stream).

Vesicle fuses with the plasma membrane.

Vesicle carrying molecules for export moves to the perimeter of the cell.

Endocytosis

Endocytosis (left) occurs by invagination (infolding) of the plasma membrane, which then forms vesicles or vacuoles that become detached and enter the cytoplasm. There are two main types of endocytosis:

Phagocytosis: 'cell-eating'

Phagocytosis involves the cell engulfing **solid material** to form large vesicles or vacuoles (e.g. food vacuoles). Examples: Feeding in *Amoeba*, phagocytosis of foreign material and cell debris by neutrophils and macrophages.

Pinocytosis: 'cell-drinking'

Pinocytosis involves the non-specific uptake of **liquids** or fine suspensions into the cell to form small pinocytic vesicles. Pinocytosis is used primarily for absorbing extracellular fluid. Examples: Uptake in many protozoa, some cells of the liver, and some plant cells. .

Areas of enlargement

Exocytosis

Exocytosis (left) is the reverse process to endocytosis. In multicellular organisms, various types of cells are specialised to manufacture and export products, such as proteins, from the cell to elsewhere in the body or outside it. Exocytosis occurs by fusion of the vesicle membrane and the plasma membrane, followed by release of the vesicle contents to the outside of the cell.

1. Distinguish between **phagocytosis** and **pinocytosis**: _____

   _____

2. Describe an example of phagocytosis and identify the cell type involved: _____

   _____

3. Describe an example of exocytosis and identify the cell type involved: _____

   _____

4. Explain why cytosis is affected by changes in oxygen level, whereas diffusion is not: _____

   _____

5. Identify the processes by which the following substances enter a living macrophage (for help, see page on diffusion):

   (a) Oxygen: _____   (c) Water: _____

   (b) Cellular debris: _____   (d) Glucose: _____

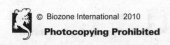
Periodicals:
What is endocytosis?

Related activities: Active and Passive Transport
Web links: Cellular Transport

RA 2

Ion Pumps

Diffusion alone cannot supply the cell's entire requirements for molecules (and ions). Some molecules (e.g. glucose) are required by the cell in higher concentrations than occur outside the cell. Others (e.g. sodium) must be removed from the cell in order to maintain fluid balance. These molecules must be moved across the plasma membrane by active transport mechanisms. **Active transport** requires the expenditure of energy because the molecules (or ions) must be moved **against** their concentration gradient. The work of active transport is performed by specific carrier proteins in the membrane. These transport proteins harness the energy of ATP to pump molecules from a low to a high concentration. When ATP transfers a phosphate group to the carrier protein, the protein changes its shape in such a way as to move the bound molecule across the membrane. Three types of membrane pump are illustrated below. The sodium-potassium pump (below, centre) is almost universal in animal cells and is common in plant cells also. The concentration gradient created by ion pumps such as this and the proton pump (left) is frequently coupled to the transport of molecules such as glucose (e.g. in the intestine) as shown below right.

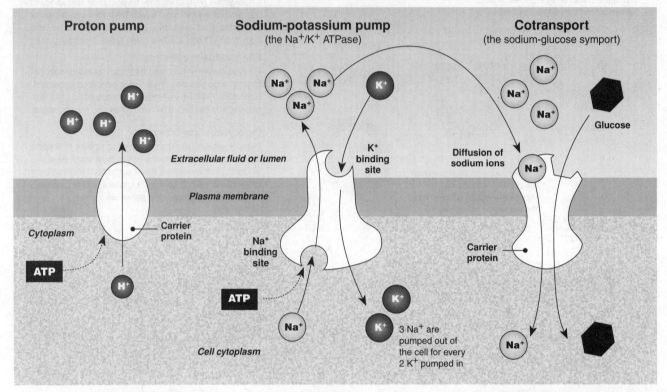

Proton pumps

ATP driven proton pumps use energy to remove hydrogen ions (H$^+$) from inside the cell to the outside. This creates a large difference in the proton concentration either side of the membrane, with the inside of the plasma membrane being negatively charged. This potential difference can be coupled to the transport of other molecules.

Sodium-potassium pump

The sodium-potassium pump is a specific protein in the membrane that uses energy in the form of ATP to exchange sodium ions (Na$^+$) for potassium ions (K$^+$) across the membrane. The unequal balance of Na$^+$ and K$^+$ across the membrane creates large concentration gradients that can be used to drive transport of other substances (e.g. cotransport of glucose).

Cotransport (coupled transport)

In intestinal epithelial cells, a gradient in sodium ions drives the active transport of **glucose**. The specific transport protein couples the return of Na$^+$ down its concentration gradient to the transport of glucose into the intestinal epithelial cell. A low intracellular concentration of Na$^+$ (and therefore the concentration gradient) is maintained by a sodium-potassium pump.

1. Explain why the ATP is required for membrane pump systems to operate: _____

_____

2. (a) Explain what is meant by cotransport: _____

_____

(b) Explain how cotransport is used to move glucose into the intestinal epithelial cells: _____

_____

_____

(c) Explain what happens to the glucose that is transported into the intestinal epithelial cells: _____

_____

3. Describe two consequences of the extracellular accumulation of sodium ions: _____

_____

_____

Related activities: Active and Passive Transport, Osmosis in Cells
Web links: Cellular Transport, Symport

Periodicals:
How biological membranes
achieve selective transport

The Human Gas Exchange System

The paired lungs of mammals, including humans, are located within the thorax and are connected to the outside air by way of a system of tubular passageways: the trachea, bronchi, and bronchioles. Ciliated, mucus secreting epithelium lines this system of tubules, trapping and removing dust and pathogens before they reach the gas exchange surfaces. Each lung is divided into a number of lobes, each receiving its own bronchus. Each bronchus divides many times, terminating in the respiratory

bronchioles from which arise 2-11 alveolar ducts and numerous **alveoli** (air sacs). These provide a very large surface area (70 m²) for the exchange of respiratory gases by diffusion between the alveoli and the blood in the capillaries. Gas exchange efficiency is also improved by the thin permeable membranes, a good blood supply, and efficient ventilation. The details of the exchange of gases across the respiratory membrane are described on the next page.

Morphology of the Gas Exchange System

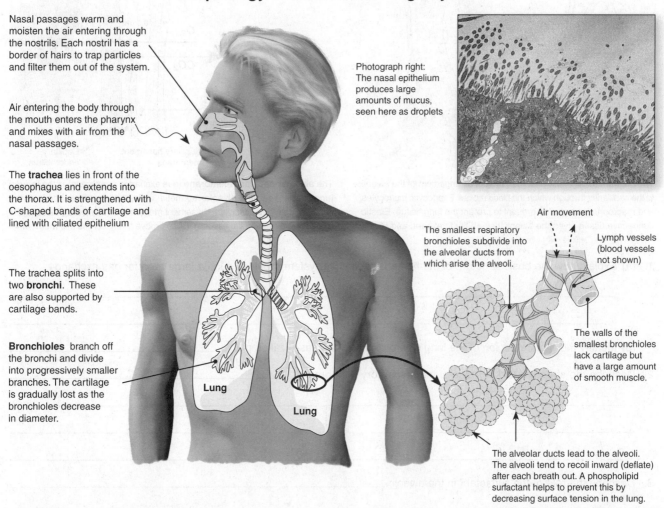

Nasal passages warm and moisten the air entering through the nostrils. Each nostril has a border of hairs to trap particles and filter them out of the system.

Air entering the body through the mouth enters the pharynx and mixes with air from the nasal passages.

The **trachea** lies in front of the oesophagus and extends into the thorax. It is strengthened with C-shaped bands of cartilage and lined with ciliated epithelium

The trachea splits into two **bronchi**. These are also supported by cartilage bands.

Bronchioles branch off the bronchi and divide into progressively smaller branches. The cartilage is gradually lost as the bronchioles decrease in diameter.

Lung

Lung

Photograph right: The nasal epithelium produces large amounts of mucus, seen here as droplets

Air movement

The smallest respiratory bronchioles subdivide into the alveolar ducts from which arise the alveoli.

Lymph vessels (blood vessels not shown)

The walls of the smallest bronchioles lack cartilage but have a large amount of smooth muscle.

The alveolar ducts lead to the alveoli. The alveoli tend to recoil inward (deflate) after each breath out. A phospholipid surfactant helps to prevent this by decreasing surface tension in the lung.

Adaptations for Rapid Gas Exchange

The human gas exchange system has several adaptations for rapid gas exchange so that it can keep up with the body's demands. Millions of tiny alveoli each act as a separate gas exchange surface. Their combined **surface area** facilitates rapid gas exchange.

The alveoli have **thin membranes** (one cell thick). The capillaries also have very thin membranes. Gas exchange between the two membranes is rapid because gases diffuse across a short distance.

Lastly, a **concentration gradient** across the membrane is required for rapid gas exchange. This is maintained by having a good blood supply that removes oxygen from the alveoli and transports it to other parts of the body, and also by a constant source of new oxygen supplied by breathing air into the lungs.

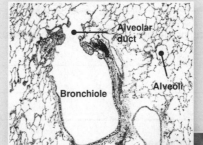

Alveolar duct

Bronchiole

Alveoli

The thin walls of the alveoli (left) and the capillaries of the lung, allow rapid diffusion of gas across between their membranes. The adult human lung contains approximately 300 million alveoli.

Breathing in air (right) brings fresh air from the outside into the lungs. This constant supply of oxygen, along with constant removal of oxygen by the capillaries, maintains a concentration gradient that promotes rapid diffusion.

An Alveolus

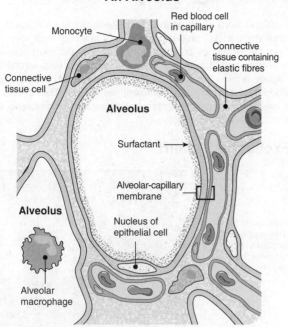

The diagram above illustrates the physical arrangement of the alveoli to the capillaries through which the blood moves. Phagocytic monocytes and macrophages are also present to protect the lung tissue. Elastic connective tissue gives the alveoli their ability to expand and recoil.

The Alveolar-Capillary Membrane

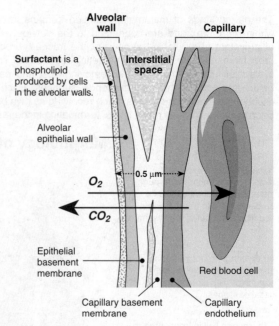

The **alveolar-capillary membrane** (gas exchange membrane) is the layered junction between the alveolar epithelial cells, the endothelial cells of the capillary, and their associated basement membranes (thin, collagenous layers underlying the epithelial tissues). Gases move freely across this membrane.

1. (a) Explain how the basic structure of the human respiratory system provides such a large area for gas exchange:

_____

_____

   (b) Identify the general region of the lung where exchange of gases takes place: _____

2. Describe the structure and purpose of the respiratory membrane: _____

_____

_____

_____

3. Describe the role of the surfactant in the alveoli: _____

_____

_____

4. Describe how the following properties improve gas exchange across the membrane surfaces of the lung:

   (a) Large surface area: _____

_____

_____

   (b) Thin, permeable membrane wall: _____

_____

_____

   (c) Concentration gradient: _____

_____

_____

KEY TERMS: Crossword

Complete the crossword below, which will test your understanding of key terms in this chapter and their meanings

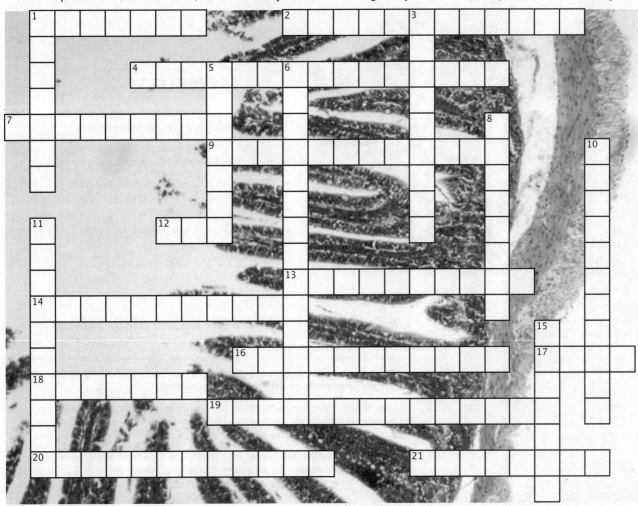

CLUES ACROSS

1. Protein embedded in the plasma membrane that provides a passage of certain molecules the cell (_____ protein).

2. This molecule has a hydrophilic end and hydrophobic end and are major components of the plasma membrane.

4. The energy-requiring movement of substances across a biological membrane against a concentration gradient (2 words 6,8).

7. The passive movement of molecules from high to low concentration.

9. The active engulfment of solid material into the cell.

12. A nucleotide comprising a purine base, a pentose sugar, and three phosphate groups, which acts as the cell's energy carrier.

13. Process by which material within a cell is enclosed in a lipid membrane and transported to the plasma membrane where it is released to the outside.

14. Active transport in which molecules are engulfed by the plasma membrane, forming a phagosome or food vacuole within the cell.

CLUES DOWN

1. Protein embedded in the plasma membrane that can transport molecules or ions across the membrane against their gradient by use of ATP or co-transport (_____ protein).

3. A solution with a lower solute concentration relative to another solution (across a membrane).

5. A transmembrane protein that moves ions across a plasma membrane against their concentration gradient (2 words: 3, 5).

6. The pressure exerted on the cell wall by the cell contents (2 words: 6, 8).

8. A term describing solutions of equal tonicity.

10. The exchange of oxygen and carbon dioxide across the gas exchange membrane (2 words: 3, 8).

11. A solution with higher solute concentration relative to another solution (across a membrane).

15. Microscopic structures in the lungs of most vertebrates that form the terminus of the bronchioles. The site of gas exchange.

Unit 1 6B101

2

Proteins, Genes and Health

KEY CONCEPTS

▶ Chains of amino acids are joined together by condensation reactions to form proteins.

▶ Protein synthesis is determined by the DNA sequence. DNA provides the blueprint for life.

▶ Mistakes in DNA replication can result in mutations. Most of these, such as cystic fibrosis are likely to be deleterious.

▶ Gene mutations can be inherited.

KEY TERMS

allele
amino acid
base pairing rule
biological catalyst
condensation reaction
cystic fibrosis
DNA
DNA polymerase
DNA replication
enzyme
fibrous protein
gene
gene therapy
genetic code
genetic screening
genotype
globular protein
hydrogen bonding
hydrolysis reaction
messenger RNA
monohybrid inheritance
mutation
nucleic acid
pedigree diagram
peptide bond
phenotype
polypeptide
prenatal testing
primary structure
protein
protein synthesis
RNA
transcription
transfer RNA
translation
triplet code

OBJECTIVES

☐ 1. Use the **KEY TERMS** to help you understand and complete these objectives.

Amino Acids, Proteins, and Enzymes   page 87-93

☐ 2. Describe the basic structure of an **amino acid**. Understand how amino acids are joined by **peptide bonds** to form **polypeptides**.

☐ 3. Explain how the **primary structure** of a **protein** determines its **three dimensional structure.** Distinguish between the structural and functional roles of **globular** or **fibrous** proteins.

☐ 4. Explain that **enzymes** are **biological catalysts**. Describe their mode of action and explain how their structure influences their specificity. Explain how enzyme concentration levels affect **reaction rates**. †

Molecular Genetics   page 94-104

☐ 5. Describe the basic structure of a **nucleic acid**.

☐ 6. Describe the **double helix model** for **DNA**, including the **base pairing rule**, and **hydrogen bonding** between **complementary strands**. Describe the structure of **RNA**.

☐ 7. Explain how **Meselson and Stahl's experiment** provided supporting evidence for the mechanism of **DNA replication**.

☐ 8. Describe how the **genetic code** is a **triplet code**.

Gene Expression, Inheritance, and Mutation   page 105-120

☐ 9. Describe **protein synthesis**, including **transcription**, **translation**, **messenger RNA**, **transfer RNA** and the role of the **template DNA strand**.

☐ 10. Explain how mistakes in DNA replication can cause **mutations**. Explain how the **cystic fibrosis mutation** can arise. Explain how the CF mutation affects functioning of the gas exchange, digestive, and reproductive systems.

☐ 11. Demonstrate and understanding of the terms; **gene**, **allele**, **genotype**, **phenotype**, **recessive**, **dominant**, **homozygote**, and **heterozygote**.

☐ 12. Explain **monohybrid inheritance** and interpret genetic **pedigree diagrams** relating to inheritance.

☐ 13. Describe the principles of **gene therapy**, differentiating between **somatic** and **germ line** therapy.

Genetic Screening   page 121-123

☐ 14. Explain how **genetic screening** can be used to detect mutations. Include **preimplantation genetic diagnosis,** and prenatal testing (**amniocentesis, chorionic villus sampling**). Appreciate that postnatal tests are also used for a variety of genetics tests for metabolic disorders.

☐ 15. Discuss the social and ethical issues associated with genetic screening.

Periodicals:
listings for this chapter are on page 229

Weblinks:
www.biozone.co.uk/ weblink/Edx-AS-2542.html

Teacher Resource CD-ROM:
Enzyme Cofactors and Inhibitors

Amino Acids

Amino acids are the basic units from which proteins are made. Plants can manufacture all the amino acids they require from simpler molecules, but animals must obtain a certain number of ready-made amino acids (called **essential amino acids**) from their diet. Which amino acids are essential varies from species to species, as different metabolisms are able to synthesize different substances. The distinction between essential and non-essential amino acids is somewhat unclear though, as some amino acids can be produced from others and some are interconvertible by the urea cycle. Amino acids can combine to form peptide chains in a **condensation reaction**. The reverse reaction, which breaks up peptide chains uses water and is called **hydrolysis**.

The Structure of Amino Acids

There are over 150 amino acids found in cells, but only 20 occur commonly in proteins. The remaining, non-protein amino acids have roles as intermediates in metabolic reactions, or as neurotransmitters and hormones. All amino acids have a common structure. The only difference between them lies with the **'R' group** in the general formula. The 'R' group is variable, meaning it is different in each kind of amino acid, and the different groups can have quite different chemical properties.

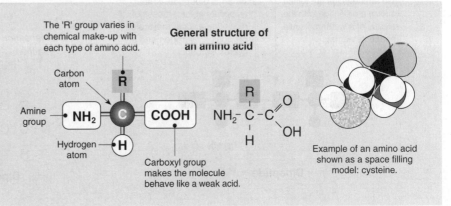

The 'R' group varies in chemical make-up with each type of amino acid.

General structure of an amino acid

Carbon atom

Amine group • **NH₂**   **C**   **COOH**

Hydrogen atom • **H**

Carboxyl group makes the molecule behave like a weak acid.

Example of an amino acid shown as a space filling model: cysteine.

The Properties of Amino Acids

Three examples of amino acids with different chemical properties are shown right, with their specific 'R' groups outlined.

All amino acids, apart from the simplest one (glycine) show **optical isomerism**, i.e. two mirror image forms exist. With a very few minor exceptions, only the **L-forms** are found in living organisms.

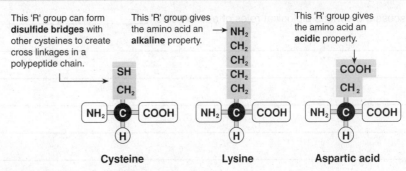

This 'R' group can form **disulfide bridges** with other cysteines to create cross linkages in a polypeptide chain.

SH
CH₂

NH₂ C COOH

H

Cysteine

This 'R' group gives the amino acid an **alkaline** property.

NH₂
CH₂
CH₂
CH₂
CH₂

NH₂ C COOH

H

Lysine

This 'R' group gives the amino acid an **acidic** property.

COOH
CH₂

NH₂ C COOH

H

Aspartic acid

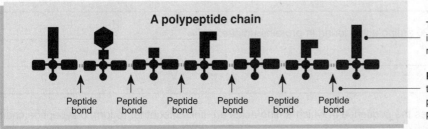

A polypeptide chain

Peptide bond · Peptide bond · Peptide bond · Peptide bond · Peptide bond · Peptide bond

The order of amino acids in a protein is directed by the order of nucleotides in DNA and mRNA.

Peptide bonds link amino acids together in long polymers called polypeptide chains. These may form part or all of a protein.

The amino acids are linked together by peptide bonds to form long chains of up to several hundred amino acids (called polypeptide chains). These chains may be functional units (complete by themselves) or they may need to be joined to other polypeptide chains before they can carry out their function. In humans, not all amino acids can be manufactured by our body: ten must be taken in with our diet (eight in adults). These are the 'essential amino acids'. They are indicated by the symbol ◆ on the right. Those indicated with as asterisk are also required by infants.

Amino acids occurring in proteins

Alanine	Glycine	Proline
Arginine *	Histidine *	Serine
Asparagine	Isoleucine ◆	Threonine ◆
Aspartic acid	Leucine ◆	Tryptophan ◆
Cysteine	Lysine ◆	Tyrosine
Glutamine	Methionine ◆	Valine ◆
Glutamic acid	Phenylalanine ◆	

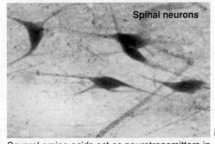

Spinal neurons

Several amino acids act as neurotransmitters in the central nervous system, Glutamic acid and ABA (gamma amino butyric acid) are the most common neurotransmitters in the brain. Others, such as glycine, are restricted to the spinal cord.

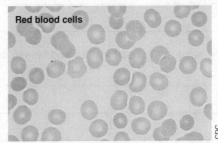

Red blood cells

Amino acids tend to stabilise the pH of solutions in which they are present (e.g. blood and tissue fluid) because they will remove excess H⁺ or OH⁻ ions. They retain this buffer capacity even when incorporated into peptides and proteins.

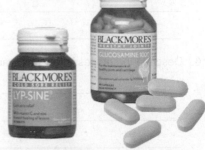

Amino acids are widely available as dietary supplements for specific purposes. Lysine is sold as relief for herpes infections and glucosamine supplements are used for alleviating the symptoms of arthritis and other joint disorders.

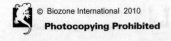

Condensation and Hydrolysis Reactions

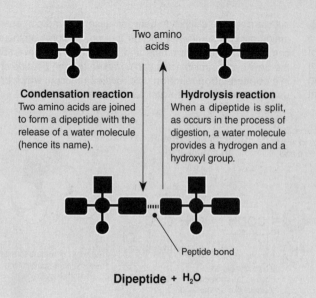

Condensation reaction
Two amino acids are joined to form a dipeptide with the release of a water molecule (hence its name).

Two amino acids

Hydrolysis reaction
When a dipeptide is split, as occurs in the process of digestion, a water molecule provides a hydrogen and a hydroxyl group.

Peptide bond

Dipeptide + H$_2$O

Amino acid   Amino acid

Condensation reaction   Hydrolysis reaction

Dipeptide + H$_2$O

1. Discuss the various biological roles of amino acids: _____

_____

_____

_____

_____

_____

_____

2. Describe what makes each of the 20 amino acids found in proteins unique: _____

_____

_____

3. Describe the process that determines the sequence in which amino acids are linked together to form polypeptide chains:

_____

_____

4. Explain how the chemistry of amino acids enables them to act as buffers in biological tissues: _____

_____

_____

5. Giving examples, explain what is meant by an **essential amino acid**: _____

_____

_____

6. Describe the processes by which amino acids are joined together and broken down: _____

_____

_____

_____

_____

_____

Proteins

The precise folding of a protein into its **tertiary structure** creates a three dimensional arrangement of the active 'R' groups. It is this structure that gives a protein its unique chemical properties. If a protein loses this precise structure (through **denaturation**), it is usually unable to carry out its biological function. Proteins can be classified on the basis of structure (e.g. globular vs fibrous) or function, as described overleaf. The entire collection of proteins in a particular cell type is termed the **cellular proteome**, while the complete proteome for an organism comprises all the various cellular proteomes. An organism's proteome is larger than its genome in the sense that there are more proteins than genes. This is the result of alternative splicing of genes and modifications made to proteins such as phosphorylation and glycosylation made after they are translated.

Primary Structure - 1° *(amino acid sequence)*

Strings of hundreds of amino acids link together with peptide bonds to form molecules called polypeptide chains. There are 20 different kinds of amino acids that can be linked together in a vast number of different combinations. This sequence is called the **primary structure**. It is the arrangement of attraction and repulsion points in the amino acid chain that determines the higher levels of organisation in the protein and its biological function.

Secondary Structure - 2° *(α-helix or β pleated sheet)*

Polypeptides become folded in various ways, referred to as the secondary (2°) structure. The most common types of 2° structures are a coiled α-**helix** and a β-**pleated sheet**. Secondary structures are maintained with hydrogen bonds between neighbouring CO and NH groups. H-bonds, although individually weak, provide considerable strength when there are a large number of them. The example, right, shows the two main types of secondary structure. In both, the **'R' side groups** (not shown) project out from the structure. Most globular proteins contain regions of α-helices together with β-sheets. Keratin (a fibrous protein) is composed almost entirely of α-helices. Fibroin (silk protein), is another fibrous protein, almost entirely in β-sheet form.

Tertiary Structure - 3° *(folding)*

Every protein has a precise structure formed by the folding of the secondary structure into a complex shape called the **tertiary structure**. The protein folds up because various points on the secondary structure are attracted to one another. The strongest links are caused by bonding between neighbouring **cysteine** amino acids which form disulfide bridges. Other interactions that are involved in folding include weak ionic and hydrogen bonds as well as hydrophobic interactions.

Quaternary Structure - 4°

Some proteins (such as enzymes) are complete and functional with a tertiary structure only. However, many complex proteins exist as aggregations of polypeptide chains. The arrangement of the polypeptide chains into a functional protein is termed the **quaternary structure**. The example (right) shows a molecule of haemoglobin, a globular protein composed of 4 polypeptide sub-units joined together; two identical **beta chains** and two identical **alpha chains**. Each has a haem (iron containing) group at the centre of the chain, which binds oxygen. Proteins containing non-protein material are **conjugated proteins**. The non-protein part is the **prosthetic group**.

Denaturation of Proteins

Denaturation refers to the loss of the three-dimensional structure (and usually also the biological function) of a protein. Denaturation is often, although not always, permanent. It results from an alteration of the bonds that maintain the secondary and tertiary structure of the protein, even though the sequence of amino acids remains unchanged. Agents that cause denaturation are:

- **Strong acids and alkalis**: Disrupt ionic bonds and result in coagulation of the protein. Long exposure also breaks down the primary structure of the protein.
- **Heavy metals**: May disrupt ionic bonds, form strong bonds with the carboxyl groups of the R groups, and reduce protein charge. The general effect is to cause the precipitation of the protein.
- **Heat and radiation** (e.g. UV): Cause disruption of the bonds in the protein through increased energy provided to the atoms.
- **Detergents and solvents**: Form bonds with the non-polar groups in the protein, thereby disrupting hydrogen bonding.

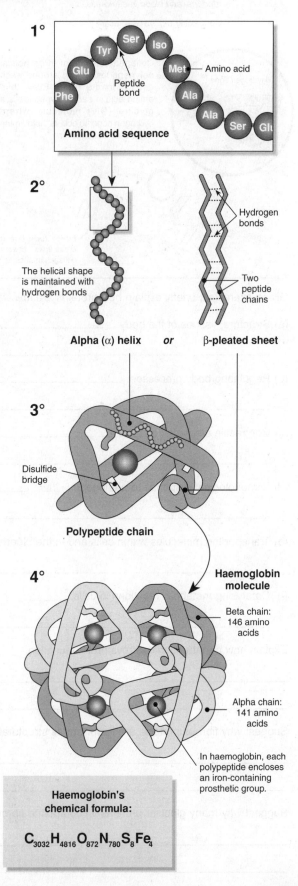

1°

Tyr — Ser — Iso — Glu — Met — Amino acid — Phe — Peptide bond — Ala — Ala — Ser — Glu

Amino acid sequence

2°

The helical shape is maintained with hydrogen bonds

Hydrogen bonds

Two peptide chains

Alpha (α) helix *or* **β-pleated sheet**

3°

Disulfide bridge

Polypeptide chain

Haemoglobin molecule

4°

Beta chain: 146 amino acids

Alpha chain: 141 amino acids

In haemoglobin, each polypeptide encloses an iron-containing prosthetic group.

Haemoglobin's chemical formula:

$$C_{3032}H_{4816}O_{872}N_{780}S_8Fe_4$$

© Biozone International 2010

Periodicals: What is tertiary structure? Modelling protein folding

Related activities: Amino Acids, The Simplest Case: Genes to Proteins
Web links: Amino Acids and Proteins

RA 2

Proteins, Genes & Health

Structural Classification of Proteins

Fibrous Proteins

Properties
- Water insoluble
- Very tough physically; may be supple or stretchy
- Parallel polypeptide chains in long fibres or sheets

Function
- Structural role in cells and organisms *e.g. collagen found in connective tissue, cartilage, bones, tendons, and blood vessel walls.*
- Contractile *e.g. myosin, actin*

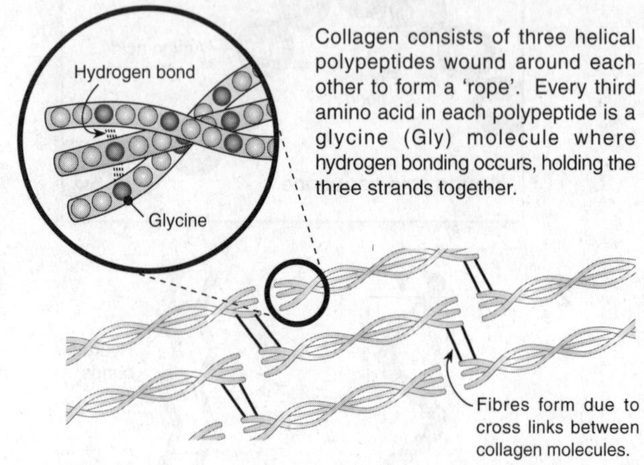

Globular Proteins

Properties
- Easily water soluble
- Tertiary structure critical to function
- Polypeptide chains folded into a spherical shape

Function
- Catalytic *e.g. enzymes*
- Regulatory *e.g. hormones (insulin)*
- Transport *e.g. haemoglobin*
- Protective *e.g. antibodies*

Collagen consists of three helical polypeptides wound around each other to form a 'rope'. Every third amino acid in each polypeptide is a glycine (Gly) molecule where hydrogen bonding occurs, holding the three strands together.

Hydrogen bond

Glycine

Fibres form due to cross links between collagen molecules.

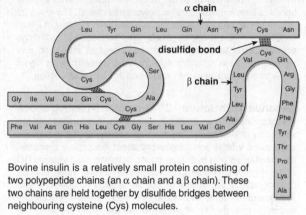

Bovine insulin is a relatively small protein consisting of two polypeptide chains (an α chain and a β chain). These two chains are held together by disulfide bridges between neighbouring cysteine (Cys) molecules.

1. Giving examples, briefly explain how proteins are involved in the following functional roles:

(a) Structural tissues of the body: _____

_____

(b) Regulating body processes: _____

_____

(c) Contractile elements: _____

_____

(d) Immunological response to pathogens: _____

_____

(e) Transporting molecules within cells and in the bloodstream: _____

_____

(f) Catalysing metabolic reactions in cells: _____

_____

2. Explain how denaturation destroys protein function: _____

_____

_____

3. Suggest why fibrous proteins are important as structural molecules in cells: _____

_____

_____

4. Suggest why many globular proteins, in contrast to fibrous proteins, have a catalytic or regulatory role:

_____

_____

Enzymes

Most enzymes are proteins. They are capable of catalysing (speeding up) biochemical reactions and are therefore called biological **catalysts**. Enzymes act on one or more compounds (called the **substrate**). They may break a single substrate molecule down into simpler substances, or join two or more substrate molecules chemically together. The enzyme itself is unchanged in the reaction; its presence merely allows the reaction to take place more rapidly. When the substrate attains the required **activation energy** to enable it to change into the product, there is a 50% chance that it will proceed forward to form the product, otherwise it reverts back to a stable form of

the reactant again. The part of the enzyme's surface into which the substrate is bound and undergoes reaction is known as the **active site**. This is made of different parts of polypeptide chain folded in a specific shape so they are closer together. For some enzymes, the complexity of the binding sites can be very precise, allowing only a single kind of substrate to bind to it. Some other enzymes have lower **specificity** and will accept a wide range of substrates of the same general type (e.g. lipases break up any fatty acid chain length of lipid). This is because the enzyme is specific for the type of chemical bond involved and not an exact substrate.

Enzyme Structure

The model on the right is of an enzyme called *Ribonuclease S*, that breaks up RNA molecules. It is a typical enzyme, being a globular protein and composed of up to several hundred atoms. The darkly shaded areas are called **active sites** and make up the **cleft**; the region into which the substrate molecule(s) are drawn. The correct positioning of these sites is critical for the catalytic reaction to occur. The substrate (RNA in this case) is drawn into the cleft by the active sites. By doing so, it puts the substrate molecule under stress, causing the reaction to proceed more readily.

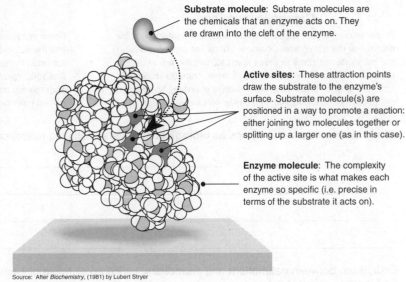

Substrate molecule: Substrate molecules are the chemicals that an enzyme acts on. They are drawn into the cleft of the enzyme.

Active sites: These attraction points draw the substrate to the enzyme's surface. Substrate molecule(s) are positioned in a way to promote a reaction: either joining two molecules together or splitting up a larger one (as in this case).

Enzyme molecule: The complexity of the active site is what makes each enzyme so specific (i.e. precise in terms of the substrate it acts on).

Source: After *Biochemistry*, (1981) by Lubert Stryer

How Enzymes Work

The **lock and key** model proposed earlier last century suggested that the substrate was simply drawn into a closely matching cleft on the enzyme molecule. More recent studies have revealed that the process more likely involves an **induced fit** (see diagram on the right), where the enzyme or the reactants change their shape slightly. The reactants become bound to enzymes by weak chemical bonds. This binding can weaken bonds within the reactants themselves, allowing the reaction to proceed more readily.

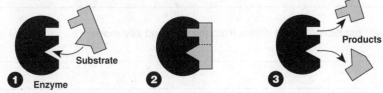

The presence of an enzyme simply makes it easier for a reaction to take place. All **catalysts** speed up reactions by influencing the stability of bonds in the reactants. They may also provide an alternative reaction pathway, thus lowering the activation energy needed for a reaction to take place (see the graph below).

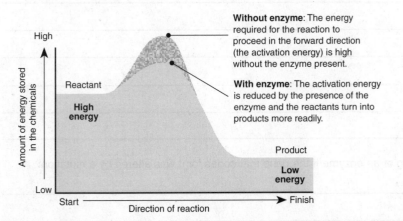

Without enzyme: The energy required for the reaction to proceed in the forward direction (the activation energy) is high without the enzyme present.

With enzyme: The activation energy is reduced by the presence of the enzyme and the reactants turn into products more readily.

Induced Fit Model

An enzyme fits to its substrate somewhat like a lock and key. The shape of the enzyme changes when the substrate fits into the cleft (called the **induced fit**):

1 Two substrate molecules are drawn into the cleft of the enzyme.

2 The enzyme changes shape, forcing the substrate molecules to combine.

3 The resulting end product is released by the enzyme which returns to its normal shape, ready to receive more.

Proteins, Genes & Health

© Biozone International 2010
Photocopying Prohibited

Periodicals: Enzymes: nature's catalytic machines

Related activities: Enzyme Reaction Rates
Web links: How Enzymes Work, Enzyme Cofactors and Inhibitors

RA 2

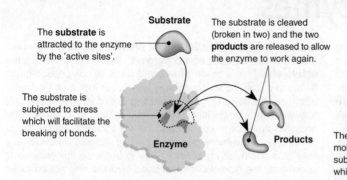

The **substrate** is attracted to the enzyme by the 'active sites'.

The substrate is subjected to stress which will facilitate the breaking of bonds.

The substrate is cleaved (broken in two) and the two **products** are released to allow the enzyme to work again.

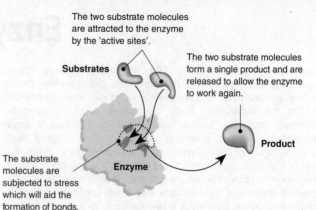

The two substrate molecules are attracted to the enzyme by the 'active sites'.

The two substrate molecules form a single product and are released to allow the enzyme to work again.

The substrate molecules are subjected to stress which will aid the formation of bonds.

Catabolic reactions

Some enzymes can cause a single substrate molecule to be drawn into the active site. Chemical bonds are broken, causing the substrate molecule to break apart to become two separate molecules. Catabolic reactions break down complex molecules into simpler ones and involve a net release of energy, so they are called exergonic. **Examples**: *hydrolysis, cellular respiration*.

Anabolic reactions

Some enzymes can cause two substrate molecules to be drawn into the active site. Chemical bonds are formed, causing the two substrate molecules to form bonds and become a single molecule. Anabolic reactions involve the net use of energy (they are endergonic) and build more complex molecules and structures from simpler ones. **Examples**: *protein synthesis, photosynthesis*.

1. Give a brief account of enzymes as **biological catalysts**, including reference to the role of the **active site**:

_____

_____

_____

_____

2. Distinguish between **catabolism** and **anabolism**, giving an example of each and identifying each reaction as **endergonic** or **exergonic**:

_____

_____

_____

3. Outline the key features of the '**lock and key**' model of enzyme action: _____

_____

_____

4. Outline the '**induced fit**' model of enzyme action, explaining how it differs from the lock and key model:

_____

_____

_____

5. Describe two factors that could cause enzyme denaturation, explaining how they exert their effects:

(a) _____

_____

_____

_____

(b) _____

_____

_____

_____

6. Explain what might happen to the functioning of an enzyme if the gene that codes for it was altered by a mutation:

_____

_____

Enzyme Reaction Rates

Enzymes are sensitive molecules. They often have a narrow range of conditions under which they operate properly. For most of the enzymes associated with plant and animal metabolism, there is little activity at low temperatures. As the temperature increases, so too does the enzyme activity, until the point is reached where the temperature is high enough to damage the enzyme's structure. At this point, the enzyme ceases to function; a phenomenon called enzyme or protein **denaturation**.

Extremes in acidity (pH) can also cause the protein structure of enzymes to denature. Poisons often work by denaturing enzymes or occupying the enzyme's active site so that it does not function. In some cases, enzymes will not function without cofactors, such as vitamins or trace elements. In the four graphs below, the rate of reaction or degree of enzyme activity is plotted against each of four factors that affect enzyme performance. Answer the questions relating to each graph:

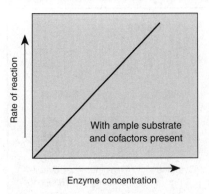

1. **Enzyme concentration**

   (a) Describe the change in the rate of reaction when the enzyme concentration is increased (assuming there is plenty of the substrate present):

   _____

   _____

   (b) Suggest how a cell may vary the amount of enzyme present in a cell:

   _____

   _____

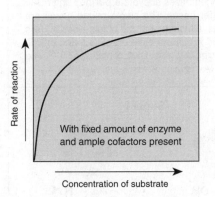

2. **Substrate concentration**

   (a) Describe the change in the rate of reaction when the substrate concentration is **increased** (assuming a fixed amount of enzyme and ample cofactors):

   _____

   _____

   (b) Explain why the rate changes the way it does: _____

   _____

   _____

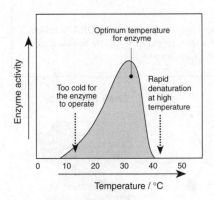

3. **Temperature**

   Higher temperatures speed up all reactions, but few enzymes can tolerate temperatures higher than 50–60°C. The rate at which enzymes are **denatured** (change their shape and become inactive) increases with higher temperatures.

   (a) Describe what is meant by an optimum temperature for enzyme activity:

   _____

   _____

   (b) Explain why most enzymes perform poorly at low temperatures:

   _____

   _____

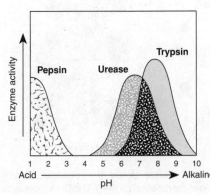

4. **pH (acidity/alkalinity)**

   Like all proteins, enzymes are **denatured** by extremes of **pH** (very acid or alkaline). Within these extremes, most enzymes are still influenced by pH. Each enzyme has a preferred pH range for optimum activity.

   (a) State the optimum pH for each of the enzymes:

   Pepsin: _____ Trypsin: _____ Urease: _____

   (b) Pepsin acts on proteins in the stomach. Explain how its optimum pH is suited to its working environment:

   _____

   _____

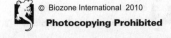
Proteins, Genes & Health

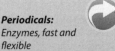

Nucleic Acids

Nucleic acids are a special group of chemicals in cells concerned with the transmission of inherited information. They have the capacity to store the information that controls cellular activity. The central nucleic acid is called **deoxyribonucleic acid** (DNA). DNA is a major component of chromosomes and is found primarily in the nucleus, although a small amount is found in mitochondria and chloroplasts. Other **ribonucleic acids** (RNA) are involved in the 'reading' of the DNA information. All nucleic acids are made up of simple repeating units called **nucleotides**, linked together to form chains or strands, often of great length. The strands vary in the sequence of the bases found on each nucleotide. It is this sequence which provides the 'genetic code' for the cell. In addition to nucleic acids, certain nucleotides and their derivatives are also important as suppliers of energy (**ATP**) or as hydrogen ion and electron carriers in respiration and photosynthesis (NAD, NADP, and FAD).

Chemical Structure of a Nucleotide

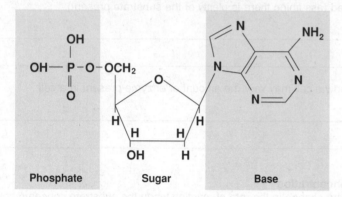

Phosphate   Sugar   Base

Symbolic Form of a Nucleotide

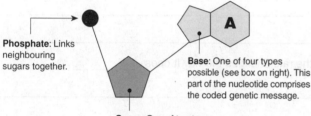

Phosphate: Links neighbouring sugars together.

Base: One of four types possible (see box on right). This part of the nucleotide comprises the coded genetic message.

Sugar: One of two types possible: ribose in RNA and deoxyribose in DNA.

Nucleotides are the building blocks of DNA. Their precise sequence in a DNA molecule provides the genetic instructions for the organism to which it governs. Accidental changes in nucleotide sequences are a cause of mutations, usually harming the organism, but occasionally providing benefits.

Bases

Purines:

A — Adenine   G — Guanine

Pyrimidines:

C — Cytosine   T — Thymine *(DNA only)*   U — Uracil *(RNA only)*

The two-ringed bases above are **purines** and make up the longer bases. The single-ringed bases are **pyrimidines**. Although only one of four kinds of base can be used in a nucleotide, **uracil** is found only in RNA, replacing **thymine**. DNA contains: A, T, G, and C, while RNA contains A, U, G, and C.

Sugars

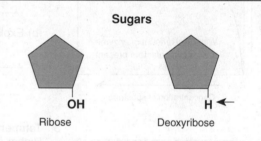

Ribose   Deoxyribose

Deoxyribose sugar is found only in DNA. It differs from **ribose** sugar, found in RNA, by the lack of a single oxygen atom (arrowed).

RNA Molecule

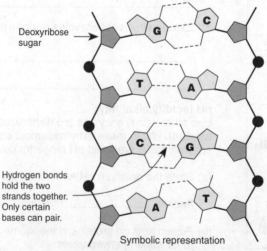

In RNA, uracil replaces thymine in the code.

Ribose sugar

Ribonucleic acid (RNA) comprises a *single strand* of nucleotides linked together.

DNA Molecule

Deoxyribose sugar

Hydrogen bonds hold the two strands together. Only certain bases can pair.

Symbolic representation

Deoxyribonucleic acid (DNA) comprises a *double strand* of nucleotides linked together. It is shown unwound in the symbolic representation (left). The DNA molecule takes on a twisted, double helix shape as shown in the space filling model on the right.

DNA Molecule

Space filling model

Related activities: *Creating a DNA Molecule*
Web links: *The Structure of DNA*

Periodicals:
DNA: 50 years of the double helix

Formation of a nucleotide

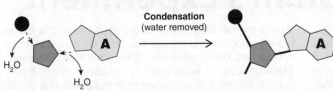

Condensation
(water removed)

A nucleotide is formed when phosphoric acid and a base are chemically bonded to a sugar molecule. In both cases, water is given off, and they are therefore condensation reactions. In the reverse reaction, a nucleotide is broken apart by the addition of water (**hydrolysis**).

Formation of a dinucleotide

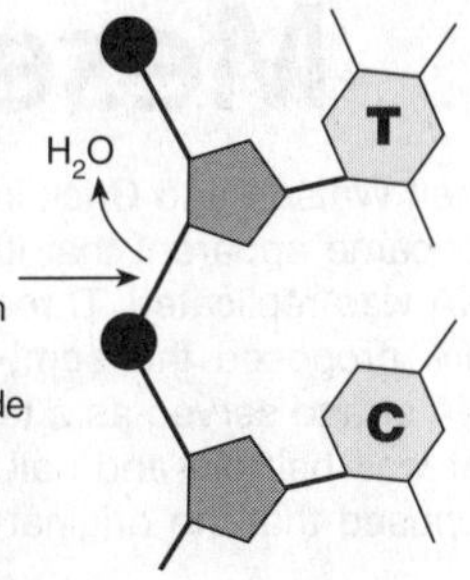

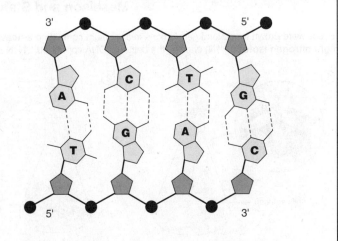

Two nucleotides are linked together by a condensation reaction between the phosphate of one nucleotide and the sugar of another.

Double-Stranded DNA

The **double-helix** structure of DNA is like a ladder twisted into a corkscrew shape around its longitudinal axis. It is 'unwound' here to show the relationships between the bases.

- The way the correct pairs of bases are attracted to each other to form hydrogen bonds is determined by the number of bonds they can form and the shape (length) of the base.

- The **template strand** the side of the DNA molecule that stores the information that is transcribed into mRNA. The template strand is also called the **antisense strand**.

- The other side (often called the **coding strand**) has the same nucleotide sequence as the mRNA except that T in DNA substitutes for U in mRNA. The coding strand is also called the **sense strand**.

1. The diagram above depicts a double-stranded DNA molecule. Label the following parts on the diagram:
   (a) **Sugar** (deoxyribose)
   (b) **Phosphate**
   (c) **Hydrogen bonds** (between bases)
   (d) **Purine** bases
   (e) **Pyrimidine** bases

2. (a) Explain the **base-pairing rule** that applies in double-stranded DNA: _____

   _____

   (b) Explain how this differs in mRNA: _____

   (c) Describe the purpose of the hydrogen bonds in double-stranded DNA: _____

   _____

3. Describe the functional role of nucleotides: _____

   _____

4. Distinguish between the **template strand** and **coding strand** of DNA, identifying the functional role of each:

   _____

   _____

   _____

5. Complete the following table summarising the differences between DNA and RNA molecules:

	DNA	RNA
Sugar present		
Bases present		
Number of strands		
Relative length		

Meselson and Stahl's Experiment

When Watson and Crick identified the structure of DNA in 1953 it became apparent that its structure could help to explain how DNA was replicated. Three models were proposed. Watson and Crick proposed the **semi-conservative model** in which each DNA strand served as a template, forming a new DNA molecule that was half old and half new DNA. The **conservative model** proposed that the original DNA served as a complete template

so that the resulting DNA was completely new. The **dispersive model** proposed that the two new DNA molecules had part new and part old DNA interspersed throughout them. **Meselson and Stahl** devised an experiment to determine which theory was correct. By using *E. coli* grown in differing isotopes of nitrogen, Meselson and Stahl provided evidence that confirmed Watson and Crick's semi-conservative model of DNA replication (below).

Meselson and Stahl's Experiment

E. coli were grown for several generations in a medium containing a **heavy nitrogen isotope** (^{15}N) and transferred to a medium containing a **light nitrogen isotope** (^{14}N) once all the bacterial DNA contained ^{15}N. Newly synthesised DNA would contain ^{14}N, old DNA would contain ^{15}N.

①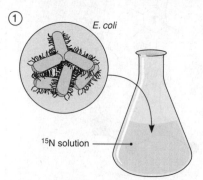

E. coli were grown in a nutrient solution containing ^{15}N. After 14 generations, all the bacterial DNA contained ^{15}N. A sample is removed. This is **generation 0**.

②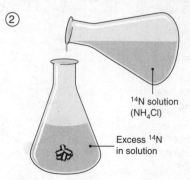

Generation 0 is added to a solution containing excess ^{14}N (as NH_4Cl). During replication, new DNA will incorporate ^{14}N and be 'lighter' than the original DNA (which contains only ^{15}N).

③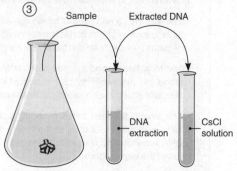

Every generation (~ 20 minutes), a sample is taken and treated to release the DNA. The DNA is placed in a CsCl solution which provides a density gradient for separation of the DNA.

④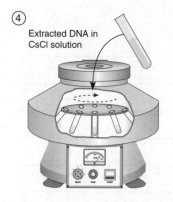

Samples are spun in a high speed ultracentrifuge at 140,000 *g* for 20 hours. Heavier ^{15}N DNA moves closer to the bottom of the test tube than light ^{14}N DNA or $^{14}N/^{15}N$ intermediate DNA.

⑤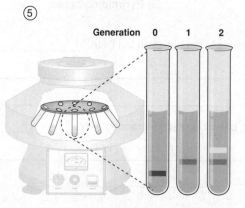

All the DNA in the generation 0 sample moved to the bottom of the test tube. All the DNA in the generation 1 sample moved to an intermediate position. At generation 2 half the DNA was at the intermediate position and half was near the top of the test tube. In subsequent generations, more DNA was near the top and less was in the intermediate position.

⑥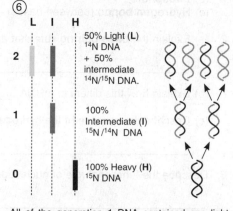

All of the generation 1 DNA contained one light strand (^{14}N) and one heavy (^{15}N) strand to produce an intermediate density. At generation 2, 50% of the DNA was light and 50% was intermediate DNA. This combination of light and intermediate (hybrid) DNA confirmed the semi conservative replication model.

1. Explain why Meselson and Stahl's experiment supports the semi-conservative replication model: _____

_____

_____

_____

2. Identify the replication model that fits the following hypothetical data:

(a) 100% of generation 0 is "heavy DNA", 50% of generation 1 is "heavy" and 50% is "light", and 25% of generation 2 is "heavy" and 75% is "light":

_____

(b) 100% of generation 0 is "heavy DNA", 100% of generation 1 is "intermediate DNA", and 100% generation 2 lies between the "intermediate" and "light" DNA regions:

_____

Related activities: DNA Replication
Web links: Meselson and Stahl Animation

DNA Replication

The replication of DNA is a necessary preliminary step for cell division (both mitosis and meiosis). This process creates the **two chromatids** that are found in chromosomes that are preparing to divide. By this process, the whole chromosome is essentially duplicated, but is still held together by a common centromere. Enzymes are responsible for all of the key events. The diagram below shows the essential steps in the process. The diagram on the facing page shows how enzymes are involved at each stage.

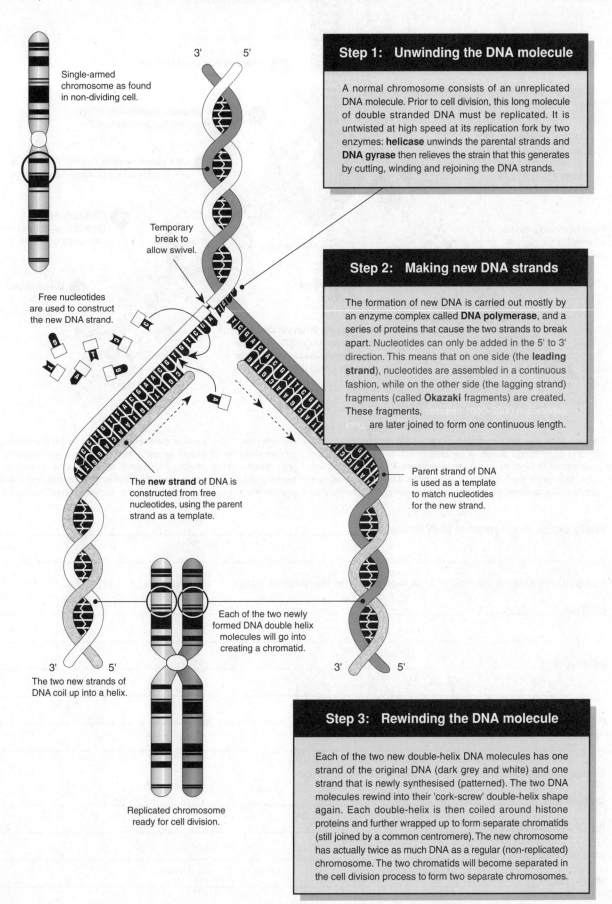

Single-armed chromosome as found in non-dividing cell.

3' 5'

Temporary break to allow swivel.

Free nucleotides are used to construct the new DNA strand.

The **new strand** of DNA is constructed from free nucleotides, using the parent strand as a template.

Parent strand of DNA is used as a template to match nucleotides for the new strand.

Each of the two newly formed DNA double helix molecules will go into creating a chromatid.

3' 5'

The two new strands of DNA coil up into a helix.

3' 5'

Replicated chromosome ready for cell division.

Step 1:   Unwinding the DNA molecule

A normal chromosome consists of an unreplicated DNA molecule. Prior to cell division, this long molecule of double stranded DNA must be replicated. It is untwisted at high speed at its replication fork by two enzymes: **helicase** unwinds the parental strands and **DNA gyrase** then relieves the strain that this generates by cutting, winding and rejoining the DNA strands.

Step 2:   Making new DNA strands

The formation of new DNA is carried out mostly by an enzyme complex called **DNA polymerase**, and a series of proteins that cause the two strands to break apart. Nucleotides can only be added in the 5' to 3' direction. This means that on one side (the **leading strand**), nucleotides are assembled in a continuous fashion, while on the other side (the lagging strand) fragments (called **Okazaki** fragments) are created. These fragments, between 1000–2000 nucleotides long, are later joined to form one continuous length.

Step 3:   Rewinding the DNA molecule

Each of the two new double-helix DNA molecules has one strand of the original DNA (dark grey and white) and one strand that is newly synthesised (patterned). The two DNA molecules rewind into their 'cork-screw' double-helix shape again. Each double-helix is then coiled around histone proteins and further wrapped up to form separate chromatids (still joined by a common centromere). The new chromosome has actually twice as much DNA as a regular (non-replicated) chromosome. The two chromatids will become separated in the cell division process to form two separate chromosomes.

Proteins, Genes & Health

Periodicals:
DNA polymerase:
replication and application

Related activities: Mitosis and the Cell Cycle,
Web links: DNA Replication

DA 3

Enzyme Control of DNA Replication

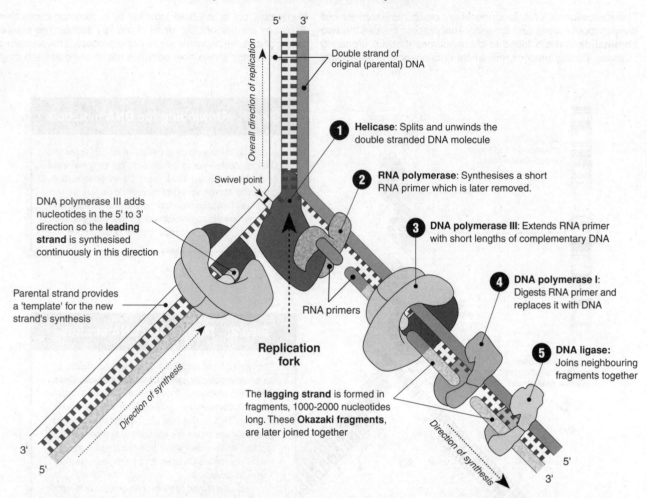

5' 3'

Overall direction of replication

Double strand of original (parental) DNA

1 **Helicase**: Splits and unwinds the double stranded DNA molecule

Swivel point

2 **RNA polymerase**: Synthesises a short RNA primer which is later removed.

DNA polymerase III adds nucleotides in the 5' to 3' direction so the **leading strand** is synthesised continuously in this direction

3 **DNA polymerase III**: Extends RNA primer with short lengths of complementary DNA

Parental strand provides a 'template' for the new strand's synthesis

RNA primers

4 **DNA polymerase I**: Digests RNA primer and replaces it with DNA

Replication fork

5 **DNA ligase**: Joins neighbouring fragments together

Direction of synthesis

The **lagging strand** is formed in fragments, 1000-2000 nucleotides long. These **Okazaki fragments**, are later joined together

Direction of synthesis

3'
5'

5'
3'

The sequence of enzyme controlled events in DNA replication is shown above (1-5). Although shown as separate, many of the enzymes are found clustered together as enzyme complexes. These enzymes are also able to 'proof-read' the new DNA strand as it is made and correct mistakes. The polymerase enzyme can only work in one direction, so that one new strand is constructed as a continuous length (the leading strand) while the other new strand is made in short segments to be later joined together (the lagging strand). **NOTE** that the nucleotides are present as deoxynucleoside triphosphates. When hydrolysed, these provide the energy for incorporating the nucleotide into the strand.

1. Briefly explain the purpose of DNA replication: _____

_____

2. Summarise the steps involved in DNA replication (on the previous page):

   (a) Step 1: _____

   _____

   (b) Step 2: _____

   _____

   (c) Step 3: _____

   _____

3. Explain the role of the following enzymes in DNA replication:

   (a) Helicase: _____

   (b) DNA polymerase I: _____

   (c) DNA polymerase III: _____

   (d) Ligase: _____

4. Determine the time it would take for a bacteria to replicate its DNA (see note in diagram above): _____

Creating a DNA Model

Although DNA molecules can be enormous in terms of their molecular size, they are made up of simple repeating units called **nucleotides**. A number of factors control the way in which these nucleotide building blocks are linked together. These factors cause the nucleotides to join together in a predictable way. This is referred to as the **base pairing rule** and can be used to construct a complementary DNA strand from a template strand, as illustrated in the exercise below:

DNA Base Pairing Rule			
Adenine	is always attracted to	**Thymine**	A ⟷ T
Thymine	is always attracted to	**Adenine**	T ⟷ A
Cytosine	is always attracted to	**Guanine**	C ⟷ G
Guanine	is always attracted to	**Cytosine**	G ⟷ C

1. Cut out page 101 and separate each of the 24 nucleotides by cutting along the columns and rows (see arrows indicating two such cutting points). Although drawn as geometric shapes, these symbols represent chemical structures.

2. Place one of each of the four kinds of nucleotide on their correct spaces below:

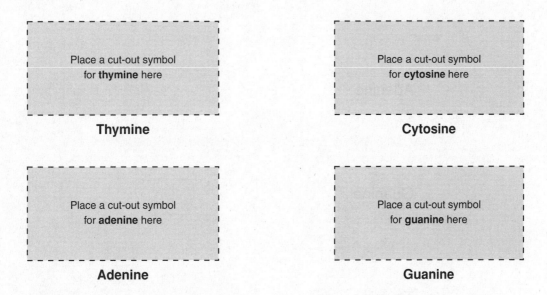

Place a cut-out symbol for **thymine** here

Thymine

Place a cut-out symbol for **cytosine** here

Cytosine

Place a cut-out symbol for **adenine** here

Adenine

Place a cut-out symbol for **guanine** here

Guanine

3. Identify and **label** each of the following features on the *adenine* nucleotide immediately above: **phosphate**, **sugar**, **base**, **hydrogen bonds**

4. Create one strand of the DNA molecule by placing the 9 correct 'cut out' nucleotides in the labelled spaces on the following page (DNA molecule). Make sure these are the right way up (with the **P** on the left) and are aligned with the left hand edge of each box. Begin with thymine and end with guanine.

5. Create the complementary strand of DNA by using the base pairing rule above. Note that the nucleotides have to be arranged upside down.

6. Under normal circumstances, it is not possible for adenine to pair up with guanine or cytosine, nor for any other mismatches to occur. Describe the two factors that prevent a mismatch from occurring:

(a) Factor 1: _____

_____

(b) Factor 2: _____

_____

7. Once you have checked that the arrangement is correct, you may glue, paste or tape these nucleotides in place.

NOTE:	There may be some value in keeping these pieces loose in order to practise the base pairing rule. For this purpose, *removable tape* would be best.

Proteins, Genes & Health

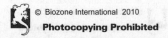
© Biozone International 2010
Photocopying Prohibited

Periodicals:
DNA: 50 years of the double helix

Related activities: Nucleic Acids

PA 2

DNA Molecule

Put the named nucleotides on the left hand side to create the template strand

Put the matching **complementary** nucleotides opposite the template strand

Thymine

Cytosine

Adenine

Adenine

Guanine

Thymine

Thymine

Cytosine

Guanine

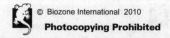

© Biozone International 2010

Photocopying Prohibited

Nucleotides

Tear out this page along the perforation and separate each of the 24 nucleotides by cutting along the columns and rows (see arrows indicating the cutting points).

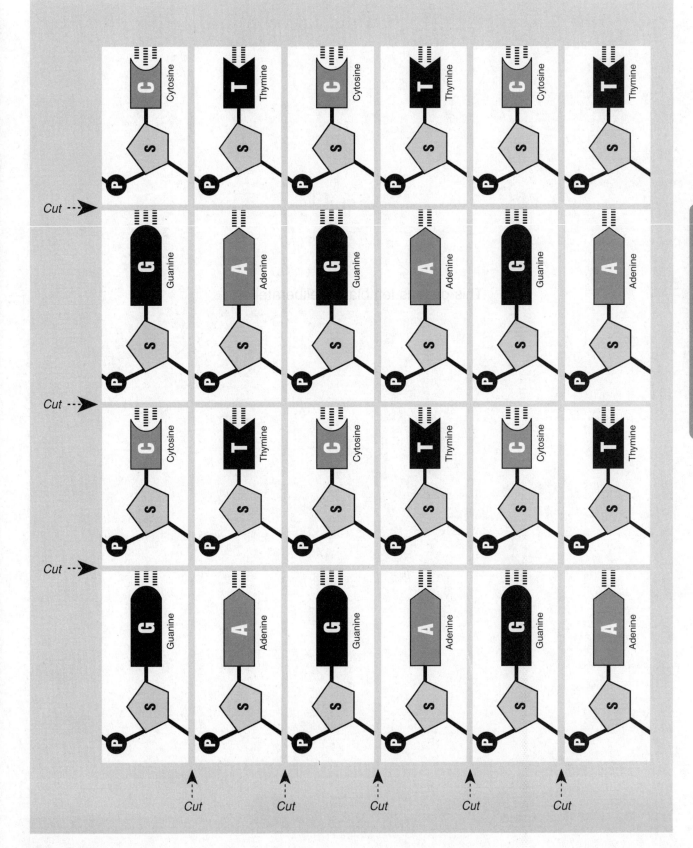

Cut ➤

Cut ➤

Cut ➤

Cut   Cut   Cut   Cut   Cut

Proteins, Genes & Health

This page is left blank deliberately

Review of DNA Replication

The diagram below summarises the main steps in DNA replication. You should use this activity to test your understanding of the main features of DNA replication, using the knowledge gained in the previous activity to fill in the missing information. You should attempt this from what you have learned, but refer to the previous activity if you require help.

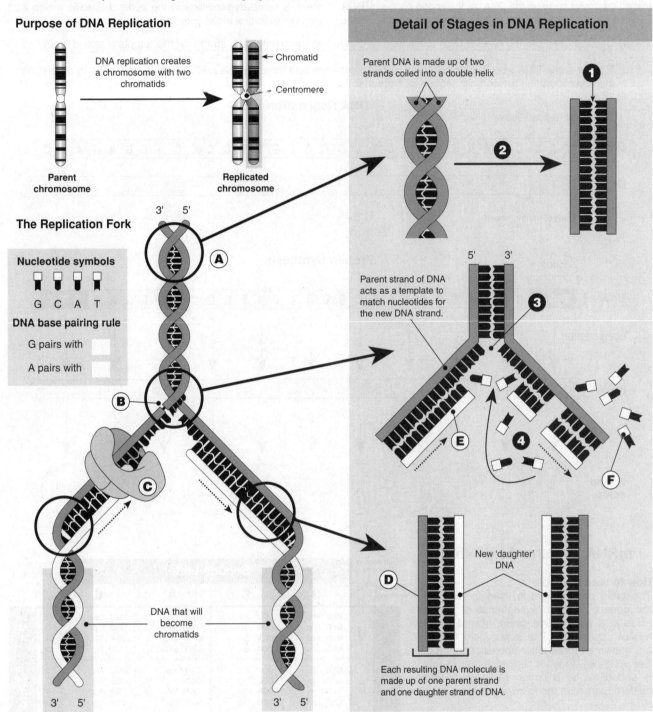

Purpose of DNA Replication

DNA replication creates a chromosome with two chromatids

← Chromatid

← Centromere

Parent chromosome

Replicated chromosome

Detail of Stages in DNA Replication

Parent DNA is made up of two strands coiled into a double helix

Parent strand of DNA acts as a template to match nucleotides for the new DNA strand.

New 'daughter' DNA

Each resulting DNA molecule is made up of one parent strand and one daughter strand of DNA.

The Replication Fork

3'   5'

Nucleotide symbols

G C A T

DNA base pairing rule

G pairs with

A pairs with

DNA that will become chromatids

3'   5'   3'   5'

Proteins, Genes & Health

1. In the white boxes provided in the diagram, state the base pairing rule for making a strand of DNA:

2. Identify each of the structures marked with a letter. (A-F):

   A: _____   C: _____   E: _____

   B: _____   D: _____   F: _____

3. Match each of the processes (1-4) to the correct summary of the process provided below:

   ☐ Unwinding of parent DNA double helix   ☐ Unzipping of parent DNA

   ☐ Free nucleotides occupy spaces alongside exposed bases   ☐ DNA strands are joined by base pairing

The Genetic Code

The genetic code consists of the sequence of bases arranged along the DNA molecule. It consists of a four-letter alphabet (from the four kinds of bases) and is read as three-letter words (called the triplet code on the DNA, or the codon on the mRNA). Each of the different kinds of triplet codes for a specific amino

acid. This code is represented at the bottom of the page in the **mRNA-amino acid table**. There are 64 possible combinations of the four kinds of bases making up three-letter words. As a result, there is some **degeneracy** in the code; a specific amino acid may have several triplet codes.

1. (a) Use the base-pairing rule for DNA replication to create the complementary strand for the template strand below.

   (b) For the same DNA strand, determine the mRNA sequence and then use the mRNA–amino acid table to determine the corresponding amino acid sequence. Note that in mRNA, uracil (U) replaces thymine (T) and pairs with adenine.

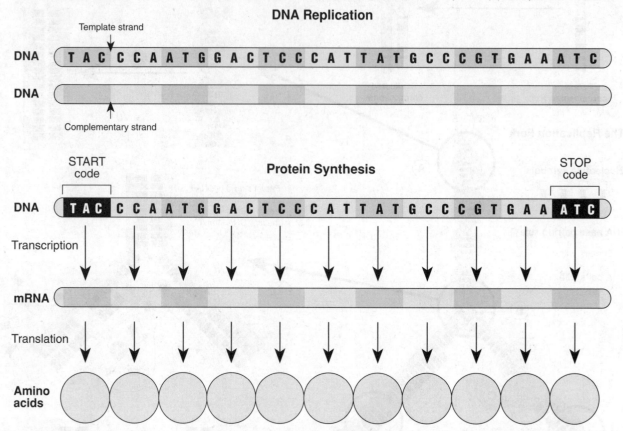

mRNA - Amino Acid Table

How to read the table

The table on the right is used to 'decode' the genetic code as a sequence of amino acids in a polypeptide chain, from a given mRNA sequence. The amino acid names are shown as three letter abbreviations (e.g. Ser = serine). To work out which amino acid is coded for by a codon (3 bases in the mRNA), carry out the following steps:

i   Look for the first letter of the codon in the row on the left hand side of the table.

ii Look for the column that intersects the same row from above that matches the second base.

iii Locate the third base in the codon by looking along the row on the right hand side that matches your codon.

	Second Letter				
First Letter	**U**	**C**	**A**	**G**	**Third Letter**
U	UUU Phe UUC Phe UUA Leu UUG Leu	UCU Ser UCC Ser UCA Ser UCG Ser	UAU Tyr UAC Tyr UAA STOP UAG STOP	UGU Cys UGC Cys UGA STOP UGG Try	U C A G
C	CUU Leu CUC Leu CUA Leu CUG Leu	CCU Pro CCC Pro CCA Pro CCG Pro	CAU His CAC His CAA Gln CAG Gln	CGU Arg CGC Arg CGA Arg CGG Arg	U C A G
A	AUU Iso AUC Iso AUA Iso AUG Met	ACU Thr ACC Thr ACA Thr ACG Thr	AAU Asn AAC Asn AAA Lys AAG Lys	AGU Ser AGC Ser AGA Arg AGG Arg	U C A G
G	GUU Val GUC Val GUA Val GUG Val	GCU Ala GCC Ala GCA Ala GCG Ala	GAU Asp GAC Asp GAA Glu GAG Glu	GGU Gly GGC Gly GGA Gly GGG Gly	U C A G

Read first letter here • Read second letter here • Read third letter here

Example: **GAU** codes for Asp (asparagine)

2. (a) State the mRNA START and STOP codons: _____

   (b) Describe the function of the START and STOP codons in a mRNA sequence: _____

   _____

Related activities: DNA Replication

The Simplest Case: Genes to Proteins

The traditionally held view of genes was as sections of DNA coding only for protein. This view has been revised in recent years with the discovery that much of the nonprotein-coding DNA encodes functional RNAs; it is not all non-coding "junk" DNA as was previously assumed. In fact, our concept of what constitutes a gene is changing rapidly and now encompasses all those segments of DNA that are transcribed (to RNA). This activity considers only the simplest scenario: one in which the gene codes for a functional protein. **Nucleotides**, the basic unit

of genetic information, are read in groups of three (**triplets**). Some triplets have a special controlling function in the making of a polypeptide chain. The equivalent of the triplet on the mRNA molecule is the **codon**. Three codons can signify termination of the amino acid chain (UAG, UAA and UGA in the mRNA code). The codon AUG is found at the beginning of every gene (on mRNA) and marks the starting point for reading the gene. The genes required to form a functional end-product (in this case, a functional protein) are collectively called a **transcription unit**.

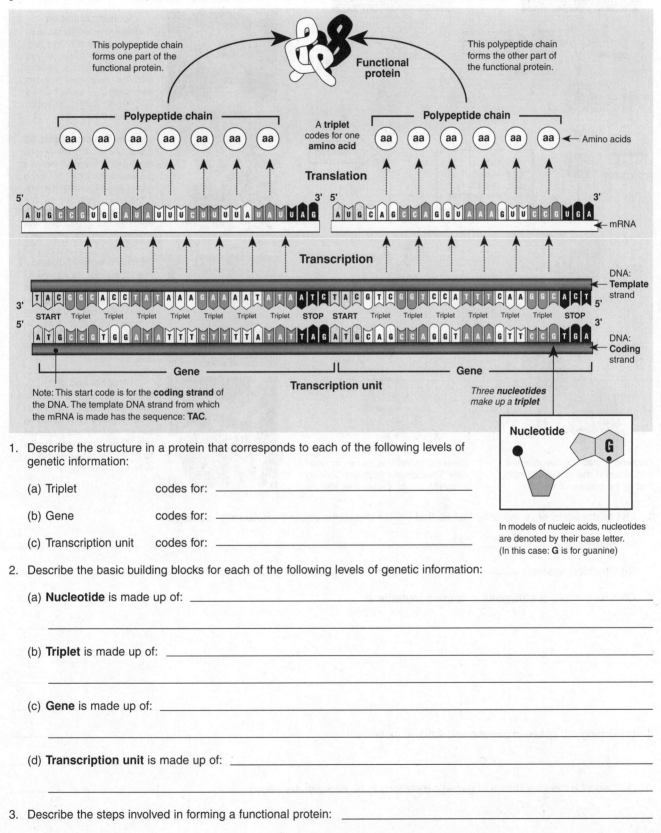

In models of nucleic acids, nucleotides are denoted by their base letter. (In this case: **G** is for guanine)

Note: This start code is for the **coding strand** of the DNA. The template DNA strand from which the mRNA is made has the sequence: **TAC**.

Three **nucleotides** make up a **triplet**

Proteins, Genes & Health

1. Describe the structure in a protein that corresponds to each of the following levels of genetic information:

   (a) Triplet   codes for: _____

   (b) Gene   codes for: _____

   (c) Transcription unit   codes for: _____

2. Describe the basic building blocks for each of the following levels of genetic information:

   (a) **Nucleotide** is made up of: _____

   _____

   (b) **Triplet** is made up of: _____

   _____

   (c) **Gene** is made up of: _____

   _____

   (d) **Transcription unit** is made up of: _____

   _____

3. Describe the steps involved in forming a functional protein: _____

   _____

Causes of Mutations

A **mutation** is a permanent change in the DNA sequence of a gene. Mutations occur in all organisms spontaneously. The natural rate at which a gene will undergo a change is normally very low, but this rate can be increased by the actions of **mutagens**. A mutagen is a physical or chemical agent that changes the genetic material. Only those mutations taking place in cells producing gametes will be inherited. These are termed **gametic mutations.** If mutations occur in a body cell after the organism has begun to develop beyond the zygote (fertilised egg cell) stage, they are called **somatic mutations**. Such mutations usually give rise to **chimaeras** (mixture of cell types, some with the mutation and some without, in the same organism). In some cases a mutation may trigger the onset of **cancer**, if the normal controls over gene regulation and expression are disrupted.

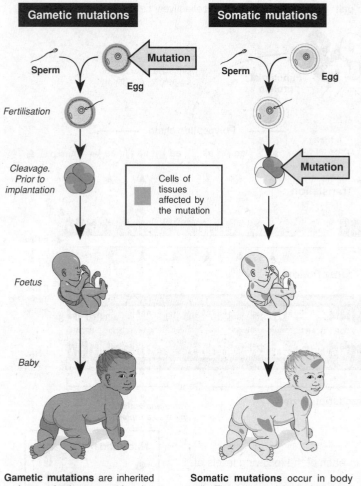

Gametic mutations are inherited and occur in the testes of males and in the ovaries of females.

Somatic mutations occur in body cells. They are not inherited but may affect the person during their lifetime.

Mutagens and their Effects

Ionising radiation
Nuclear and ultraviolet radiation, gamma rays, and X-rays (e.g. from medical treatment) are forms of high energy, ionising radiation associated with the development of some cancers (e.g. thyroid cancer, leukemia, and skin cancer).

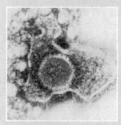

Viruses & microorganisms
Some viruses, e.g. the herpes virus (left), integrate into the human chromosome, disrupting genes and triggering cancers. Aflatoxins, produced by the fungus *Aspergillus flavus,* can contaminate grain, and are potent inducers of liver cancer.

Environmental poisons
Many chemicals are mutagenic. Synthetic and natural examples include organic solvents such as carbon tetrachloride (used in dry cleaning fluid) and formaldehyde, tobacco and coal tars, benzene, asbestos, vinyl chlorides, some dyes, and nitrites.

Alcohol and diet
High alcohol intake increases the risk of some cancers. Diets high in fat, especially those containing burned meat or meats preserved with nitrates, slow gut passage time, allowing mutagenic irritants to form in the bowel.

1. List examples of environmental factors that induce mutations under the following headings:

   (a) Radiation: _____

   (b) Chemical agents: _____

2. Discuss the role of **mutagens** in causing **mutations**: _____
   _____
   _____
   _____
   _____

3. (a) Distinguish between gametic and somatic mutations: _____
   _____

   (b) Describe the importance of gametic mutations in an evolutionary sense: _____
   _____
   _____

Related activities: The Effects of Mutations
Web links: Radiation and DNA
Periodicals:
What is a mutation?
© Biozone International 2010
Photocopying Prohibited

The Effects of Mutations

It is not correct to assume that all mutations are harmful. There are many documented cases where mutations conferring a survival advantage have arisen in a population. Such **beneficial mutations** occur most often among viruses and bacteria but occur in multicellular organisms also (e.g. insects). Sometimes, a mutation may be neutral and have no immediate effect. If there is no selective pressure against it, a mutation may be carried in the population and be of benefit (or harm) at some future time.

Harmful Mutations

There are many well-documented examples of mutations that cause harmful effects. Examples are the mutations giving rise to **cystic fibrosis** (CF) and **sickle cell disease**. The sickle cell mutation involves a change to only one base in the DNA sequence, whereas the CF mutation involves the loss of three nucleotides. The malformed proteins that result from these mutations cannot carry out their normal biological functions.

Albinism is caused by a mutation in the gene that produces an enzyme in the metabolic pathway to produce melanin. It occurs in a large number of animals (photos, right). Albinos are not common in the wild because they tend to be more vulnerable to predation.

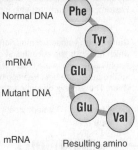

Albinism is widespread in the animal kingdom. A mutation in the metabolic pathway to melanin results in a lack of pigment.

Neutral Mutations

Some mutations are neutral because a change in the gene sequence does not result in a change to protein product of that gene. Such silent mutations are neither harmful nor beneficial at the time they occur, but may be important in an evolutionary sense if they later become subject to selection pressure. This example shows how a change to the DNA sequence (normal vs. mutant) can be silenced if there is no change to the amino acid sequence. The redundancy in the code means that both the original (GAA) and the mutated (GAG) triplet still code for glutamic acid.

Normal DNA

mRNA

Mutant DNA

mRNA

Phe
Tyr
Glu
Glu
Val

Resulting amino acid chain

Beneficial Mutations

Bacteria reproduce asexually by binary fission. They are susceptible to antibiotics (substances that harm them or inhibit their growth) but are well-known for acquiring **antibiotic resistance** through mutation. The genes for bacterial resistance can be transferred within or even between bacterial species. New, multi-resistant bacterial superbugs have arisen in this way.

Viruses, including HIV and *Influenzavirus*, have membrane envelopes coated with glycoproteins. These are used by the host to identify the virus so that it can be destroyed. The genes coding for these glycoproteins are constantly mutating. The result is that each new viral 'strain' goes undetected by the immune system until well after the infection is established.

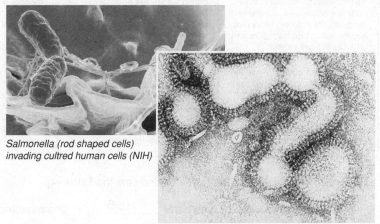

Salmonella (rod shaped cells) invading cultred human cells (NIH)

Influenzavirus showing glycoprotein soikes (Photo: DS)

1. Explain the evolutionary significance of **neutral mutations**: _____

_____

_____

2. Giving examples, explain the difference between **harmful** and **beneficial mutations**: _____

_____

_____

_____

Inherited Metabolic Disorders

Humans have more than 6000 physiological diseases attributed to mutations in single genes and over one hundred syndromes known to be caused by chromosomal abnormality. The number of genetic disorders identified increases every year. Rapid progress of the Human Genome Project is enabling the identification of the genetic basis of these disorders. This will facilitate the development of new drug therapies and gene therapies. Four genetic disorders are summarised below.

Sickle Cell Disease	β-Thalassaemia	Cystic Fibrosis	Huntington Disease
Synonym: Sickle cell anaemia	**Synonyms**: Cooley anaemia, Mediterranean anaemia	**Synonyms**: Mucoviscidosis, CF	**Synonyms**: Huntington's chorea, HD (abbreviated)
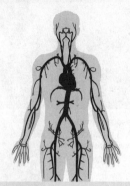	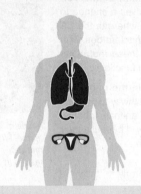		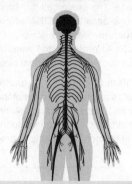
Incidence: Most common in people of African, Mediterranean, Middle Eastern, and Indian descent.	**Incidence**: More common in people from Mediterranean and Middle Eastern descent.	**Incidence**: 1:25 people in the UK carry the CF gene. About 8 000 people in the UK have CF.	**Incidence**: An uncommon disease present in 1 in 20 000. An estimated 6 500 - 8 000 people in the UK have Huntington disease.
Gene type: Autosomal mutation which results in the substitution of a single nucleotide in the HBB gene that codes for the beta haemoglobin chain. The allele is codominant.	**Gene type**: Autosomal recessive mutation of the HBB gene coding for the haemoglobin beta chain. It may arise through a gene deletion or a nucleotide deletion or insertion.	**Gene type**: Autosomal recessive. Over 500 different recessive mutations (deletions, missense, nonsense, terminator codon) of the CFTR gene have been identified.	**Gene type**: An autosomal dominant mutation of the HD gene (IT15) caused by an increase in the length (36-125) of a CAG repeat region (normal range is 11-30 repeats).
Gene location: Chromosome 11 HBB	**Gene location**: Chromosome 11 HBB	**Gene location**: Chromosome 7 CFTR 	**Gene location**: Chromosome 4 IT15
Symptoms: Include: anaemia; mild to severe pain in the chest, joints, back, or abdomen; jaundice; kidney failure; repeated infections; in particular pneumonia or meningitis; eye problems including blindness; swollen hands and feet; gallstones (at an early age); strokes.	**Symptoms**: The result of haemoglobin with few or no beta chains, causes a severe anaemia during the first few years of life. People with this condition are tired and pale because not enough oxygen reaches the cells.	**Symptoms**: Disruption of glands: the *pancreas*, *intestinal glands*, *biliary tree* (biliary cirrhosis), *bronchial glands* (chronic lung infections), and *sweat glands* (high salt content of which becomes depleted in a hot environment). *Infertility* occurs in males/females.	**Symptoms**: Mutant gene forms defective protein: **huntingtin**. Progressive, selective *nerve cell death* associated with chorea (jerky, involuntary movements), *psychiatric disorders*, and *dementia* (memory loss, disorientation, impaired ability to reason, and personality changes).
Treatment and outlook: Patients are given folic acid. Acute episodes may require oxygen therapy, intravenous infusions of fluid, and antibiotic drugs. Experimental therapies include bone marrow transplants and gene therapy.	**Treatment and outlook**: Patients require frequent blood transfusions. This causes iron build-up in the organs, which is treated with drugs. Bone marrow transplants and gene therapy hold promise and are probable future treatments.	**Treatment and outlook:** Conventional: chest physiotherapy, a modified diet, and the use of TOBI antibiotic to control lung infections. Outlook: Gene transfer therapy inserting normal CFTR gene using adenovirus vectors and liposomes.	**Treatment and outlook**: Surgical treatment may be possible. Research is underway to discover drugs that interfere with *huntingtin* protein. Genetic counselling coupled with genetic screening of embryos may be developed in the future.

1. For each of the genetic disorder below, indicate the following:

   (a) Sickle cell disease: Gene name: <u>HBB</u> Chromosome: <u>11</u> Mutation type: <u>Substitution</u>

   (b) β-thalassaemia: Gene name: _____ Chromosome: ____ Mutation type: _____

   (c) Cystic fibrosis: Gene name: _____ Chromosome: ____ Mutation type: _____

   (d) Huntington disease: Gene name: _____ Chromosome: ____ Mutation type: _____

2. Explain the cause of the symptoms for people suffering from β-thalassaemia: _____

   _____

3. Suggest a reason for the differences in the country-specific incidence rates for some genetic disorders:

   _____

   _____

Related activities: Cystic Fibrosis Mutation

Cystic Fibrosis Mutation

Cystic fibrosis an inherited disorder caused by a mutation of the **CF gene**. It is one of the most common lethal autosomal recessive conditions affecting caucasians, with an incidence of 1 in 2500 live births and a **carrier frequency** of 4%. It is uncommon in Asians and Africans. The CF gene's protein product, **CFTR**, is a membrane-based protein with a function in regulating the transport of chloride across the membrane. A faulty gene in turn codes for a faulty CFTR. More than 500 mutations of the CF gene have been described, giving rise to disease symptoms of varying severity. One mutation is particularly common and accounts for more than 70% of all defective CF genes. This mutation, called δ(delta)F508, leads to the absence of CFTR from its proper position in the membrane. This mutation is described below. Another CF mutation, R117H, which is also relatively common, produces a partially functional CFTR protein. The DNA sequence below is part of the transcribing sequence for the **normal** CF gene.

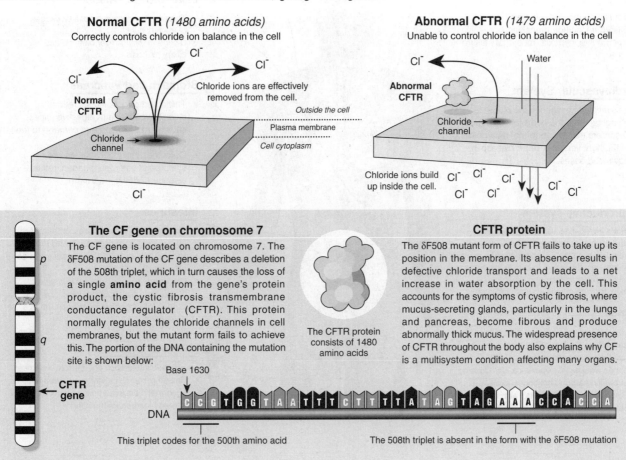

Normal CFTR *(1480 amino acids)*
Correctly controls chloride ion balance in the cell

Abnormal CFTR *(1479 amino acids)*
Unable to control chloride ion balance in the cell

The CF gene on chromosome 7

The CF gene is located on chromosome 7. The δF508 mutation of the CF gene describes a deletion of the 508th triplet, which in turn causes the loss of a single **amino acid** from the gene's protein product, the cystic fibrosis transmembrane conductance regulator (CFTR). This protein normally regulates the chloride channels in cell membranes, but the mutant form fails to achieve this. The portion of the DNA containing the mutation site is shown below:

The CFTR protein consists of 1480 amino acids

CFTR protein

The δF508 mutant form of CFTR fails to take up its position in the membrane. Its absence results in defective chloride transport and leads to a net increase in water absorption by the cell. This accounts for the symptoms of cystic fibrosis, where mucus-secreting glands, particularly in the lungs and pancreas, become fibrous and produce abnormally thick mucus. The widespread presence of CFTR throughout the body also explains why CF is a multisystem condition affecting many organs.

Base 1630

DNA C C G T G G T A A T T T C T T T T A T A G T A G A A A C C A C C A

This triplet codes for the 500th amino acid

The 508th triplet is absent in the form with the δF508 mutation

1. (a) Write the mRNA sequence for the transcribing DNA strand above: _____

    _____

   (b) Use the mRNA-amino acid table earlier in this workbook to determine the amino acid sequence coded by the mRNA for the fragment of the normal protein we are studying here:

    _____

2. (a) Rewrite the mRNA sequence for the mutant DNA strand: _____

    _____

   (b) State what kind of mutation δF508 is: _____

   (c) Determine the amino acid sequence coded by the mRNA for the fragment of the δF508 mutant protein:

    _____

   (d) Identify the amino acid that has been removed from the protein by this mutation: _____

3. Suggest why cystic fibrosis is a disease with varying degrees of severity: _____

    _____

    _____

Periodicals:
Tertiary structure

Related activities: The Genetic Code, Inherited Metabolic Disorder
Web links: Cystic Fibrosis Interactive Tutorial

RA 3

Proteins, Genes & Health

Possible Expressions of Cystic Fibrosis

Gas Exchange System

- Airways clog with mucus, causing repeated lung infections.
- Mucus causes excessive coughing and inflammation, resulting in structural damage to the lung tissue.
- Sufferers have difficulty breathing.
- CF sufferers are often hypoxic and ventilation may be required.
- Disorders include bronchitis, pneumonia, hypoxia, respiratory failure

Cardiovascular System

- Low levels of oxygen in the blood (hypoxia) force the heart to work harder.
- Disorders include abnormal enlargement of the right ventricle, heart disease in advanced cases.

Reproductive system

- In males, absence of the vas deferens (the ducts that transport sperm) results in infertility even though sperm may be normal.
- In females, thickened mucus in the oviducts and uterus may disrupt ovulation and menstruation may stop (amenorrhoea).

Bones

- Low nutrient absorption results in poor absorption of vitamin D and calcium, resulting in poor bone development and bones weakness. Hypoxia causes clubbing, or swelling around the finger joints.
- Disorders include clubbing, arthritis, and osteoporosis.

Skin

- Failure of chloride transport in the sweat glands leads to excessively salty sweat. This was the basis of a diagnostic sweat test for cystic fibrosis before genetic screening was available.

Liver and Gall Bladder

- Thickened secretions cause blockages of the bile duct.
- Disorders include liver damage and biliary cirrhosis.

Stomach and Pancreas

- Thickened pancreatic secretions block the passage of digestive juices causing malnutrition and damage to the pancreatic islet cells.
- Outcomes include: Diabetes-like symptoms, gastroesophageal reflux disease (GERD), heartburn.

Intestines

- Thickened mucus reduces uptake of the fat-soluble vitamins A, D, E and K.
- Stool is foul-smelling and greasy (fat not absorbed).
- Disorders include internal intestinal blockages, constipation, blockage and rupture of the bowel by meconium in newborns.

Muscular system

- Exercise intolerance and muscle weakness.
- Recent research indicates sufferers have abnormal muscle aerobic metabolism and altered bicarbonate transport.

Cystic fibrosis is a multi-organ disorder that mainly affects mucus lined organs, particularly the **respiratory**, **digestive** and **reproductive** systems. Secondary abnormalities in bone development and heart disorders may also arise. The majority of these abnormalities result from thickened mucus that blocks narrow passages such as those in the intestines or bronchioles. CF patients also suffer frequent respiratory tract infections as a result of bacteria proliferating in the thick mucus of the lungs.

4. Describe the effect cystic fibrosis on mucus lined organs: _____

_____

5. Explain why osteoporosis and heart disease are secondary results of cystic fibrosis: _____

_____

_____

_____

6. Explain how the lungs are affected by CF : _____

_____

_____

_____

_____

Alleles

Sexually reproducing organisms in nearly all cases have paired sets of chromosomes, one set coming from each parent. The equivalent chromosomes that form a pair are termed **homologues**. They contain equivalent sets of genes on them. But there is the potential for different versions of a gene to exist in a population and these are termed **alleles**.

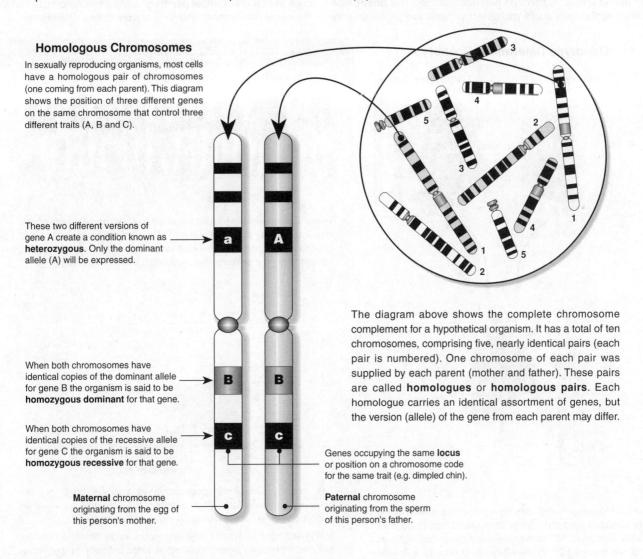

Homologous Chromosomes

In sexually reproducing organisms, most cells have a homologous pair of chromosomes (one coming from each parent). This diagram shows the position of three different genes on the same chromosome that control three different traits (A, B and C).

These two different versions of gene A create a condition known as **heterozygous**. Only the dominant allele (A) will be expressed.

When both chromosomes have identical copies of the dominant allele for gene B the organism is said to be **homozygous dominant** for that gene.

When both chromosomes have identical copies of the recessive allele for gene C the organism is said to be **homozygous recessive** for that gene.

Maternal chromosome originating from the egg of this person's mother.

The diagram above shows the complete chromosome complement for a hypothetical organism. It has a total of ten chromosomes, comprising five, nearly identical pairs (each pair is numbered). One chromosome of each pair was supplied by each parent (mother and father). These pairs are called **homologues** or **homologous pairs**. Each homologue carries an identical assortment of genes, but the version (allele) of the gene from each parent may differ.

Genes occupying the same **locus** or position on a chromosome code for the same trait (e.g. dimpled chin).

Paternal chromosome originating from the sperm of this person's father.

1. Define the following terms used to describe the allele combinations in the genotype for a given gene:

   (a) Heterozygous: _____

   (b) Homozygous dominant: _____

   (c) Homozygous recessive: _____

2. For a gene given the symbol 'A', describe the alleles present in an organism that is identified as:

   (a) Heterozygous: _____ (b) Homozygous dominant: _____ (c) Homozygous recessive: _____

3. Explain what is meant by a homologous pair of chromosomes: _____

   _____

   _____

4. Discuss the significance of genes existing as **alleles**: _____

   _____

   _____

   _____

   _____

Related activities: Meiosis

A 1

Proteins, Genes & Health

Phenotype and Genotype

The distinction between phenotype and genotype is an important one in genetics. The **genotype** is the complete set of genetic information for an individual, and is determined by the combination of alleles it inherits from the parental gametes. The **phenotype** refers to the individual's expressed or observed characteristics and is a result of interactions between its genotype and its internal and external environment. A number of environmental factors may contribute towards phenotype, including obvious influences such as nutrition and (for plants) growing environment, as well as hormonal environment and temperature during development.

Genotype Determines Phenotype

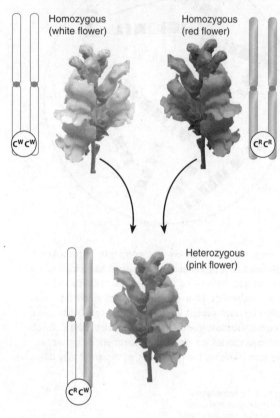

In the snapdragon (*Antirrhinum*), flower colour is determined by the alleles of both parents. When plants homozygous for white flowers (C^WC^W) are crossed with plants homozygous for red flowers (C^RC^R) the offspring all have pink flowers (C^RC^W). The alleles for flower colour show incomplete dominance and both maternal and paternal alleles contribute equally to the final intermediate phenotype.

Same Genotype, Different Phenotype

These calves were produced by cloning. Their genetic profiles are identical but they all exhibit different coat patterns (phenotypes). This is because pigment cells migrate randomly in the embryos causing variations in coat patterns that distinguish different individuals.

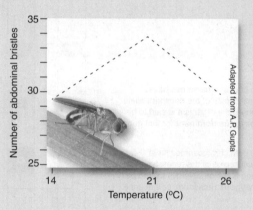

The fruit fly *Drosophila pseudoobscura* can be bred to have homozygous genes for a single chromosome. This enables the effect of environmental factors on the phenotype to be studied.

In the example above, *D. pseudoobscura* eggs and the resulting hatched individuals were incubated at three different temperatures. The number of abdominal bristles (phenotype) was influenced by the environment (temperature).

1. Explain the difference between genotype and phenotype: _____

   _____

   _____

2. In some pairs of identical twins, one will develop a disease (e.g. Parkinson's) and the other will not. A recent study has revealed that identical twins have small variations in their genomes. Explain how this may account for the difference in disease susceptibility:

   _____

   _____

   _____

3. Explain why identical twins have different fingerprint patterns. _____

   _____

   _____

Related activities: Alleles, Gene Interactions and Gene Inactivations

Mendel's Pea Plant Experiments

Gregor Mendel (1822-1884), pictured on the right, was an Austrian monk who is regarded as the 'father of genetics'. He carried out some pioneering work using pea plants to study the inheritance patterns of a number of **traits** (characteristics). Mendel observed that characters could be masked in one generation of peas but could reappear in later generations. He showed that inheritance involved the passing on to offspring of discrete units of inheritance; what we now call genes. Mendel examined a number of phenotypic traits and found that they were inherited in predictable ratios, depending on the phenotype of the parents. Below are some of his results from crossing heterozygous plants (e.g. tall plants that were the offspring of tall and dwarf parent plants: Tt x Tt). The numbers in the results column represent how many offspring had those phenotypic features.

1. Study the **results** for each of the six experiments below. Determine which of the two phenotypes is the dominant one, and which is the recessive. Place your answers in the spaces in the **dominance** column in the table below.

2. Calculate the ratio of dominant phenotypes to recessive phenotypes (to two decimal places). The first one (for seed shape) has been done for you (5474 ÷ 1850 = 2.96). Place your answers in the spaces provided in the table below:

Trait	Possible Phenotypes		Results		Dominance	Ratio
Seed shape	*Wrinkled*	*Round*	Wrinkled Round **TOTAL**	1850 5474 **7324**	Dominant: Round Recessive: Wrinkled	2.96 : 1
Seed color	*Green*	*Yellow*	Green Yellow **TOTAL**	2001 6022 **8023**	Dominant: Recessive:	
Pod color	*Green*	*Yellow*	Green Yellow **TOTAL**	428 152 **580**	Dominant: Recessive:	
Flower position	*Axial*	*Terminal*	Axial Terminal **TOTAL**	651 207 **858**	Dominant: Recessive:	
Pod shape	*Constricted*	*Inflated*	Constricted Inflated **TOTAL**	299 882 **1181**	Dominant: Recessive:	
Stem length	*Tall*	*Dwarf*	Tall Dwarf **TOTAL**	787 277 **1064**	Dominant: Recessive:	

3. Mendel's experiments identified that two heterozygous parents should produce offspring in the ratio of three times as many dominant offspring to those showing the recessive phenotype.

   (a) State which three of Mendel's experiments provided ratios closest to the theoretical 3:1 ratio:

   _____

   (b) Suggest a possible reason why these results deviated less from the theoretical ratio than the others:

   _____

   _____

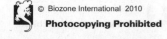
© Biozone International 2010
Photocopying Prohibited

Periodicals:
Mendel's legacy

 Related activities: Alleles, Mendel's Laws of Inheritance
Web links: Basic Principles of Genetics, interactive Mendelian Genetics

DA 2

Proteins, Genes & Health

Monohybrid Cross

The study of **single-gene inheritance** is achieved by performing **monohybrid crosses**. The six basic types of matings possible among the three genotypes can be observed by studying a pair of alleles that govern coat colour in the guinea pig. A dominant allele: given the symbol **B** produces **black** hair, and its recessive allele: **b**, produces white. Each of the parents can produce two types of gamete by the process of **meiosis** (in reality there are four, but you get identical pairs). Determine the **genotype** and **phenotype frequencies** for the crosses below (enter the frequencies in the spaces provided). For crosses 3 to 6, you must also determine gametes produced by each parent (write these in the circles), and offspring (F₁) genotypes and phenotypes (write in the genotype inside the offspring and state if black or white).

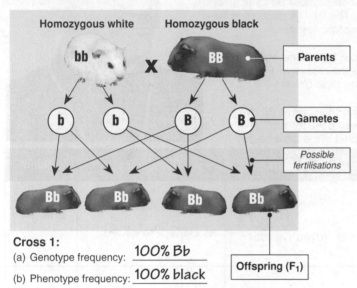

Homozygous white   Homozygous black

Parents

Gametes

Possible fertilisations

Cross 1:

(a) Genotype frequency: __100% Bb__

(b) Phenotype frequency: __100% black__

Offspring (F₁)

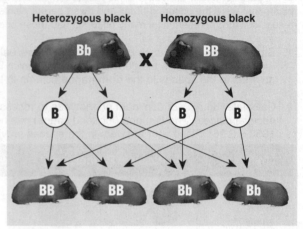

Heterozygous black   Homozygous black

Cross 2:

(a) Genotype frequency: _____

(b) Phenotype frequency: _____

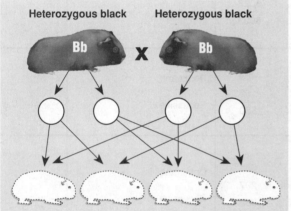

Heterozygous black   Heterozygous black

Cross 3:

(a) Genotype frequency: _____

(b) Phenotype frequency: _____

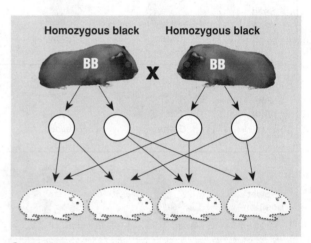

Homozygous black   Homozygous black

Cross 4:

(a) Genotype frequency: _____

(b) Phenotype frequency: _____

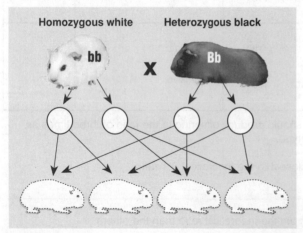

Homozygous white   Heterozygous black

Cross 5:

(a) Genotype frequency: _____

(b) Phenotype frequency: _____

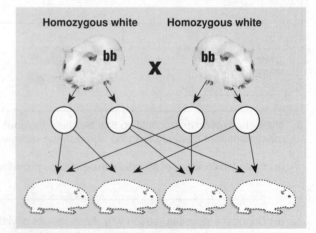

Homozygous white   Homozygous white

Cross 6:

(a) Genotype frequency: _____

(b) Phenotype frequency: _____

A 1

Related activities: Alleles
Web links: Drag and Drop Genetics

Pedigree Analysis

It is socially and ethically unacceptable to manipulate the mating patterns in people, so to understand how human traits are inherited, geneticists analyse the results of matings that have already occurred. In order to do this, as much information as possible must be collected about a family's history for the inheritance of a particular trait so that a family **pedigree** can be produced. The pedigree analysis looks a little like a family tree. It describes the relationships of parents and their children across a number of generations. Standardised symbols are used in pedigree analysis trees (below). A pedigree helps geneticists understand how particular traits are inherited, and provides a way to predict if and how that trait will be passed on to subsequent generations. This is particularly important for lethal traits or for genetic traits that produce severe disabilities.

Sample Pedigree Chart

Pedigree charts are a way of graphically illustrating inheritance patterns over a number of generations. They are used to study the inheritance of genetic disorders. The key (below the chart) should be consulted to make sense of the various symbols. Particular individuals are identified by their generation number and their order number in that generation. For example, **II-6** is the sixth person in the second row. The arrow indicates the **propositus**; the person through whom the pedigree was discovered (i.e. who reported the condition).

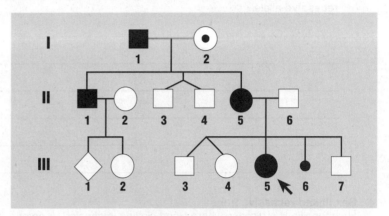

If the chart on the right were illustrating a human family tree, it would represent three generations: grandparents (I-1 and I-2) with three sons and one daughter. Two of the sons (II-3 and II-4) are identical twins, but did not marry or have any children. The other son (II-1) married and had a daughter and another child (sex unknown). The daughter (II-5) married and had two sons and two daughters (plus a child that died in infancy).

For the particular trait being studied, the grandfather was expressing the phenotype (showing the trait) and the grandmother was a carrier. One of their sons and one of their daughters also show the trait, together with one of their granddaughters.

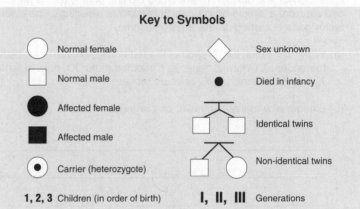

Key to Symbols

○ Normal female

□ Normal male

● Affected female

■ Affected male

◉ Carrier (heterozygote)

1, 2, 3 Children (in order of birth)

◇ Sex unknown

● Died in infancy

⊤ Identical twins

⊤ Non-identical twins

I, II, III Generations

---

1. **Autosomal dominant traits**

   An unusual trait found in some humans is woolly hair (not to be confused with curly hair). Each affected individual will have at least one affected parent.

   (a) Write the genotype for each of the individuals on the chart using the following letter codes:
   WW woolly hair; **Ww** woolly hair (heterozygous); **W-** woolly hair, but unknown if homozygous; **ww** normal hair

   (b) Describe a feature of this inheritance pattern that suggests the trait is the result of a **dominant** allele:

   _____

2. **Sex linked dominant traits**

   A rare form of rickets is inherited on the X chromosome. All daughters of affected males will be affected. More females than males will show the trait.

   (a) Write the genotype for each of the individuals on the chart using the following letter codes:
   XY normal male; **$X_R Y$** affected male; **XX** normal female; **X_{R-}** female (unknown if homozygous); **$X_R X_R$** affected female.

   (b) Explain why more females than males will be affected:

   _____

   _____

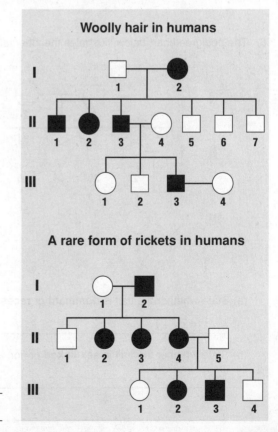

Woolly hair in humans

A rare form of rickets in humans

Proteins, Genes & Health

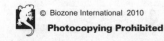

3. **Autosomal recessive traits**

Albinos lack pigment in the hair, skin and eyes. This trait is inherited as an autosomal recessive allele (i.e. it is not carried on the sex chromosome).

(a) Write the genotype for each of the individuals on the chart using the following letter codes: **PP** normal skin colour; **P-** normal, but unknown if homozygous; **Pp** carrier; **pp** albino.

(b) Explain why the parents (II-3) and (II-4) must be **carriers** of a **recessive** allele:

_____

_____

_____

_____

_____

_____

Albinism in humans

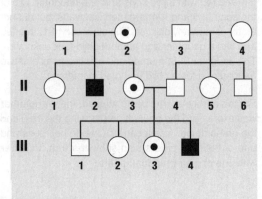

4. **Sex linked recessive traits**

Haemophilia is a disease where blood clotting is affected. A person can die from a simple bruise (which is internal bleeding). The clotting factor gene is carried on the X chromosome.

(a) Write the genotype for each of the individuals on the chart using the codes: **XY** normal male; **X$_h$Y** affected male; **XX** normal female; **X$_h$X** female carrier; **X$_h$X$_h$** affected female:

(b) Explain why males can never be carriers:

_____

_____

_____

_____

Haemophilia in humans

6. The pedigree chart below illustrates the inheritance of a trait (darker symbols) in two families joined in marriage.

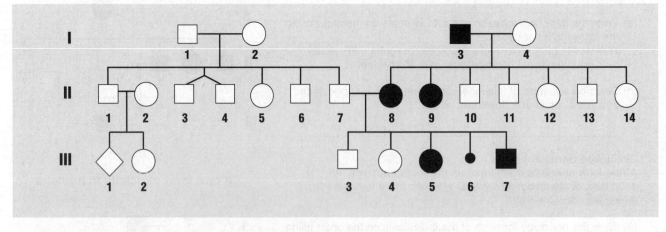

(a) State whether the trait is **dominant** or **recessive**, and explain your reasoning: _____

_____

(b) State whether the trait is **sex linked** or not, and explain your reasoning: _____

_____

Gene Therapy

Gene therapy refers to the application of gene technology to treat disease by correcting or replacing faulty genes. It was first envisioned as a treatment, or even a cure, for genetic disorders, but it could also be used to treat a wide range of diseases, including those that resist conventional treatments. Although varying in detail, all gene therapies are based around the same technique. Normal (non-faulty) DNA containing the correct gene is inserted into a vector, which is able to transfer the DNA into the patient's cells in a process called **transfection**. The vector may be a virus, liposome, or any one of a variety of other molecular transporters. The vector is introduced into a sample of the patient's cells and these are cultured to amplify the correct gene. The cultured cells are then transferred back to the patient. The use of altered stem cells instead of mature somatic cells has so far achieved longer lasting results in many patients. The treatment of somatic cells or stem cells may be **therapeutic** but the changes are not inherited. **Germline therapy** would enable genetic changes to be passed on. To date there have been limited successes with gene therapy because transfection of targeted cells is inefficient and side effects can be severe or even fatal. However there have been a small number of successes, including the treatment of one form of SCID, a genetic disease affecting the immune system.

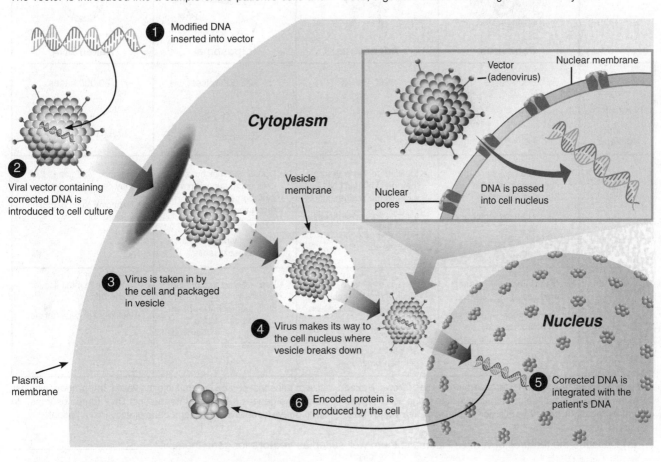

1 Modified DNA inserted into vector

2 Viral vector containing corrected DNA is introduced to cell culture

Cytoplasm

Vesicle membrane

3 Virus is taken in by the cell and packaged in vesicle

4 Virus makes its way to the cell nucleus where vesicle breaks down

Plasma membrane

6 Encoded protein is produced by the cell

Vector (adenovirus)

Nuclear membrane

Nuclear pores

DNA is passed into cell nucleus

Nucleus

5 Corrected DNA is integrated with the patient's DNA

Proteins, Genes & Health

1. (a) Describe the general principle of gene therapy: _____

_____

_____

(b) Describe the medical areas where gene therapy might be used:_____

_____

2. Explain the significance of transfecting **germline cells** rather than **somatic cells**: _____

_____

_____

3. Describe the purpose of **gene amplification** in gene therapy: _____

_____

4. Explain why genetically modified stem cells offer greater potential in gene therapy treatments than GM somatic cells:

_____

_____

_____

© Biozone International 2010
Photocopying Prohibited

Periodicals: Genes can come true

Related activities: Vectors for Gene Therapy, Gene Delivery Systems
Web links: Gene Therapy, Gene Therapy Primer

RA 2

Vectors for Gene Therapy

Gene therapy usually requires a **vector** (carrier) to introduce the DNA. The majority of approved clinical gene therapy protocols (63%) employ **retroviral vectors** to deliver the selected gene to the target cells, although there is considerable risk in using these vectors (below). Other widely used vectors include adenoviral vectors (16%), and liposomes (13%). The remaining 8% employ a variety of vector systems, the majority of which include injection of naked plasmid DNA.

Vectors That Can Be Used For Gene Therapy

	Retrovirus	**Adenovirus**	**Liposome**	**Naked DNA**
Insert size:	8000 bases	8000 bases	>20 000 bases	>20 000 bases
Integration:	Yes	No	No	No
***In vivo* delivery:**	Poor	High	Variable	Poor
Advantages	• Integrate genes into the chromosomes of the human host cell. • Offers chance for long-term stability.	• Modified for gene therapy, they infect human cells and express the normal gene. • Most do not cause disease. • Have a large capacity to carry foreign genes.	• Liposomes seek out target cells using sugars in their membranes that are recognised by cell receptors. • Have no viral genes that may cause disease.	• Have no viral genes that may cause disease. • Expected to be useful for vaccination.
Disadvantages	• Many infect only cells that are dividing. • Genes integrate randomly into chromosomes, so might disrupt useful genes in the host cell.	• Viruses may have poor survival due to attack by the host's immune system. • Genes may function only sporadically because they are not integrated into host cell's chromosome.	• Less efficient than viruses at transferring genes into cells, but recent work on using sugars to aid targeting have improved success rate.	• Unstable in most tissues of the body. • Inefficient at gene transfer.

In the table above, the following terms are defined as follows: **Naked DNA**: the genes are applied by ballistic injection (firing using a gene gun) or by regular hypodermic injection of plasmid DNA. **Insert size**: size of gene that can be inserted into the vector. **Integration**: whether or not the gene is integrated into the host DNA (chromosomes). ***In vivo* delivery**: ability to transfer a gene directly into a patient.

1. (a) Describe the features of viruses that make them well suited as **vectors** for gene therapy: _____

_____

_____

(b) Identify two problems with using viral vectors for gene therapy: _____

_____

_____

2. (a) Suggest why it may be beneficial for a (therapeutic) gene to integrate into the patient's chromosome: _____

_____

_____

(b) Explain why this has the potential to cause problems for the patient: _____

_____

3. (a) Suggest why naked DNA is likely to be unstable within a patient's tissues: _____

_____

(b) Suggest why enclosing the DNA within liposomes might provide greater stability: _____

_____

Related activities: Gene Therapy, Gene Delivery Systems
Web links: Gene Therapy Primer

Periodicals:
Tools you can trust

Gene Delivery Systems

The mapping of the human genome has improved the feasibility of gene therapy as a option for treating an increasingly wide range of diseases, but it remains technically difficult to deliver genes successfully to a patient. Even after a gene has been identified, cloned, and transferred to a patient, it must be expressed normally. To date, the success of gene therapy has been generally poor, and improvements have been short-lived or counteracted by adverse side effects. Inserted genes may reach only about 1% of target cells and those that reach their destination may work inefficiently and produce too little protein, too slowly to be of benefit. In addition, many patients react immunologically to the vectors used in gene transfer. Much of the current research is focussed on improving the efficiency of gene transfer and expression. One of the first gene therapy trials was for **cystic fibrosis** (CF). CF was an obvious candidate for gene therapy because, in most cases, the disease is caused by a single, known gene mutation. However, despite its early promise, gene therapy for this disease has been disappointing (below right). Another candidate for gene therapy, again because the disease is caused by single, known mutations, is Severe Combined Immune Deficiency (SCID) (below left). Gene therapies developing for this disease have so far proved promising.

Treating SCID using Gene Therapy

The most common form of **SCID** (Severe Combined Immune Deficiency) is **X-linked SCID**, which results from mutations to a gene on the X chromosome encoding the **common gamma chain**, a protein forming part of a receptor complex for numerous types of leukocytes. A less common form of the disease, (**ADA-SCID**) is caused by a defective gene that codes for the enzyme adenosine deaminase (ADA).

Both of these types of SCID lead to immune system failure. A common treatment for SCID is bone marrow transplant, but this is not always successful and runs the risks of infection from unscreened viruses. **Gene therapy** appears to hold the best chances of producing a cure for SCID because the mutation affects only one gene whose location is known. DNA containing the corrected gene is placed into a **gutted retrovirus** and introduced to a sample of the patient's **bone marrow.** The treated cells are then returned to the patient.

In some patients with ADA-SCID, treatment was so successful that supplementation with purified ADA was no longer required. The treatment carries risks though. In early trials, two of ten treated patients developed leukemia when the corrected gene was inserted next to a gene regulating cell growth.

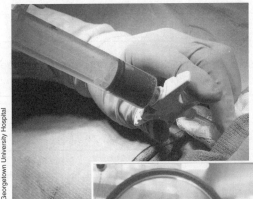

Samples of bone marrow being extracted prior to treatment with gene therapy.

Georgetown University Hospital

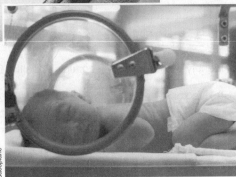

Jacoplane

Detection of SCID is difficult for the first months of an infant's life due to the mother's antibodies being present in the blood. Suspected SCID patients must be kept in sterile conditions at all times to avoid infection.

Proteins, Genes & Health

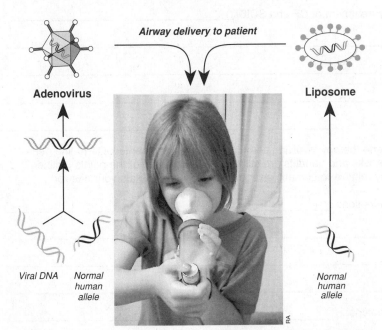

Airway delivery to patient

Adenovirus

Liposome

Viral DNA   *Normal human allele*

Normal human allele

RA

An **adenovirus** that normally causes colds is genetically modified to make it safe and to carry the normal (unmutated) CFTR ('cystic fibrosis') gene.

Liposomes are tiny fat globules. Normal CF genes are enclosed in liposomes, which fuse with plasma membranes and deliver the genes into the cells.

Gene Therapy - Potential Treatment for Cystic Fibrosis?

Cystic fibrosis (CF) is caused by a mutation to the gene coding for a chloride ion channel important in creating sweat, digestive juices, and mucus. The dysfunction results in abnormally thick, sticky mucus that accumulates in the lungs and intestines. The identification and isolation of the CF gene in 1989 meant that scientists could look for ways in which to correct the genetic defect rather than just treating the symptoms using traditional therapies.

The main target of CF gene therapy is the lung, because the progressive lung damage associated with the disease is eventually lethal.

In trials, normal genes were isolated and inserted into patients using vectors such as **adenoviruses** and **liposomes**, delivered via the airways (left). The results of trials were disappointing: on average, there was only a 25% correction, the effects were short lived, and the benefits were quickly reversed. Alarmingly, the adenovirus used in one of the trials led to the death of one patient.

Source: Cystic Fibrosis Trust, UK.

Related activities*: Gene Therapy, Vectors for Gene Therapy, Cystic Fibrosis Mutation*
Web links*: Gene Therapy Case Study: Cystic Fibrosis*

A 2

1. A great deal of current research is being devoted to discovering a gene therapy solution to treat **cystic fibrosis** (CF):

   (a) Describe the symptoms of CF: _____

   _____

   _____

   (b) Explain why this genetic disease has been so eagerly targeted by gene therapy researchers: _____

   _____

   _____

   _____

   _____

   (c) Outline some of the problems so far encountered with gene therapy for CF: _____

   _____

   _____

   _____

   _____

   _____

2. Identify two vectors for introducing healthy CFTR genes into CF patients.

   (a) Vector 1: _____

   (b) Vector 2: _____

3. (a) Describe the difference between X-linked SCID and ADA-SCID: _____

   _____

   (b) Identify the vector used in the treatment of SCID: _____

   _____

4. Briefly outline the differences in the gene therapy treatment of CF and SCID:_____

   _____

   _____

   _____

   _____

5. Changes made to chromosomes as a result of gene therapy involving somatic cells are not inherited. Germ-line gene therapy has the potential to cure disease, but the risks and benefits are still not clear. For each of the points outlined below, evaluate the risk of germ-line gene therapy relative to somatic cell gene therapy and explain your answer:

   (a) Chance of interfering with an essential gene function: _____

   _____

   _____

   _____

   (b) Misuse of the therapy to selectively alter phenotype: _____

   _____

   _____

   _____

Genetic Testing

Genetic testing involves the direct analysis of DNA to determine if any mutations exist which may cause genetic diseases (e.g. Down syndrome, CF). Currently about 900 different tests are available. Genetic testing is carried out for a number of reasons. **Preimplantation genetic diagnosis** (embryo screening) is performed on embryos prior to implantation. The developing embryos are tested and the parents select a healthy embryo for implantation into the woman's uterus. **Prenatal testing** of the developing foetus in the womb can be used to detect genetic abnormalities after development has begun. **Postnatal testing** of newborns routinely screens for a range of diseases. Early detection enables rapid treatment, and this can improve newborn health, and prevent severe disability or even death. Genetic testing is also available to adults. Reasons to test include to diagnose a genetic disease or determine if they are a disease carrier. This enables the carrier to determine the risk of passing on the disease to future children. All genetic tests have social and ethical issues associated with their use.

Genetic Counselling

Genetic counselling analyses the risk of producing offspring with a known gene defect. Counsellors identify families at risk, investigate the problem present in the family, interpret information about the disorder, analyse inheritance patterns and risks of recurrence, and review available options with the family.

Increasingly, there are DNA tests for the identification of specific defective genes. People usually consider genetic counselling if they have a family history of a genetic disorder, or if a routine prenatal screening test yields an unexpected result.

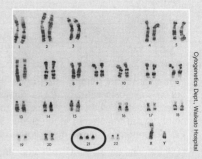

Cytogenetics Dept., Waikato Hospital

Genetic counselling provides information to families who have members with birth defects or genetic disorders, and to families who may be at risk of a variety of inherited conditions.

Most pregnant women in developed countries will have a prenatal test to detect chromosomal abnormalities such as Down syndrome (karyotype above), and developmental abnormalities such as neural tube defects.

Amniocentesis

Performed at: 14-16 weeks into the pregnancy. The amniotic fluid (which naturally contains some foetal cells) is centrifuged, and the cells are cultured, examined biochemically, and karyotyped.

Used for: Detection of nearly 300 chromosomal disorders, such as Down syndrome, neural tube defects (e.g. spina bifida), and inborn errors of metabolism.

Recommended: A maternal age nearing or over 40, when parents are carriers of an inherited disorder or already have a child with a chromosomal disorder.

Associated risks: Risk of miscarriage through damage to foetus or placenta. In women younger than 35, the risk of miscarriage through the procedure is greater than the risk of carrying a child with chromosomal abnormalities.

Chorionic Villus Sampling (CVS)

Performed at: 8-10 weeks gestation. Using ultrasound guidance, a narrow tube is inserted through the cervix and a sample of the foetal chorionic villi is taken from the placenta. Compared with amniocentesis, more foetal cells are obtained so analysis can be completed earlier and more quickly.

Used for: As for amniocentesis: detection of chromosomal and metabolic disorders.

Recommended: Recommendations as for amniocentesis.

Associated risks: Risk of miscarriage is higher than for amniocentesis but, if abortion is recommended, this can be performed sooner. Note that both amniocentesis and CVS rely on the ultrasound to determine the position of the foetus and placenta in the uterus.

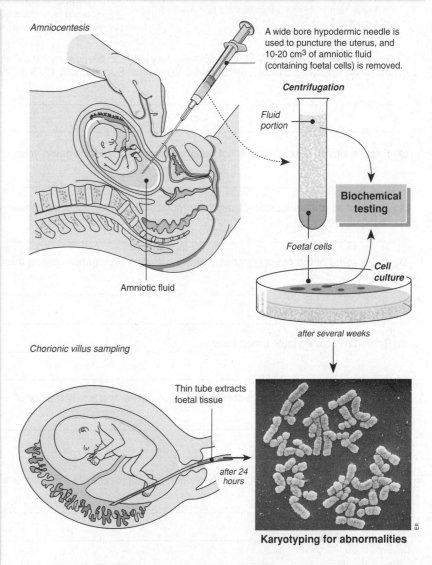

Amniocentesis

A wide bore hypodermic needle is used to puncture the uterus, and 10-20 cm³ of amniotic fluid (containing foetal cells) is removed.

Centrifugation

Fluid portion

Biochemical testing

Foetal cells

Cell culture

Amniotic fluid

after several weeks

Chorionic villus sampling

Thin tube extracts foetal tissue

after 24 hours

Karyotyping for abnormalities

Preimplantation Genetic Diagnosis

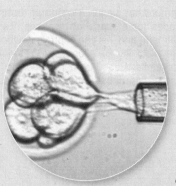

IVF techniques are used to produce embryos for genetic testing. A single cell is removed at the 8-cell stage (left), and analysed for genetic disorders.

The embryo is not damaged by the removal of the cell because at this early stage the embryonic cells are not specialised.

Preimplantation genetic diagnosis (PGD) is carried out on embryos created by *in vitro* fertilisation. It aims to detect genetic defects in embryos, so that only 'healthy' embryos are transferred to the uterus for development. PGD is performed on two main groups of patients:

(1) those who have a specific disorder and want to ensure that it is not transferred to their children.

(2) IVF patients who want their embryos screened for chromosomal aneuploidies.

PGD decreases the occurrence of heritable genetic disease, and provides an alternative to prenatal testing. Testing prior to implantation is advantageous because it removes the ethical dilemma of having to decide whether to terminate the pregnancy if a genetic disorder is detected.

The Ethics of Genetic Testing

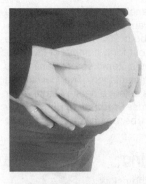

For some people, the discovery of a genetic defect in their unborn child provides them with an opportunity to come to terms with the situation and prepare for the delivery of a special needs baby. For others, the information will lead to the decision to terminate the pregnancy, an action some people believe is morally wrong because they feel it devalues human life.

There is concern that parents will be able to pick certain characteristics resulting in "designer babies". This is seen in countries where more value is placed on the birth of a boy child than a girl. China has a high male/female ratio in the population because of sex selection based on a preference for boys. Unwanted female foetuses are terminated.

In the UK, genetic testing is regulated by the human fertilisation embryology authority (HFEA). They control who can carry out testing and the suitability of the testing. However, in some countries genetic testing is run as a commercial venture, with fewer regulations.

1. Chorionic villus sampling (CVS), if performed very early in pregnancy (at 5-7 weeks) may cause limb abnormalities, probably via upsetting critical sites of foetal blood flow. Suggest why CVS might be performed at such an early stage:

_____

_____

2. (a) Explain why amniocentesis is not usually recommended for women younger than 35: _____

_____

(b) Suggest when amniocentesis might be recommended for younger women, in spite of the risk: _____

_____

3. Describe some of the ethical concerns of the following information gained through prenatal diagnoses:

(a) Gender determination: _____

_____

(b) Termination of a viable pregnancy: _____

_____

4. Outline the benefits of **carrier screening** to a couple with a family history of a genetic disorder:

_____

_____

5. What is the main advantage of preimplantation genetic diagnosis over prenatal screening: _____

_____

_____

_____

Postnatal Testing

There are a number of genetically inherited metabolic disorders in humans that involve interruption of metabolic pathways. Most are very rare with incidence rates of one in millions. However, some are common enough to warrant testing of all newborn babies (five days after birth). A sample of baby's blood is taken from a heel prick and blotted onto an absorbent card (below, right). The sample is sent away for genetic tests that could save the baby's life, or prevent the development of serious physical or mental problems. Most babies born in Britain are normal. A very few have rare, but serious disorders that are caused by a defective gene that gives rise to a defective protein. This protein is usually an enzyme that is unable to carry out its vital step in a metabolic pathway. In the UK, five disorders are currently tested for (below left).

Metabolic Disorders Tested in Newborns

Congenital Hypothyroidism (sporadic)

Caused by:	Not enough normal thyroid gland.
Leads to:	Slowed growth and mental development.
Probability:	1 in **4000** newborn babies

Sickle Cell Disorder (autosomal recessive)

Caused by:	A faulty enzyme which produces sickle haemoglobin.
Leads to:	Anaemia, painful clots in small vessels.
Probability:	1 in **1900** newborn babies

Phenylketonuria (PKU) (autosomal recessive)

Caused by:	A faulty gene means an enzyme is missing from the liver. Without this enzyme, the essential amino acid phenylalanine rises to harmful levels.
Leads to:	Brain damage.
Probability:	1 in **10000** newborn babies

Cystic fibrosis (autosomal recessive)

Caused by:	Abnormal chloride transport, abnormally thick mucus secretions.
Leads to:	Poor growth, chest infections, shortened life.
Probability:	1 in **2500** newborn babies

Medium Chain Acyl-CoA Dehydrogenase Deficiency (MCADD) (autosomal recessive)

Caused by:	A faulty gene which prevents production of MCADD
Leads to:	Cannot break down fats for energy. Serious illness or death if untreated.
Probability:	1 in **10000** newborn babies

Results of the test

- The vast majority of results are negative for all of the conditions described above.
- The results of the test should be available within a few days of the test being taken.
- If a test is positive, a paediatrician is consulted to discuss the results and treatment.

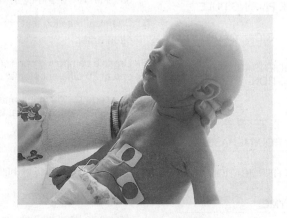

Newborn Baby Blood Spot Test

The card below shows the front of the blood test collecting paper (a kind of blotting paper). Blood is obtained from a heel prick from the baby's foot on the 5th day after birth and applied to each of the three circles at the bottom of the card.

PKU BLOOD TEST
PLEASE PRINT USING A BALLPOINT PEN

Name _____
Address and Full Post Code _____

District Health Board _____
Place of Birth _____
G.P. Name _____
Address _____

Date of Birth _____ Sex _____
Date of First Milk Feed _____

Mother's DOB _____
Type of Feeding – Bottle ☐ Breast ☐ Other _____
Tick if Baby is Premature ☐ On Antibiotics ☐ In Hospital ☐

Date Specimen Taken _____
Name of person taking the specimen _____

FILL ALL CIRCLES THROUGH TO THE BACK WITH BLOOD

○ ⊙ ⊙ ○

Please ensure that ALL information requested on the card is filled in correctly

Blood is applied here, filling each of the 4 circles

Proteins, Genes & Health

---

1. Explain what is meant by a **metabolic disorder**: _____
_____

2. Explain briefly the purpose of the **newborn baby blood test**: _____
_____

3. Suggest why the blood samples are not taken until the 5th day after birth: _____
_____
_____

4. Define the term **congenital**: _____

Related activities: Genetic Testing, Inherited Metabolic Disorders

ERA 1

KEY TERMS: Mix and Match

INSTRUCTIONS: Test your vocab by matching each term to its correct definition, as identified by its preceding letter code.

ALLELE

AMINO ACID

BASE PAIRING RULE

BIOLOGICAL CATALYST

CONDENSATION REACTION

DNA

DNA POLYMERASE

DNA REPLICATION

ENZYME

GENE

GENE THERAPY

GENOTYPE

HYDROGEN BONDING

HYDROLYSIS REACTION

MESSENGER RNA

MONOHYBRID INHERITANCE

MUTATION

NUCLEIC ACID

PEDIGREE DIAGRAM

PEPTIDE BOND

PHENOTYPE

POLYPEPTIDE

PROTEIN

PROTEIN SYNTHESIS

RNA

TRANSCRIPTION

TRANSFER RNA

TRANSLATION

TRIPLET CODE

A Ribonucleic acid molecule used to transfer the information encoded in the DNA to the cytoplasm for translation.

B Observable characteristics in an organism.

C Pairing of bases in nucleic acids in which adenine always pairs with thymine (in DNA) or uracil (in RNA) and cytosine always pairs with guanine (forming a base-pair).

D Chemical reaction that combines two molecules by the elimination of a smaller molecule, often water.

E Organic compound consisting of a carboxyl, an amine and an R group (where R may be one of 20 different atomic groupings). Polymerise by peptide bonds to form proteins.

F The code by which information in genetic material (DNA or RNA) is encoded in a three nucleotide sequence that is translated into amino acid sequences by living cells.

G Enzyme that joins together free nucleotides to form a strand of DNA complementary to the one that is being copied.

H Chemical reaction in which a molecule is split into two parts by water (as H^+ and OH^-).

I A globular protein which acts as a catalyst to speed up a specific biological reaction.

J Molecule used by biological systems to increase reaction rates. In biological systems these are commonly enzymes.

K The application of gene technology to treat disease by correcting or replacing faulty genes.

L Process by which the DNA molecule is copied to form two new, identical daughter DNA molecules.

M A diagram showing analysis of inheritance patterns over a number of generations.

N Organic polymer formed from the assembly of amino acids joined by peptide bonds.

O Intermolecular bond that forms between hydrogen and either oxygen, nitrogen or fluorine.

P A cross between two organisms in which the inheritance patterns of one gene is studied.

Q A change to the DNA sequence of an organism. This may be a deletion, insertion, duplication, inversion or translocation of DNA in a gene or chromosome.

R The process by which mRNA is decoded to produce a specific polypeptide.

S Covalent bond formed when a carboxyl group and an amine group react and release a molecule of water and in biological systems leads to the formation of proteins.

T Ribonucleic acid. Single stranded nucleic acid that consists of nucleotides that contain ribose sugar.

U Process by which DNA is transcribed into mRNA, which is then translated into proteins (via codon-anticodon matching with tRNA carrying amino acids).

V Macromolecule that forms from the joining of multiple peptide subunits.

W The allele combination of an organism. The genotype determines the genetic potential of the individual (e.g. potential to grow to a certain height).

X Ribonucleic acid molecule that transfers amino acids to the growing polypeptide chain during translation.

Y The process by which the code contained in the DNA molecule is transcribed into a mRNA molecule.

Z A polynucleotide molecule that occurs in two forms, DNA and RNA

AA Deoxyribonucleic acid. Macromolecule consisting of many millions of units containing a phosphate group, sugar and a base (A, T, C or G). Stores the genetic information of the cell.

BB The basic unit of heredity.

CC Sequences of DNA occupying the same gene locus (position) on different, but homologous, chromosomes. The forms of a gene.

Important in this section...
- *Understand how cell ultrastructure is related to function.*
- *Describe between plant and animal cells.*
- *Describe the factors contributing to genetic variation.*
- *Recognise the importance of plants to humans.*
- *Understand the factors influencing biodiversity.*

Variation & Heredity

Heredity
- Meiosis
- Mendel's law of inheritance
- Gametes and gametogenesis
- Fertilisation in mammals and plants

Cell specialisation
- Stem cells
- Stem cell technologies
- Plant tissue culture
- Cell specialisation

Genetic variation
- Sources of genetic variation
- Gene-enivronment interactions
- Variation in phenotype
- Continuous variation

Cells & Microscopy

Cells
- Prokaryotic and eukaryotic cells
- Animal cells and organelles
- Levels of organisation
- Mitosis and the cell cycle
- Protein production

Microscopy
- Techniques and applications

Microscopy provides information about cells, and how cell structure is related to function.

Variation arises through genetic and environmental factors. Cells become specialised for function.

Unit 2

Development, Plants and the Environment

Understanding cell structure provides information on cellular function. Humans exploit the functional properties of cells to manufacture useful products. Environmental factors and genetics produce variation and biodiversity.

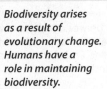

Biodiversity arises as a result of evolutionary change. Humans have a role in maintaining biodiversity.

The properties of plants are exploited by humans, including to develop new drugs.

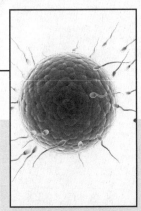

Biodiversity
- What is biodiversity?
- Britain's biodiversity
- Measuring biodiversity
- Conserving biodiversity

Speciation and evolution
- What is a species
- Adaptations to niche
- Darwin's theory
- Natural selection

Classifying organisms
- New evidence for classification
- Molecular phylogeny
- Classification systems

Plant cells and structure
- Plant cells and organelles
- Structure and function of starch
- Structure and function of cellulose
- Plant transport mechanisms
- Plant mineral deficiencies

Putting plants to use
- How humans use plants
- Sustainable use of plant products
- Tensile strength

Drug development
- William Withering's digitalis soup
- Drug development protocols
- Antimicrobial properties of plants

Biodiversity & Evolution

Plants as Resources

Unit 2
6BI02

3

Cells and Microscopy

KEY CONCEPTS

▶ Describe the structure and ultrastructure of eukaryotic and prokaryotic cells.

▶ Identify the main organelles of an animal cell, and describe their function.

▶ Describe how the cells, tissues and organs of multicellular organisms have a hierarchical system of arrangement.

▶ Explain mitosis, and its functions.

KEY TERMS

anaphase
animal cell
cell
cell cycle
centrioles
cytokinesis
electron microscopy
eukaryote
extracellular enzyme
Golgi apparatus
interphase
light microscopy
lysosome
metaphase
microscopy
mitochondria
mitosis
nucleolus
nucleus
organ
organelle
prokaryote
prophase
ribosomes
rough endoplasmic reticulum
scanning electron microscopy
smooth endoplasmic reticulum
telophase
tissue
transmission electron
    microscopy

OBJECTIVES

☐ 1. Use the **KEY TERMS** to help you understand and complete these objectives.

Types of Cells and Organelles
page 127-132, 142, 145

☐ 2. Describe the differences between **eukaryotic** and **prokaryotic** cells with reference to their size, structure, and ultrastructure.

☐ 3. Describe the ultrastructure of an **animal cell**. Identify and describe the functions of the following organelles: **nucleus, nucleolus, ribosomes, endoplasmic reticulum** (rough and smooth ER), **mitochondria, centrioles, lysosomes,** and **Golgi apparatus.**

☐ 4. Explain the role of the rough endoplasmic reticulum (rER) and the Golgi apparatus in protein transport. Explain how **extracellular enzymes** are packaged and exported from the cell.

Microscopy
page 133-138

☐ 5. Distinguish between **light microscopy** and **electron microscopy** (EM). Appreciate that electron microscopy can be used to identify cell organelles and study cellular processes such as protein synthesis and secretion.

☐ 6. Recognise cellular organelles in electron micrographs, noting their distinguishing features in each case.

Levels of Organisation
page 139-141

☐ 7. Describe the hierarchical organisation of multicellular organisms. In particular, describe the organisation of cells into tissues, tissues into organs, and organs into organ systems.

Mitosis
page 143-144

☐ 8. Describe the stages of **mitosis** as seen in a stained root tip squash. From your observations, determine the proportion of cells dividing at any one time. †

☐ 9. Use examples to describe and explain the role of mitosis in **cell growth, repair,** and **asexual reproduction**.

Periodicals:
listings for this chapter are on page 229

Weblinks:
www.biozone.co.uk/
weblink/Edx-AS-2542.html

Teacher Resource CD-ROM:
Review of eukaryotic cells

Types of Living Things

All living things can be grouped according to the type of building blocks they are made up of. Most organisms are made up of one or more **cells**. Cells can be divided into two basic types: the **prokaryotes**, which are simple cells, and the **eukaryotes**, which are much more complex in their structure. Eukaryote cells can be further categorised as protist, fungal, plant, or animal. Viruses are non-cellular and have no cell machinery of their own. All cells must secure a source of energy if they are to survive and carry out their metabolic processes. Some, called **autotrophs**, are able to obtain energy from the physical environment, by using light or chemical energy. Other types of cell, called **heterotrophs**, obtain their energy from other living organisms or their dead remains.

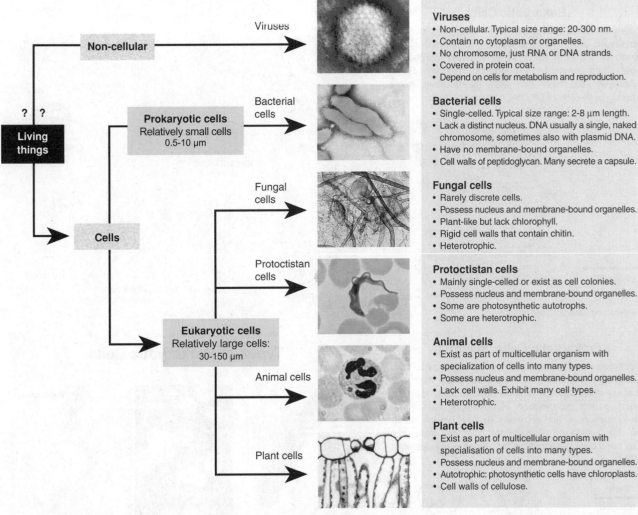

Viruses
- Non-cellular. Typical size range: 20-300 nm.
- Contain no cytoplasm or organelles.
- No chromosome, just RNA or DNA strands.
- Covered in protein coat.
- Depend on cells for metabolism and reproduction.

Bacterial cells
- Single-celled. Typical size range: 2-8 µm length.
- Lack a distinct nucleus. DNA usually a single, naked chromosome, sometimes also with plasmid DNA.
- Have no membrane-bound organelles.
- Cell walls of peptidoglycan. Many secrete a capsule.

Fungal cells
- Rarely discrete cells.
- Possess nucleus and membrane-bound organelles.
- Plant-like but lack chlorophyll.
- Rigid cell walls that contain chitin.
- Heterotrophic.

Protoctistan cells
- Mainly single-celled or exist as cell colonies.
- Possess nucleus and membrane-bound organelles.
- Some are photosynthetic autotrophs.
- Some are heterotrophic.

Animal cells
- Exist as part of multicellular organism with specialization of cells into many types.
- Possess nucleus and membrane-bound organelles.
- Lack cell walls. Exhibit many cell types.
- Heterotrophic.

Plant cells
- Exist as part of multicellular organism with specialisation of cells into many types.
- Possess nucleus and membrane-bound organelles.
- Autotrophic: photosynthetic cells have chloroplasts.
- Cell walls of cellulose.

Diagram labels: Living things → Non-cellular → Viruses; Cells → Prokaryotic cells (Relatively small cells 0.5-10 µm) → Bacterial cells; Eukaryotic cells (Relatively large cells: 30-150 µm) → Fungal cells, Protoctistan cells, Animal cells, Plant cells.

1. List the cell types above according to the way in which they obtain their energy. Include viruses in your answer as well:

   (a) Autotrophic: _____

   (b) Heterotrophic: _____

2. Consult the diagram above and describe the two main features distinguishing **eukaryotic** cells from **prokaryotic** cells:

   (a) _____

   (b) _____

3. (a) Suggest why fungi were once classified as belonging to the plant kingdom: _____

   _____

   (b) Explain why, in terms of the distinguishing features of fungi, this classification was erroneous: _____

   _____

4. Explain why the Protoctista have traditionally been a difficult group to classify: _____

   _____

   _____

Prokaryotic vs Eukaryotic Cells

Prokaryotic cells are defined by their lack of a membrane bound nucleus or any other membrane bound organelles. They are small, single celled organisms ranging in size from 0.5-10 μm, Prokaryotic cells include the **Eubacteria** and the smaller **Archaea** group, and they are often simply referred to as bacteria. Despite their simple structure, prokaryotic cells are a very metabolically diverse group. They are important in nutrient cycling, animal and plant pathogens, and have many industrial uses. **Eukaryotic cells** are characterised by a membrane bound nucleus, and the presence of other membrane bound organelles. They are relatively large cells (30-150 μm), and most eukaryotes are multicellular organisms. Eukaryotic organisms include plants, animals, fungi and protists. Eukaryotic cells are often highly specialised to fulfil specific roles within organisms.

Eukaryotic Cells

Eukaryotic cells are larger and more complex than prokaryotic cells. They are adapted to perform a wide variety of specialist roles, so have varied morphology (right) and organelle composition (below). Some eukaryotic cells have cell walls (plant, fungi, some protists), while others (animal, some protists) do not.

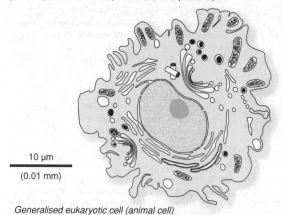

10 μm
(0.01 mm)

Generalised eukaryotic cell (animal cell)

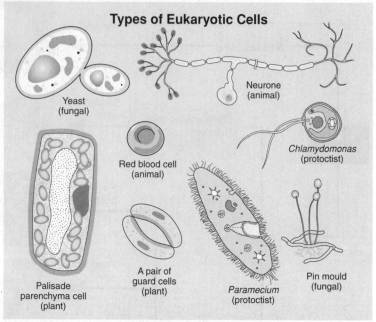

Types of Eukaryotic Cells

Yeast (fungal)

Neurone (animal)

Chlamydomonas (protoctist)

Red blood cell (animal)

Palisade parenchyma cell (plant)

A pair of guard cells (plant)

Paramecium (protoctist)

Pin mould (fungal)

Prokaryotic Cells (Bacteria)

Prokaryotes are smaller and much simpler cells than eukaryotes. Morphologically there are three main cell shapes; rods, cocci, and spirals (right). They lack the structure and organisation of eukaryotic cells (their DNA, ribosomes, and enzymes are free floating within the cell cytoplasm). Prokaryotes have cell walls, but their composition differs from the cell walls of eukaryotic cells.

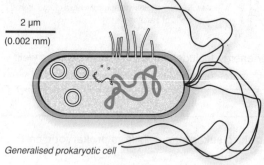

2 μm
(0.002 mm)

Generalised prokaryotic cell

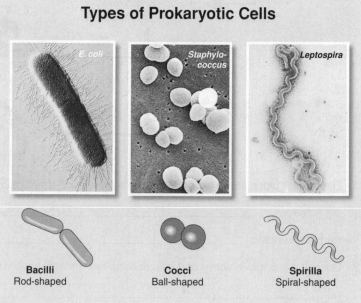

Types of Prokaryotic Cells

E. coli

Staphylo-coccus

Leptospira

Bacilli
Rod-shaped

Cocci
Ball-shaped

Spirilla
Spiral-shaped

1. Describe three features of prokaryotic cells that distinguish them from eukaryotic cells:

   (a) _____

   (b) _____

   (c) _____

2. Compare the size of prokaryotic and eukaryotic cells and relate this to their internal structure: _____

   _____

   _____

   _____

© Biozone International 2010

Related activities: Animal Cells, Plant Cells, Differentiation of Human Cells

Animal Cells

Eukaryotic cells have a similar basic structure, although they may vary tremendously in size, shape, and function. Certain features are common to almost all eukaryotic cells, including their three main regions: a **nucleus** (usually located near the centre of the cell), surrounded by a watery **cytoplasm**, which is itself enclosed by the **plasma membrane**. Animal cells do not have a regular shape, and some (such as the phagocytic white blood cells) are quite mobile. The diagram below illustrates the basic ultrastructure of an **intestinal epithelial cell**. It contains organelles common to most relatively unspecialised human cells. The intestine is lined with these columnar epithelial cells. They are taller than they are wide, with the nucleus close to the base and hairlike projections (**microvilli**) on their free surface. Microvilli increase the surface area of the cell, greatly increasing the capacity for absorption.

Structures and Organelles in an Intestinal Epithelial Cell

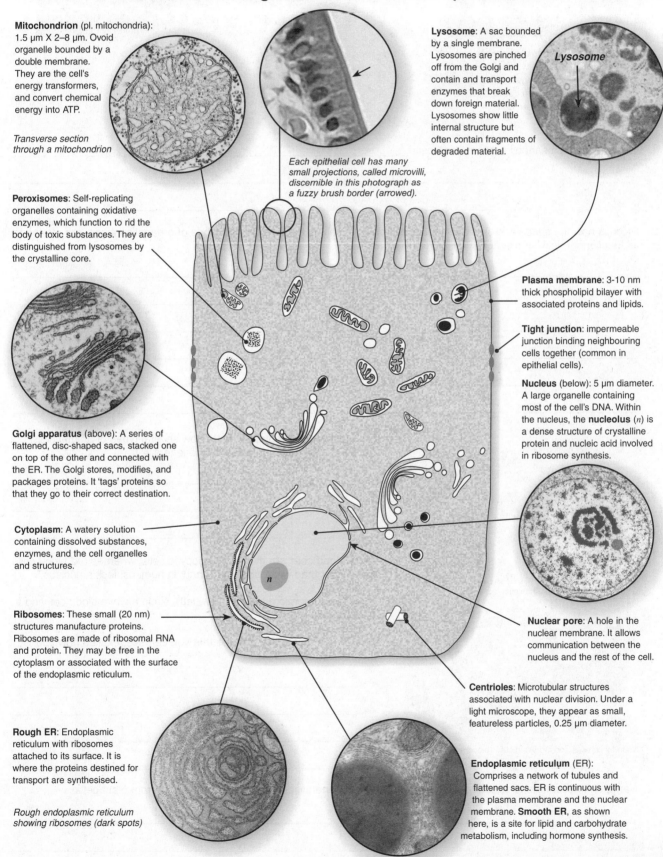

Mitochondrion (pl. mitochondria): 1.5 μm X 2–8 μm. Ovoid organelle bounded by a double membrane. They are the cell's energy transformers, and convert chemical energy into ATP.

Transverse section through a mitochondrion

Each epithelial cell has many small projections, called microvilli, discernible in this photograph as a fuzzy brush border (arrowed).

Lysosome: A sac bounded by a single membrane. Lysosomes are pinched off from the Golgi and contain and transport enzymes that break down foreign material. Lysosomes show little internal structure but often contain fragments of degraded material.

Lysosome

Peroxisomes: Self-replicating organelles containing oxidative enzymes, which function to rid the body of toxic substances. They are distinguished from lysosomes by the crystalline core.

Plasma membrane: 3-10 nm thick phospholipid bilayer with associated proteins and lipids.

Tight junction: impermeable junction binding neighbouring cells together (common in epithelial cells).

Nucleus (below): 5 μm diameter. A large organelle containing most of the cell's DNA. Within the nucleus, the **nucleolus** (*n*) is a dense structure of crystalline protein and nucleic acid involved in ribosome synthesis.

Golgi apparatus (above): A series of flattened, disc-shaped sacs, stacked one on top of the other and connected with the ER. The Golgi stores, modifies, and packages proteins. It 'tags' proteins so that they go to their correct destination.

Cytoplasm: A watery solution containing dissolved substances, enzymes, and the cell organelles and structures.

Ribosomes: These small (20 nm) structures manufacture proteins. Ribosomes are made of ribosomal RNA and protein. They may be free in the cytoplasm or associated with the surface of the endoplasmic reticulum.

Nuclear pore: A hole in the nuclear membrane. It allows communication between the nucleus and the rest of the cell.

Centrioles: Microtubular structures associated with nuclear division. Under a light microscope, they appear as small, featureless particles, 0.25 μm diameter.

Rough ER: Endoplasmic reticulum with ribosomes attached to its surface. It is where the proteins destined for transport are synthesised.

Rough endoplasmic reticulum showing ribosomes (dark spots)

Endoplasmic reticulum (ER): Comprises a network of tubules and flattened sacs. ER is continuous with the plasma membrane and the nuclear membrane. **Smooth ER**, as shown here, is a site for lipid and carbohydrate metabolism, including hormone synthesis.

Related activities: *Prokaryote vs Eukaryote Cells, Cell Structures & Organelles*
Web links: *Eukaryotic Cells Interactive Animation, Review of Eukaryotic Cells*

A 2

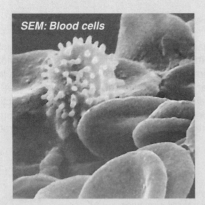

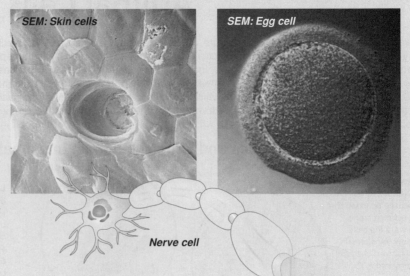

SEM: Blood cells

SEM: Skin cells

SEM: Egg cell

Many animal cells are specialised to carry out specific functions within the body. As a result, the morphology and physiology of animal cells are highly varied. Some examples are presented here.

Nerve cell

1. Explain what is meant by a generalised cell: _____

_____

_____

2. Discuss how the shape and size of a specialised cell, as well as the number and types of organelles it has, are related to its functional role. Use examples to illustrate your answer:

_____

_____

_____

_____

_____

_____

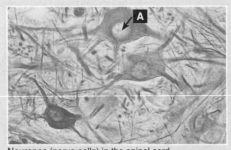

Neurones (nerve cells) in the spinal cord

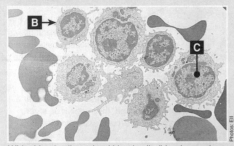

White blood cells and red blood cells (blood smear)

3. The two photomicrographs (left) show several types of animal cells. Identify the features indicated by the letters **A-C**:

A: _____

B: _____

C: _____

4. White blood cells are mobile, phagocytic cells, whereas red blood cells are smaller than white blood cells and, in humans, lack a nucleus.

(a) In the photomicrograph (below, left), circle a white blood cell and a red blood cell:

(b) With respect to the features that you can see, explain how you made your decision.

_____

_____

_____

_____

5. Name and describe one structure or organelle present in generalised animal cells but absent from plant cells:

_____

_____

Cell Structures and Organelles

Cells & Microscopy

This activity requires you to summarise information about the components of a typical eukaryotic cell. Complete the table using the list provided and by referring to other pages in this chapter. Fill in the final column with either 'YES' or 'NO'. The first has been completed for you as a guide and the log scale of measurements (next page) illustrates the relative sizes of some cells and cell structures. **List of components**: nucleus, ribosome, centrioles, mitochondrion, lysosome and vacuole (given), endoplasmic reticulum, Golgi apparatus, plasma membrane (given), cell cytoskeleton, flagella or cilia (given), cellular junctions (given).

Cell Component	Details	Visible under light microscope
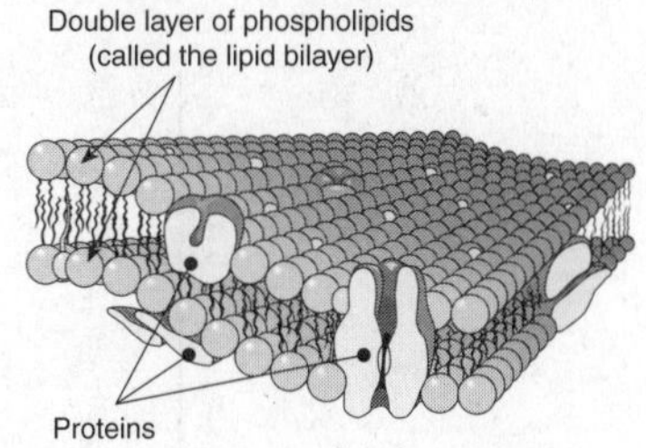 (a) Double layer of phospholipids (called the lipid bilayer) / Proteins	Name: Plasma (cell surface) membrane Location: Surrounding the cell Function: Gives the cell shape and protection. It also regulates the movement of substances into and out of the cell.	YES (but not at the level of detail shown in the diagram)
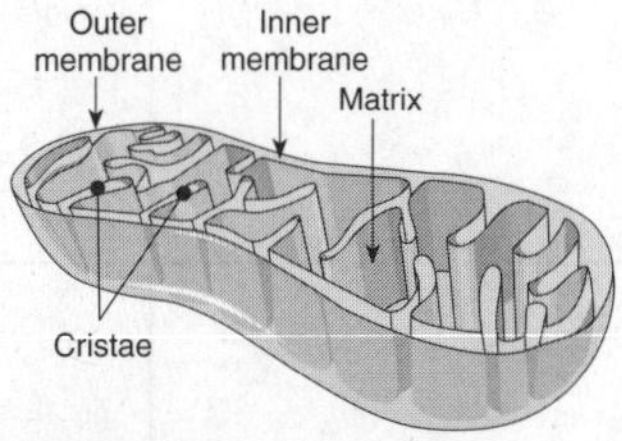 (b) Outer membrane / Inner membrane / Matrix / Cristae	Name: Location: Function:	
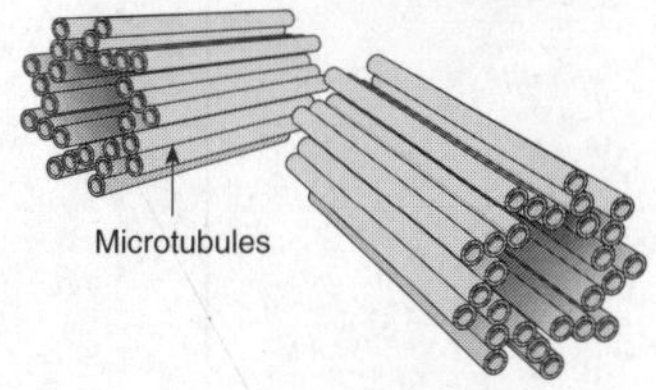 (c) Microtubules	Name: Location: Function:	
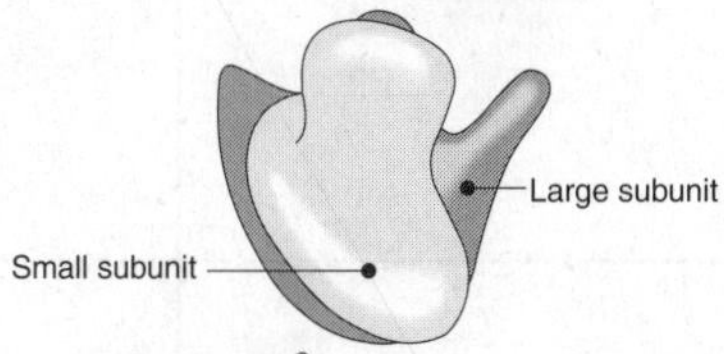 (d) Large subunit / Small subunit	Name: Location: Function:	
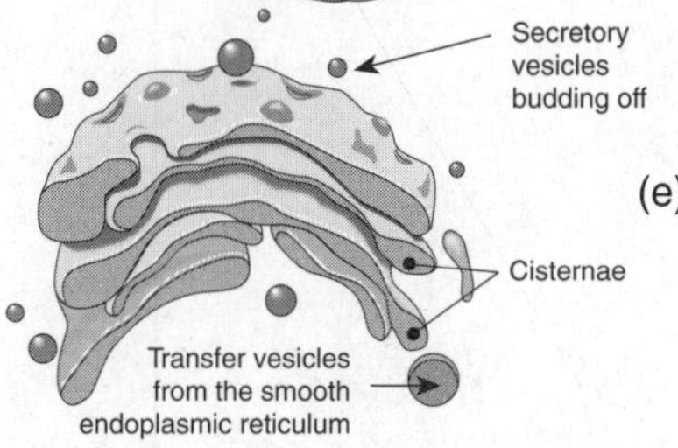 (e) Secretory vesicles budding off / Cisternae / Transfer vesicles from the smooth endoplasmic reticulum	Name: Location: Function:	
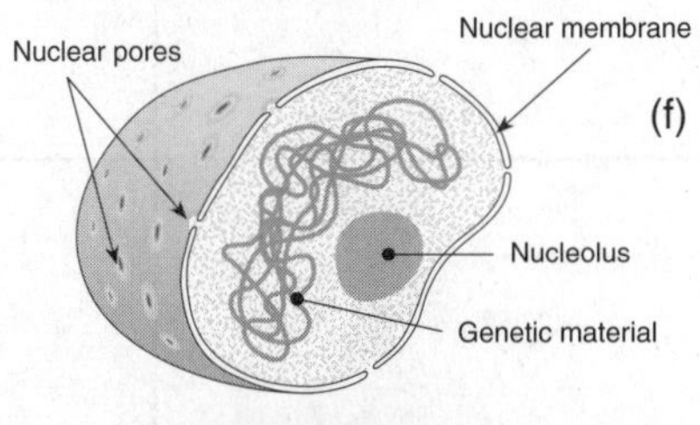 (f) Nuclear pores / Nuclear membrane / Nucleolus / Genetic material	Name: Location: Function:	
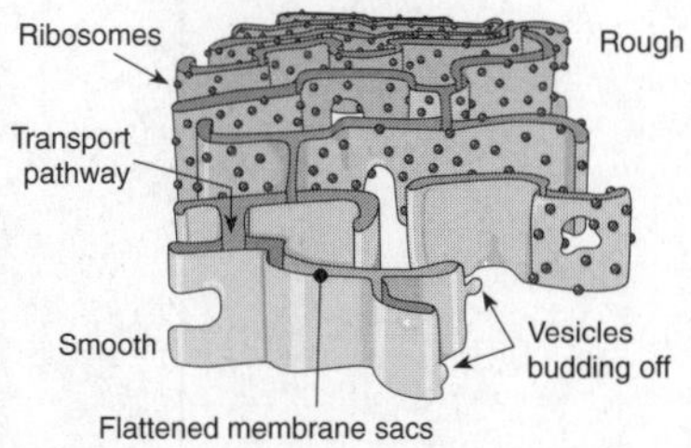 (g) Ribosomes / Rough / Transport pathway / Smooth / Vesicles budding off / Flattened membrane sacs	Name: Location: Function:	

© Biozone International 2010
Photocopying Prohibited
Periodicals: Cellular factories
Related activities: Animal Cells, Plant Cells
Web links: Eukaryotic Cells Interactive Animation
RA 2

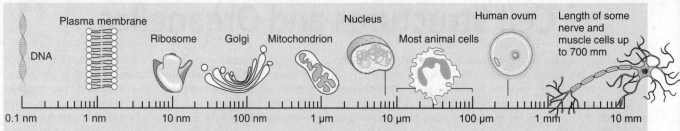

Cell Component	Details	Visible under light microscope
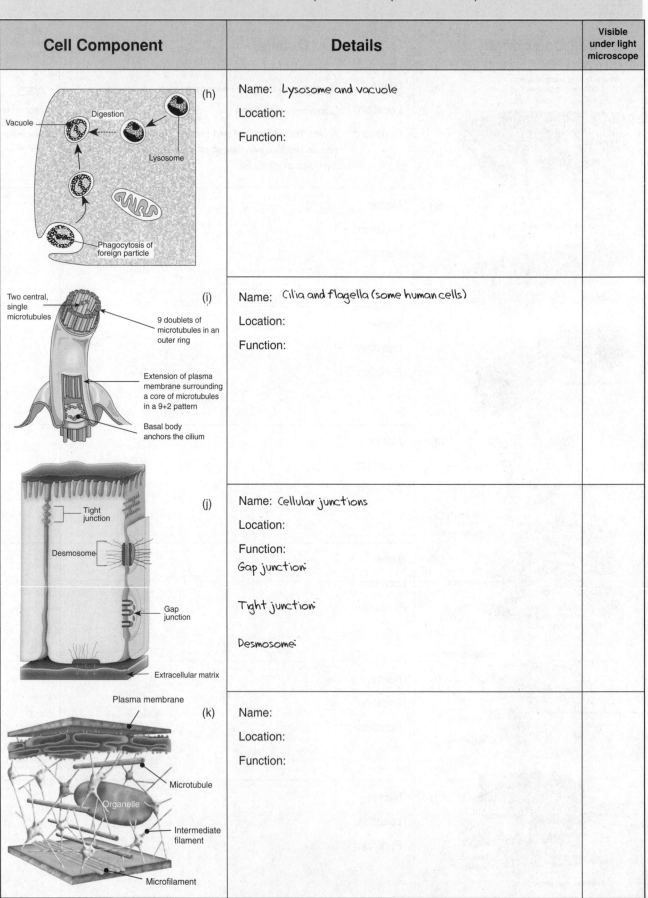 (h)	Name: Lysosome and vacuole Location: Function:	
(i)	Name: Cilia and flagella (some human cells) Location: Function:	
(j)	Name: Cellular junctions Location: Function: Gap junction: Tight junction: Desmosome:	
(k)	Name: Location: Function:	

Optical Microscopes

The light microscope is one of the most important instruments used in biology practicals, and its correct use is a basic and essential skill of biology. High power light microscopes use a combination of lenses to magnify objects up to several hundred times. They are called **compound microscopes** because there are two or more separate lenses involved. A typical compound light microscope (bright field) is shown below (top photograph). The specimens viewed with these microscopes must be thin and mostly transparent. Light is focused up through the condenser and specimen; if the specimen is thick or opaque, little or no detail will be visible. The microscope below has two eyepieces (**binocular**), although monocular microscopes, with a mirror rather than an internal light source, may still be encountered. Dissecting microscopes (lower photograph) are a type of binocular microscope used for observations at low total magnification (x4 to x50), where a large working distance between objectives and stage is required. A dissecting microscope has two separate lens systems, one for each eye. Such microscopes produce a 3-D view of the specimen and are sometimes called stereo microscopes for this reason.

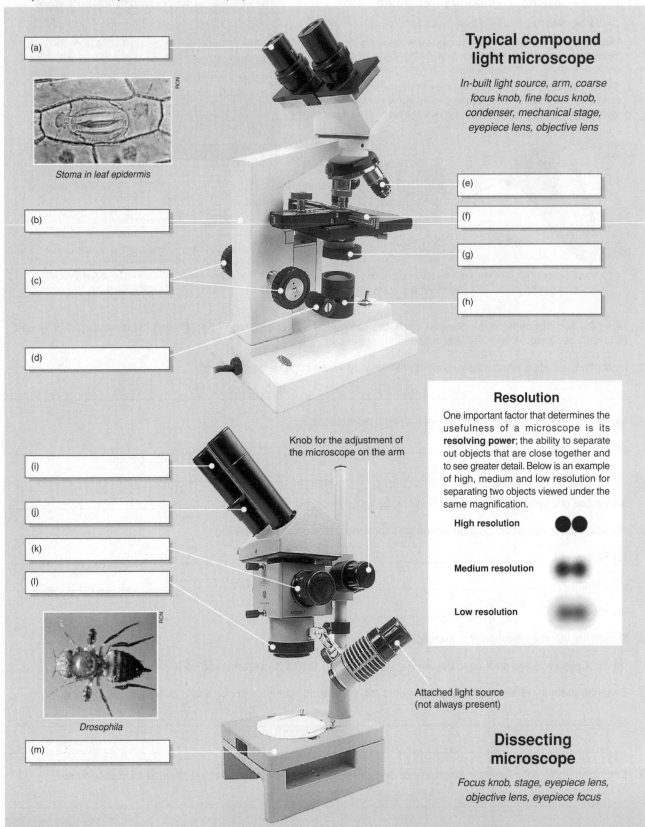

(a)

Stoma in leaf epidermis

(b)

(c)

(d)

(e)

(f)

(g)

(h)

Typical compound light microscope

In-built light source, arm, coarse focus knob, fine focus knob, condenser, mechanical stage, eyepiece lens, objective lens

Resolution

One important factor that determines the usefulness of a microscope is its **resolving power**; the ability to separate out objects that are close together and to see greater detail. Below is an example of high, medium and low resolution for separating two objects viewed under the same magnification.

High resolution

Medium resolution

Low resolution

(i)

(j)

(k)

(l)

Knob for the adjustment of the microscope on the arm

Drosophila

(m)

Attached light source (not always present)

Dissecting microscope

Focus knob, stage, eyepiece lens, objective lens, eyepiece focus

Periodicals: Light microscopy

Related activities: Plant Cells, Animal Cells
Web links: Introduction to the Microscope, LM Animation

RA 2

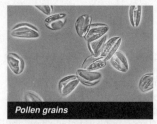

Pollen grains

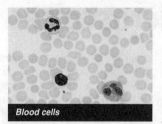

Blood cells

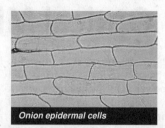

Onion epidermal cells

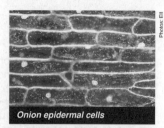

Onion epidermal cells

Photos: Eii

Phase contrast illumination increases contrast of transparent specimens by producing interference effects.

Leishman's stain is used to show red blood cells as red/pink, while staining the nucleus of white blood cells blue.

Standard **bright field** lighting shows cells with little detail; only cell walls, with the cell nuclei barely visible.

Dark field illumination is excellent for viewing near transparent specimens. The nucleus of each cell is visible.

Making a temporary wet mount

1. **Sectioning**: Very thin sections of fresh material are cut with a razorblade.

2. **Mounting**: The thin section(s) are placed in the centre of a clean glass microscope slide and covered with a drop of mounting liquid (e.g. water, glycerol or stain). A coverslip is placed on top to exclude air (below).

3. **Staining**: Dyes can be applied to stain some structures and leave others unaffected. The stains used in dyeing living tissues are called **vital stains** and they can be applied before or after the specimen is mounted.

Commonly used temporary stains

Stain	Final colour	Used for
Iodine solution	blue-black	Starch
Aniline sulfate	yellow	Lignin
Schultz's solution	blue	Starch
	blue or violet	Cellulose
	yellow	Protein, cutin, lignin, suberin
Methylene blue	blue	Nuclei

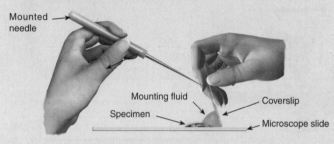

Mounted needle

Mounting fluid

Specimen

Coverslip

Microscope slide

A mounted needle is used to support the coverslip and lower it gently over the specimen. This avoids including air in the mount.

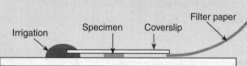

Irrigation   Specimen   Coverslip   Filter paper

If a specimen is already mounted, a drop of stain can be placed at one end of the coverslip and drawn through using filter paper (above). Water can be drawn through in the same way to remove excess stain.

1. Label the two diagrams on the previous page, the compound light microscope (a) to (h) and the dissecting microscope (i) to (m), using words from the lists supplied.

2. Describe a situation where phase contrast microscopy would improve image quality: _____

_____

3. Identify two structures that could be seen by light microscopy in:

   (a) A plant cell: _____

   (b) An animal cell: _____

4. Name one cell structure that can not be seen by light microscopy: _____

5. Identify a stain that would be appropriate for improving definition of the following:

   (a) Blood cells: _____   (d) Lignin: _____

   (b) Starch: _____   (e) Nuclei and DNA: _____

   (c) Protein: _____   (f) Cellulose: _____

6. Determine the magnification of a microscope using:

   (a) 15 X eyepiece and 40 X objective lens: _____   (b) 10 X eyepiece and 60 X objective lens: _____

7. Describe the main difference between a bright field, compound light microscope and a dissecting microscope:

_____

_____

8. Explain the difference between magnification and resolution (resolving power) with respect to microscope use:

_____

_____

Electron Microscopes

Electron microscopes (EMs) use a beam of electrons, instead of light, to produce an image. The higher resolution of EMs is due to the shorter wavelengths of electrons. There are two basic types of electron microscope: **scanning electron microscopes** (SEM) and **transmission electron microscopes** (TEM). In SEMs, the electrons are bounced off the surface of an object to produce detailed images of the external appearance. TEMs produce very clear images of specially prepared thin sections.

Transmission Electron Microscope (TEM)

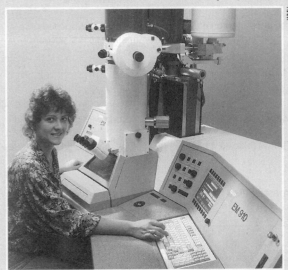

The transmission electron microscope is used to view extremely thin sections of material. Electrons pass through the specimen and are scattered. Magnetic lenses focus the image onto a fluorescent screen or photographic plate. The sections are so thin that they have to be prepared with a special machine, called an **ultramicrotome**, that can cut wafers to just 30 thousandths of a millimetre thick. It can magnify several hundred thousand times.

Scanning Electron Microscope (SEM)

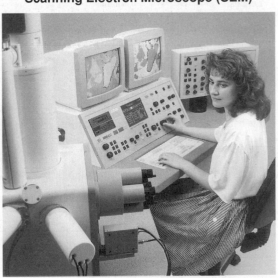

The scanning electron microscope scans a sample with a beam of primary electrons that knock electrons from its surface. These secondary electrons are picked up by a collector, amplified, and transmitted onto a viewing screen or photographic plate, producing a superb 3-D image. A microscope of this power can easily obtain clear pictures of organisms as small as bacteria and viruses. The image produced is of the outside surface only.

TEM diagram labels:
- Electron gun
- Electron beam
- Electromagnetic condenser lens
- Specimen
- Electromagnetic objective lens
- Vacuum pump
- Electromagnetic projector lens
- Eyepiece
- **TEM**
- Fluorescent screen or photographic plate

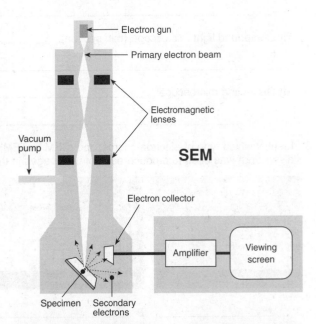

SEM diagram labels:
- Electron gun
- Primary electron beam
- Electromagnetic lenses
- Vacuum pump
- **SEM**
- Electron collector
- Amplifier
- Viewing screen
- Specimen
- Secondary electrons

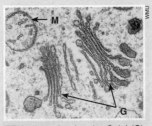

TEM photo showing the Golgi (**G**) and a mitochondrion (**M**).

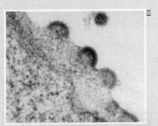

Three HIV viruses budding out of a human lymphocyte (TEM).

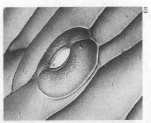

SEM photo of stoma and epidermal cells on the upper surface of a leaf.

Image of hair louse clinging to two hairs on a Hooker's sealion (SEM).

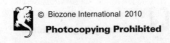
Periodicals:
TEM, SEM

Related activities: Interpreting Electron Micrographs
Web links: SEM Basics Animation

	Light Microscope	Transmission Electron Microscope (TEM)	Scanning Electron Microscope (SEM)
Radiation source:	light	electrons	electrons
Wavelength:	400-700 nm	0.005 nm	0.005 nm
Lenses:	glass	electromagnetic	electromagnetic
Specimen:	living or non-living supported on glass slide	non-living supported on a small copper grid in a vacuum	non-living supported on a metal disc in a vacuum
Maximum resolution:	200 nm	1 nm	10 nm
Maximum magnification:	1500 x	250 000 x	100 000 x
Stains:	coloured dyes	impregnated with heavy metals	coated with carbon or gold
Type of image:	coloured	monochrome (black & white)	monochrome (black & white)

1. Explain why electron microscopes are able to resolve much greater detail than a light microscope:

   _____

   _____

   _____

2. Describe two typical applications for each of the following types of microscope:

   (a) Transmission electron microscope (TEM): _____

   _____

   (b) Scanning electron microscope (SEM): _____

   _____

   (c) Compound light microscope (thin section): _____

   _____

   (d) Dissecting microscope: _____

   _____

3. Identify which type of electron microscope (SEM or TEM) or optical microscope (compound light microscope or dissecting) was used to produce each of the images in the photos below (A-H):

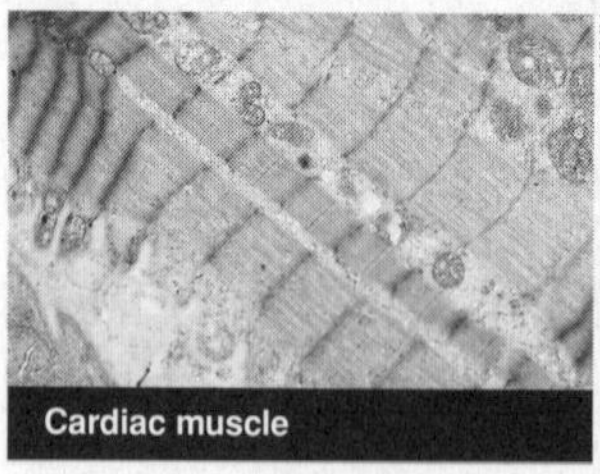

   Cardiac muscle

   A _____

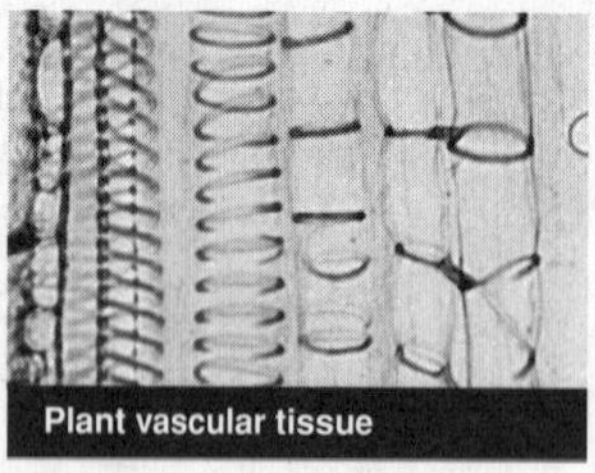

   Plant vascular tissue

   B _____

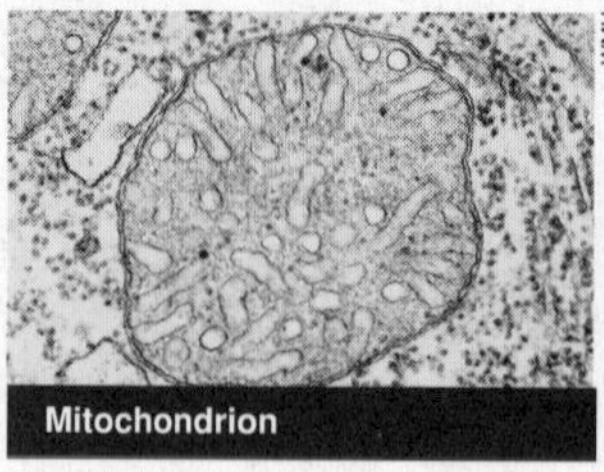

   Mitochondrion

   C _____

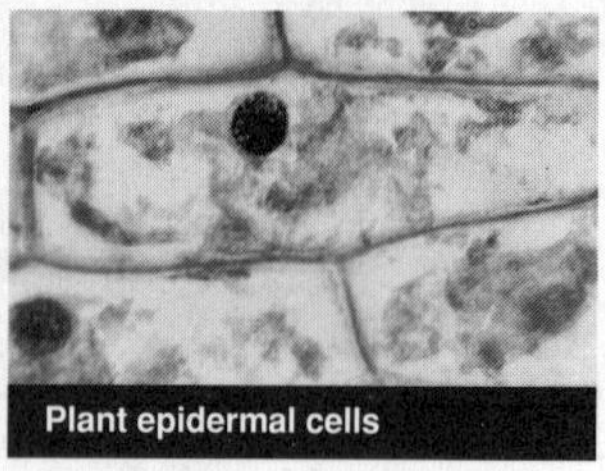

   Plant epidermal cells

   D _____

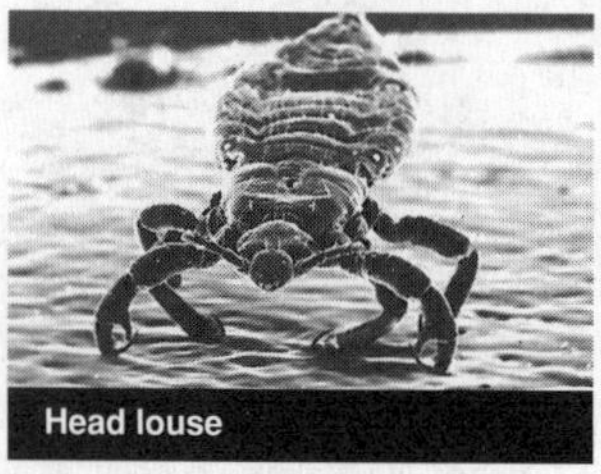

   Head louse

   E _____

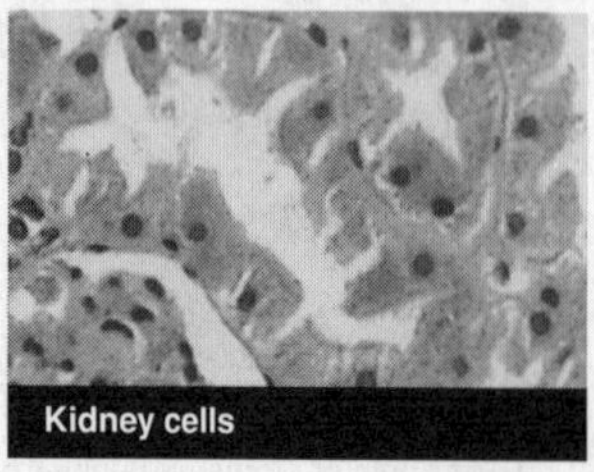

   Kidney cells

   F _____

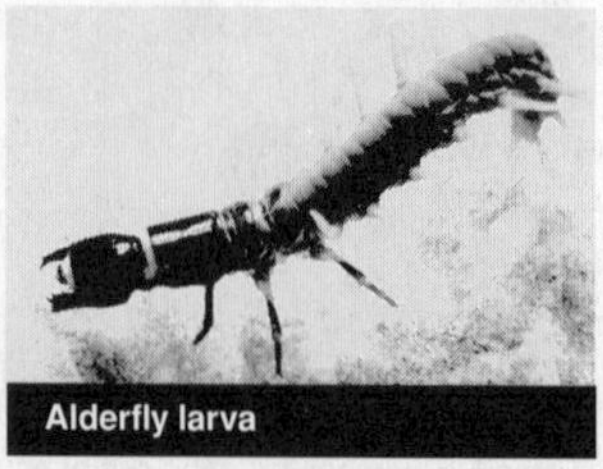

   Alderfly larva

   G _____

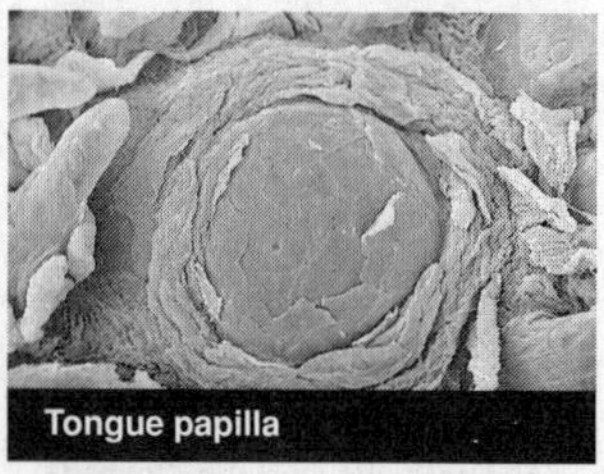

   Tongue papilla

   H _____

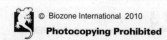

Interpreting Electron Micrographs

The photographs below were taken using a **transmission electron microscope** (TEM). They show the ultrastructure of some organelles. Remember that these photos are showing only **parts of cells**, **not whole cells**. Some of the photographs show more than one type of organelle. The questions refer to the main organelle indicated in the photo.

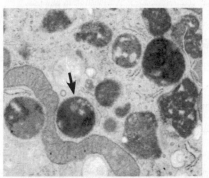
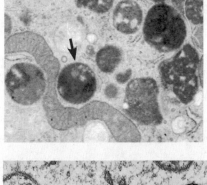

1. (a) Identify this organelle (arrowed): _____

   (b) State which kind of cell(s) this organelle would be found in:

   _____

   (c) Describe the function of this organelle: _____

   _____

   _____

   (d) Label **two** structures that can be seen inside this organelle.

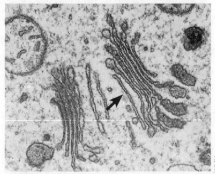

2. (a) Name this organelle (arrowed): _____

   (b) State which kind of cell(s) this organelle would be found in:

   _____

   (c) Describe the function of this organelle: _____

   _____

   _____

3. (a) Name the large, circular organelle: _____

   (b) State which kind of cell(s) this organelle would be found in:

   _____

   (c) Describe the function of this organelle: _____

   _____

   _____

   (d) Label **two** regions that can be seen **inside** this organelle.

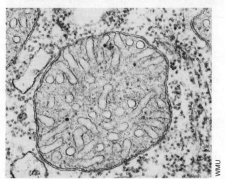

4. (a) Name and label the ribbon-like organelle in this photograph (arrowed):

   _____

   (b) State which kind of cell(s) this organelle is found in:

   _____

   (c) Describe the function of these organelles: _____

   _____

   (d) Name the dark 'blobs' attached to the organelle you have labelled:

   _____

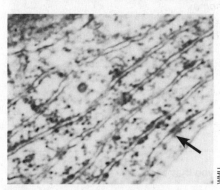
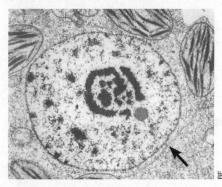

5. (a) Name this large circular structure (arrowed): _____

   (b) State which kind of cell(s) this structure would be found in: _____

   _____

   (c) Describe the function of this structure: _____

   _____

   (d) Label three features relating to this structure in the photograph.

Periodicals:
The power behind an electron microscopist

Related activities: *Electron Micrographs, Animal Cells, Plant Cells*

RA 2

Identifying Structures in an Animal Cell

Our knowledge of cell ultrastructure has been made possible by the advent of electron microscopy. Transmission electron microscopy is the most frequently used technique for viewing cellular organelles. When viewing transmission electron micrographs, the cellular organelles may appear to be quite different depending on whether they are in transverse or longitudinal section.

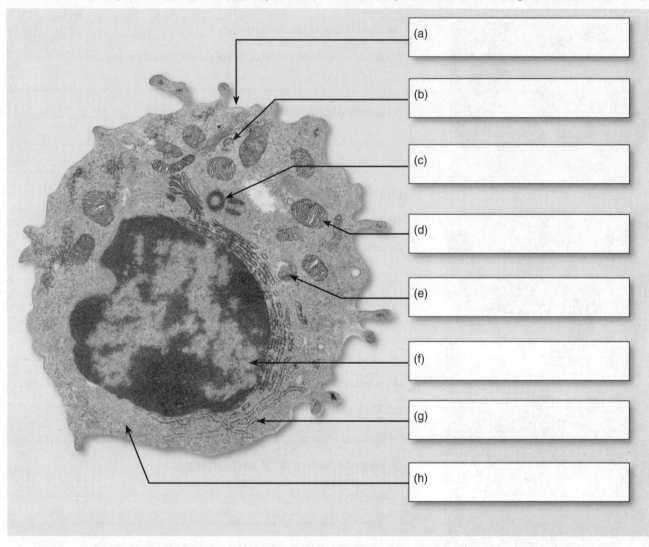

(a)

(b)

(c)

(d)

(e)

(f)

(g)

(h)

1. Identify and label the structures in the cell above using the following list of terms: *cytoplasm, plasma membrane, rough endoplasmic reticulum, mitochondrion, nucleus, centriole, Golgi apparatus, lysosome*

2. In the electron micrograph above, identify which of the organelles are shown in both transverse and longitudinal section:

    _____

3. Plants lack any of the mobile phagocytic cells typical of animals. Explain why this is the case:

    _____

    _____

4. The animal pictured above is a lymphocyte. Describe the features that suggest to you that:

    (a) It has a role in producing and secreting proteins: _____

    _____

    (b) It is metabolically very active: _____

    _____

5. Describe the features of the lymphocyte cell above that identify it as an eukaryotic cell: _____

    _____

    _____

Related activities: *Cell Structure and Organelles, Animal Cells, Prokaryotic vs Eukaryotic Cells*

Levels of Organisation

Organisation and the emergence of novel properties in complex systems are two of the defining features of living organisms. Organisms are organised according to a hierarchy of structural levels (below), each level building on the one before it. At each level, novel properties emerge that were not present at the simpler level. Hierarchical organisation allows specialised cells to group together into tissues and organs to perform a particular function. This improves efficiency of function in the organism.

In the spaces provided for each question below, assign each of the examples listed to one of the levels of organisation as indicated.

1. **Animals**: *adrenaline, blood, bone, brain, cardiac muscle, cartilage, collagen, DNA, heart, leucocyte, lysosome, mast cell, nervous system, neurone, phospholipid, reproductive system, ribosomes, Schwann cell, spleen, squamous epithelium.*

   (a) Molecular level: _____

   _____

   (b) Organelles: _____

   _____

   (c) Cells: _____

   _____

   (d) Tissues: _____

   _____

   (e) Organs: _____

   _____

   (f) Organ system: _____

   _____

   _____

2. **Plants**: *cellulose, chloroplasts, collenchyma, companion cells, DNA, epidermal cell, fibres, flowers, leaf, mesophyll, parenchyma, pectin, phloem, phospholipid, ribosomes, roots, sclerenchyma, tracheid.*

   (a) Molecular level: _____

   _____

   (b) Organelles: _____

   _____

   (c) Cells: _____

   _____

   (d) Tissues: _____

   _____

   (e) Organs: _____

   _____

   _____

MOLECULAR LEVEL

Atoms and molecules form the most basic level of organisation. This level includes all the chemicals essential for maintaining life e.g. water, ions, fats, carbohydrates, amino acids, proteins, and nucleic acids.

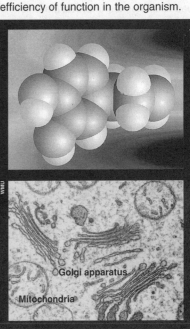

ORGANELLE LEVEL

Many diverse molecules may associate together to form complex, highly specialised structures within cells called cellular organelles e.g. mitochondria, Golgi apparatus, endoplasmic reticulum, chloroplasts.

Golgi apparatus

Mitochondria

CELLULAR LEVEL

Cells are the basic structural and functional units of an organism. Each cell type has a different structure and function; the result of cellular differentiation during development.

Animal examples include: epithelial cells, osteoblasts, muscle fibres.

Plant examples include: sclereids, xylem vessels, sieve tubes.

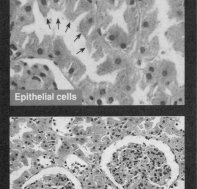

Epithelial cells

TISSUE LEVEL

Tissues are composed of groups of cells of similar structure that perform a particular, related function.

Animal examples include: epithelial tissue, bone, muscle.

Plant examples include: phloem, chlorenchyma, endodermis, xylem.

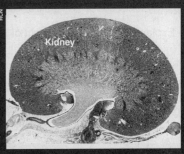

Epithelial tissue of the glomerulus

ORGAN LEVEL

Organs are structures of definite form and structure, made up of two or more tissues.

Animal examples include: heart, lungs, brain, stomach, kidney.

Plant examples include: leaves, roots, storage organs, ovary.

Kidney

ORGAN SYSTEM LEVEL

In animals, organs form parts of even larger units known as **organ systems**. An organ system is an association of organs with a common function, e.g. digestive system, cardiovascular system, and the urinary system. In all, eleven organ systems make up the **organism**.

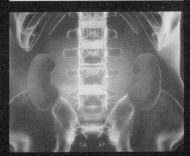

Animal Tissues

The study of tissues (plant or animal) is called **histology**. The cells of a tissue, and their associated extracellular substances, e.g. collagen, are grouped together to perform particular functions. Tissues improve the efficiency of operation because they enable tasks to be shared amongst various specialised cells. **Animal tissues** can be divided into four broad groups: **epithelial tissues**, **connective tissues**, **muscle**, and **nervous**

tissues. Organs usually consist of several types of tissue. The heart mostly consists of cardiac muscle tissue, but also has epithelial tissue, which lines the heart chambers to prevent leaking, connective tissue for strength and elasticity, and nervous tissue, in the form of neurones, which direct the contractions of the cardiac muscle. The features of some of he more familiar animal tissues are described below.

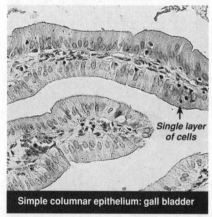

Blood

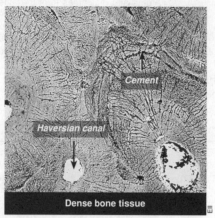

Dense bone tissue

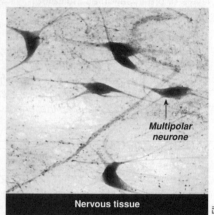

Nervous tissue

Connective tissue is the major supporting tissue of the animal body. It comprises cells, widely dispersed in a semi-fluid matrix. Connective tissues bind other structures together and provide support, and protection against damage, infection, or heat loss. Connective tissues include dentine (teeth), adipose (fat) tissue, bone (above) and cartilage, and the tissues around the body's organs and blood vessels. Blood (above, left) is a special type of liquid tissue, comprising cells floating in a liquid matrix.

Nervous tissue contains densely packed nerve cells (neurones) which are specialised for the transmission of nerve impulses. Associated with the neurones there may also be supporting cells and connective tissue containing blood vessels.

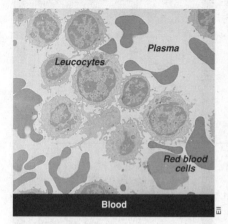

Simple columnar epithelium: gall bladder

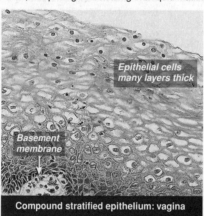

Compound stratified epithelium: vagina

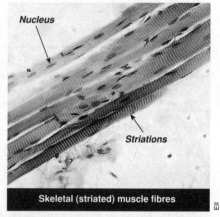

Skeletal (striated) muscle fibres

Epithelial tissue is organised into single (above, left) or layered (above) sheets. It lines internal and external surfaces (e.g. blood vessels, ducts, gut lining) and protects the underlying structures from wear, infection, and/or pressure. Epithelial cells rest on a basement membrane of fibres and collagen and are held together by a carbohydrate-based "glue". The cells may also be specialised for absorption, secretion, or excretion. Examples: stratified (compound) epithelium of vagina, ciliated epithelium of respiratory tract, cuboidal epithelium of kidney ducts, and the columnar epithelium of the intestine.

Muscle tissue consists of very highly specialised cells called fibres, held together by connective tissue. The three types of muscle in the body are cardiac muscle, skeletal muscle (above), and smooth muscle. Muscles bring about both voluntary and involuntary (unconscious) body movements.

1. Explain how the development of tissues improves functional efficiency: _____

_____

_____

2. Describe the general functional role of each of the following broad tissue types:

   (a) Epithelial tissue: _____ (c) Muscle tissue: _____

   (b) Nervous tissue: _____ (d) Connective tissue: _____

3. Identify the particular features that contribute to the particular functional role of each of the following tissue types:

   (a) Muscle tissue: _____

   _____

   (b) Nervous tissue: _____

Related activities: Levels of Organisation
Web links: Animal Tissues

Plant Tissues

Plant tissues are divided into two groups: simple and complex. **Simple tissues** contain only one cell type and form packing and support tissues. **Complex tissues** contain more than one cell type and form the conducting and support tissues of plants. Tissues are in turn grouped into tissue systems which make up the plant body. Vascular plants have three systems; the dermal, vascular, and ground tissue systems. The **dermal system** is the outer covering of the plant providing protection and reducing water loss. **Vascular tissue** provides the transport system by which water and nutrients are moved through the plant. The **ground tissue** system, which makes up the bulk of a plant, is made up mainly of simple tissues such as parenchyma, and carries out a wide variety of roles within the plant including photosynthesis, storage, and support.

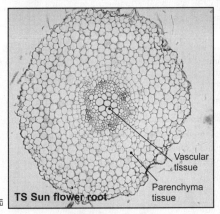
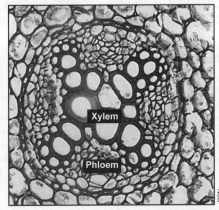
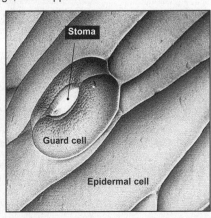

TS Sun flower root — Vascular tissue, Parenchyma tissue

Xylem, Phloem

Stoma, Guard cell, Epidermal cell

Simple Tissues

Simple tissues consists of only one or two cell types. **Parenchyma tissue** is the most common and involved in storage, photosynthesis, and secretion. **Collenchyma tissue** comprises thick-walled collenchyma cells alternating with layers of intracellular substances (pectin and cellulose) to provide flexible support. The cells of **sclerenchyma** tissue (fibres and sclereids) have rigid cell walls which provide support.

Complex Tissues

Xylem and phloem tissue (above left), which together make up the plant **vascular tissue** system, are complex tissues. Each comprises several tissue types including tracheids, vessel members, parenchyma and fibres in xylem, and sieve tube members, companion cells, parenchyma and sclerenchyma in phloem. **Dermal tissue** is also complex tissue and covers the outside of the plant. The composition of dermal tissue varies depending upon its location on the plant. Root epidermal tissue consist of epidermal cells which extend to root hairs (**trichomes**) for increasing surface area. In contrast, the epidermal tissue of leaves (above right) are covered by a waxy cuticle to reduce water loss, and specialised guard cells regulate water intake via the stomata (pores in the leaf through which gases enter and leave the leaf tissue).

1. The table below lists the major types of simple and complex plant tissue. Complete the table by filling in the role each of the tissue types plays within the plant. The first example has been completed for you.

Simple Tissue	Cell Type(s)	Role within the Plant
Parenchyma	Parenchyma cells	Involved in respiration, photosynthesis, storage and secretion.
Collenchyma		
Sclerenchyma		
Root endodermis	Endodermal cells	
Pericycle		
Complex Tissue		
Leaf mesophyll	Spongy mesophyll cells, palisade mesophyll cells	
Xylem		
Phloem		
Epidermis		

Related activities: Levels of Organisation, Xylem and Phloem
Web links: Photographic Atlas of Plant Anatomy

RA 2

Cell Sizes

Cells are extremely small and can only be seen properly when viewed through the magnifying lenses of a microscope. The diagrams below show a variety of cell types, together with a virus (non-cellular) and a multicellular microscopic animal (as a comparison). For each of these images, note the scale and relate this to the type of microscopy used.

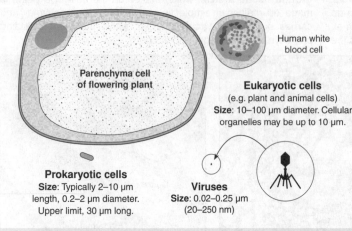

Parenchyma cell of flowering plant

Human white blood cell

Eukaryotic cells
(e.g. plant and animal cells)
Size: 10–100 µm diameter. Cellular organelles may be up to 10 µm.

Prokaryotic cells
Size: Typically 2–10 µm length, 0.2–2 µm diameter. Upper limit, 30 µm long.

Viruses
Size: 0.02–0.25 µm (20–250 nm)

Units of length (International System)

Unit	Metres	Equivalent
1 metre (m)	1 m	= 1000 millimetres
1 millimetre (mm)	10^{-3} m	= 1000 micrometres
1 micrometre (µm)	10^{-6} m	= 1000 nanometres
1 nanometre (nm)	10^{-9} m	= 1000 picometres

Micrometres are sometime referred to as **microns**. Smaller structures are usually measured in nanometres (nm) e.g. molecules (1 nm) and plasma membrane thickness (10 nm).

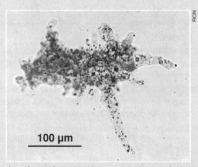

100 µm

An **Amoeba** showing extensions of the cytoplasm called pseudopodia. This protoctist changes its shape, exploring its environment.

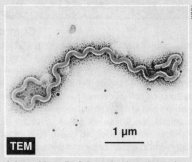

TEM

1 µm

A long thin cell of the spirochete bacterium **Leptospira pomona**, which causes the disease leptospirosis.

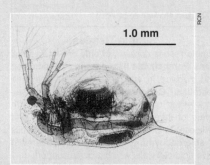

1.0 mm

Daphnia showing its internal organs. These freshwater microcrustaceans are part of the zooplankton found in lakes and ponds.

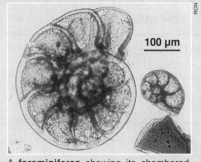

100 µm

A **foraminiferan** showing its chambered, calcified shell. These single-celled protozoans are marine planktonic amoebae.

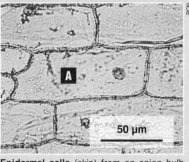

A

50 µm

Epidermal cells (skin) from an onion bulb showing the nucleus, cell walls and cytoplasm. Organelles are not visible at this resolution.

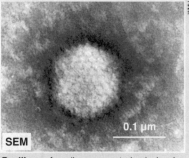

0.1 µm

SEM

Papillomavirus (human wart virus) showing its polyhedral protein coat (20 triangular faces, 12 corners) made of ball-shaped structures.

1. Using the measurement scales provided on each of the photographs above, determine the longest dimension (length or diameter) of the cell/animal/virus in µm and mm (choose the cell marked '**A**' for epidermal cells):

   (a) *Amoeba*: _____ µm _____ mm   (d) Epidermis: _____ µm _____ mm

   (b) Foraminiferan: _____ µm _____ mm   (e) *Daphnia*: _____ µm _____ mm

   (c) *Leptospira*: _____ µm _____ mm   (f) *Papillomavirus*: _____ µm _____ mm

2. List these six organisms in order of size, from the smallest to the largest: _____

   _____

3. Study the scale of your ruler and state which of these six organisms you would be able to see with your unaided eye:

   _____

4. Calculate the equivalent length in millimetres (mm) of the following measurements:

   (a) 0.25 µm: _____   (b) 450 µm: _____   (c) 200 nm: _____

Related activities: Optical Microscopes, Electron Microscopes

Mitosis and the Cell Cycle

Mitosis is part of the **cell cycle** in which an existing cell (the parent cell) divides into two (the daughter cells). Unlike meiosis, mitosis does not result in a change of chromosome numbers and the daughter cells are identical to the parent cell. Although mitosis is part of a continuous cell cycle, it is divided into stages (below) starting with prophase and ending with nuclear division or telophase. In multicellular organisms, mitosis repairs damaged cells and tissues, and produces the growth that allows an organism to reach its adult size. In unicellular organisms, and some small multicellular organisms, cell division allows organisms to reproduce asexually (as in the budding yeast shown overleaf). The example below illustrates the cell cycle in a plant cell.

The Cell Cycle and Stages of Mitosis

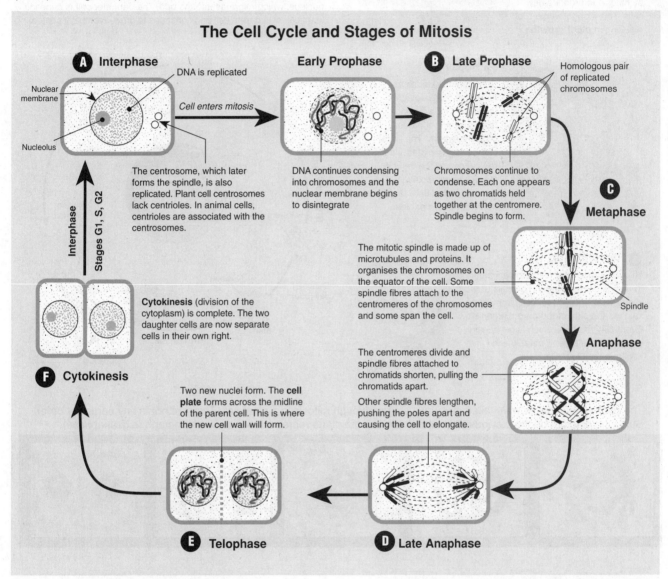

A Interphase

DNA is replicated

Nuclear membrane

Nucleolus

Cell enters mitosis

The centrosome, which later forms the spindle, is also replicated. Plant cell centrosomes lack centrioles. In animal cells, centrioles are associated with the centrosomes.

Early Prophase

DNA continues condensing into chromosomes and the nuclear membrane begins to disintegrate

B Late Prophase

Homologous pair of replicated chromosomes

Chromosomes continue to condense. Each one appears as two chromatids held together at the centromere. Spindle begins to form.

C Metaphase

The mitotic spindle is made up of microtubules and proteins. It organises the chromosomes on the equator of the cell. Some spindle fibres attach to the centromeres of the chromosomes and some span the cell.

Spindle

Anaphase

The centromeres divide and spindle fibres attached to chromatids shorten, pulling the chromatids apart.

Other spindle fibres lengthen, pushing the poles apart and causing the cell to elongate.

Interphase — Stages G1, S, G2

F Cytokinesis

Cytokinesis (division of the cytoplasm) is complete. The two daughter cells are now separate cells in their own right.

Two new nuclei form. The **cell plate** forms across the midline of the parent cell. This is where the new cell wall will form.

E Telophase

D Late Anaphase

The Cell Cycle Overview

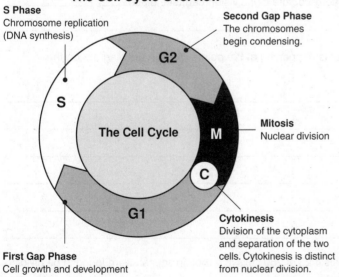

S Phase
Chromosome replication (DNA synthesis)

Second Gap Phase
The chromosomes begin condensing.

G2

S

The Cell Cycle

M

Mitosis
Nuclear division

C

Cytokinesis
Division of the cytoplasm and separation of the two cells. Cytokinesis is distinct from nuclear division.

G1

First Gap Phase
Cell growth and development

Mitosis in an Onion Root Tip Squash

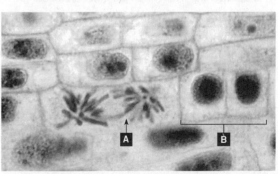

This light micrograph shows a section of the cells of an onion root tip, stained to show up the chromosomes. Cell A is in late anaphase, while in cell B the cell plate is becoming visible (cytokinesis). Cytokinesis in plant cells involves construction of a cell plate in the middle of the cell where Golgi vesicles release components for constructing a new cell wall. In animal cells, cytokinesis involves the formation of a constriction that divides the cell in two. It is usually well underway by the end of telophase and does not involve the formation of a cell plate.

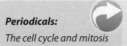

The Functions of Mitosis

❶ Growth

In plants, cell division occurs in regions of **meristematic tissue**. In the plant root tip (right), the cells in the root apical meristem are dividing by mitosis to produce new cells. This elongates the root, resulting in **plant growth**.

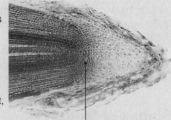

Root apical meristem

❷ Repair

Photo: AB Sheldon

Some animals, such as this skink (left), detach their limbs as a defence mechanism in a process called autotomy. The limbs can be **regenerated** via the mitotic process, although the tissue composition of the new limb differs slightly from that of the original.

❸ Reproduction

Mitotic division enables some animals to reproduce **asexually**. The cells of this Hydra (left) undergo mitosis, forming a 'bud' on the side of the parent organism. Eventually the bud, which is genetically identical to its parent, detaches to continue the life cycle.

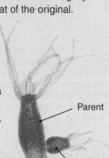

Parent

Bud

The Budding Yeast Cell Cycle

Yeasts can reproduce asexually through **budding**. In *Saccharomyces cerevisiae* (baker's yeast), budding involves mitotic division in the parent cell, with the formation of a daughter cell (or bud). As budding begins, a ring of chitin stabilises the area where the bud will appear and enzymatic activity and turgor pressure act to weaken and extrude the cell wall. New cell wall material is incorporated during this phase. The nucleus of the parent cell also divides in two, to form a daughter nucleus, which migrates into the bud. The daughter cell is genetically identical to its parent cell and continues to grow, eventually separating from the parent cell.

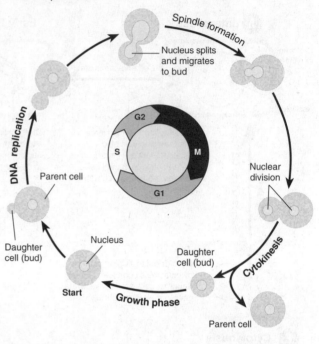

Spindle formation

Nucleus splits and migrates to bud

DNA replication

Parent cell

Nuclear division

Daughter cell (bud)

Nucleus

Cytokinesis

Daughter cell (bud)

Start

Growth phase

Parent cell

1. The photographs below were taken at various stages through mitosis in a plant cell. They are not in any particular order. Study the diagram on the previous page and determine the stage represented in each photograph (e.g. anaphase).

Photos: RCN

(a) _____   (b) _____   (c) _____   (d) _____   (e) _____

2. State two important changes that chromosomes must undergo before cell division can take place: _____

_____

3. Summarise the stages of the cell cycle by describing what is happening at the points (**A-F**) in the diagram on the previous page:

A. _____

B. _____

C. _____

D. _____

E. _____

F. _____

4. Calculate the proportion of the cell cycle occupied by mitosis if 25 of 250 cells in a root tip squash were found to be dividing:

_____

Protein Production and Transport

The **rough endoplasmic reticulum (rER)** is the primary site of protein synthesis within a cell. Some proteins are used within the cell and others are exported from the cell (e.g. antibodies produced by lymphocytes). The ER membrane keeps the newly synthesised proteins separated from proteins produced by free ribosomes in the cytoplasm. The proteins are dispatched within vesicles from a specialised region of the ER called the transitional ER, to the **Golgi apparatus**. The Golgi apparatus functions principally as a system for processing, sorting, and modifying proteins, before they are transported to their next destination. The membrane transport vesicles leaving the Golgi often have external molecules attached to them to help guide them to their destination. For example **extracellular enzymes** (such as digestive enzymes) must be transported out of the cell. They leave the Golgi in membrane bound vesicles and move to the cellular membrane where they fuse with it to release digestive enzymes. The diagram below shows a path for the production and transport of proteins between the rER and the Golgi.

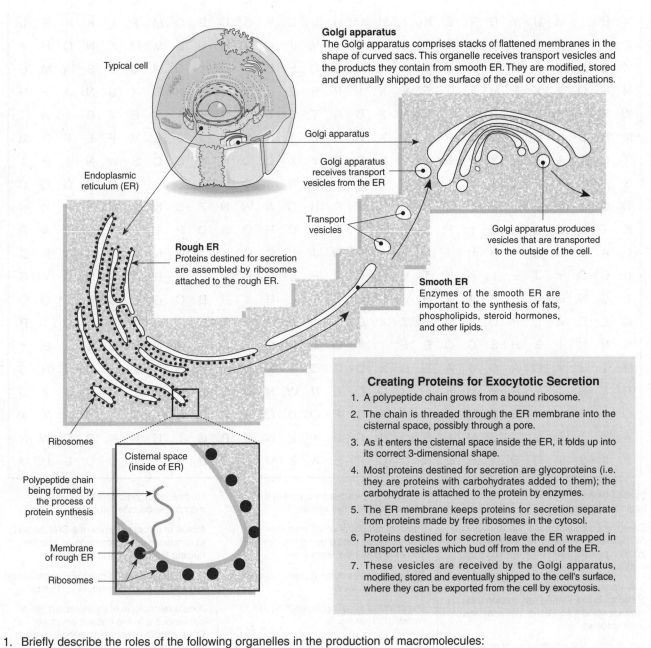

Golgi apparatus
The Golgi apparatus comprises stacks of flattened membranes in the shape of curved sacs. This organelle receives transport vesicles and the products they contain from smooth ER. They are modified, stored and eventually shipped to the surface of the cell or other destinations.

Typical cell

Golgi apparatus

Golgi apparatus receives transport vesicles from the ER

Endoplasmic reticulum (ER)

Transport vesicles

Golgi apparatus produces vesicles that are transported to the outside of the cell.

Rough ER
Proteins destined for secretion are assembled by ribosomes attached to the rough ER.

Smooth ER
Enzymes of the smooth ER are important to the synthesis of fats, phospholipids, steroid hormones, and other lipids.

Ribosomes

Cisternal space (inside of ER)

Polypeptide chain being formed by the process of protein synthesis

Membrane of rough ER

Ribosomes

Creating Proteins for Exocytotic Secretion

1. A polypeptide chain grows from a bound ribosome.

2. The chain is threaded through the ER membrane into the cisternal space, possibly through a pore.

3. As it enters the cisternal space inside the ER, it folds up into its correct 3-dimensional shape.

4. Most proteins destined for secretion are glycoproteins (i.e. they are proteins with carbohydrates added to them); the carbohydrate is attached to the protein by enzymes.

5. The ER membrane keeps proteins for secretion separate from proteins made by free ribosomes in the cytosol.

6. Proteins destined for secretion leave the ER wrapped in transport vesicles which bud off from the end of the ER.

7. These vesicles are received by the Golgi apparatus, modified, stored and eventually shipped to the cell's surface, where they can be exported from the cell by exocytosis.

1. Briefly describe the roles of the following organelles in the production of macromolecules:

   (a) Rough ER: _____

   (b) Smooth ER: _____

   (c) Golgi apparatus: _____

   (d) Transport vesicles: _____

2. Suggest why polypeptides requiring transport are synthesised by membrane-bound (rather than free) ribosomes:

   _____

   _____

© Biozone International 2010
Photocopying Prohibited

Related activities: Cell Structure and Organelles
Web links: Vesicle Transport Animation, Protein Secretion Animation

RA 2

KEY TERMS: Word Find

Use the clues below to find the relevant key terms in the WORD FIND grid

```
M  C  Q  E  U  K  A  R  Y  O  T  E  G  T  U  I  J  B  T  Y  O  T  H  A  T  U  O  Q
C  H  U  L  S  E  X  W  A  Z  S  W  I  M  U  N  Q  Z  C  P  R  H  X  S  Q  A  P  V
X  S  C  A  N  N  I  N  G  E  L  E  C  T  R  O  N  M  I  C  R  O  S  C  O  P  Y  D
P  Q  R  C  R  Y  M  Q  X  E  O  C  E  N  T  R  I  O  L  E  S  L  T  O  U  X  Z  H
K  Y  R  O  U  G  H  E  N  D  O  P  L  A  S  M  I  C  R  E  T  I  C  U  L  U  M  P
V  B  I  M  U  Y  D  Y  E  K  I  H  O  U  L  L  Y  S  O  S  O  M  E  L  K  K  S  L
E  P  G  E  Z  A  Q  A  L  S  B  H  T  L  V  J  U  J  W  U  E  V  M  K  N  Q  H  I
X  E  G  T  T  H  R  X  V  Y  A  A  X  S  O  E  S  W  Q  L  Q  M  Q  Y  S  W  M  G
N  S  Q  A  X  E  M  I  T  E  R  X  I  P  S  Z  L  A  F  W  H  B  X  U  N  J  F  H
W  R  S  P  C  D  P  U  L  A  N  S  B  A  T  E  L  O  P  H  A  S  E  V  B  I  W  T
M  P  D  H  E  C  O  C  P  S  O  X  H  B  E  T  A  L  G  J  J  L  V  F  L  F  P  M
Z  O  E  A  F  V  V  P  G  T  V  P  R  S  Y  X  O  Q  R  W  C  Q  S  W  N  I  R  I
S  I  F  S  X  C  A  T  I  B  O  L  A  U  E  N  J  Q  M  U  O  D  U  C  O  G  O  C
O  Q  B  E  L  I  J  M  O  R  G  H  Z  P  H  C  A  W  N  Z  U  N  S  N  R  F  K  R
K  T  B  L  G  N  L  I  P  K  P  M  I  T  O  C  H  O  N  D  R  I  A  T  G  I  A  O
C  A  E  L  A  T  R  H  T  A  I  J  G  W  J  L  C  C  K  T  N  N  Z  I  A  B  R  S
U  C  O  Y  J  E  I  H  N  T  V  N  N  D  F  S  G  P  N  V  H  B  A  S  N  G  Y  C
L  G  N  M  H  R  B  A  V  H  J  D  E  T  P  L  R  A  R  B  C  G  C  S  E  T  O  O
Q  E  B  A  L  P  O  K  M  V  Z  X  C  S  W  F  G  S  T  C  P  U  I  U  L  G  T  P
K  N  U  I  B  H  S  G  Q  E  F  I  T  O  I  R  O  T  H  N  Z  N  A  E  L  A  E  Y
M  I  J  V  W  A  O  J  A  D  S  X  D  D  O  S  K  U  G  F  W  T  I  W  E  X  N  S
R  L  H  I  Y  S  M  M  E  F  J  O  O  L  N  W  N  U  C  L  E  O  L  U  S  J  Z  D
O  A  B  J  Q  E  E  M  R  H  G  T  B  F  O  D  Q  Q  P  F  R  I  Q  X  X  A  Y  X
A  Y  Q  F  S  T  S  M  L  S  G  R  R  L  I  L  W  Z  P  G  I  G  N  K  R  O  W  N
F  K  S  M  O  O  T  H  E  N  D  O  P  L  A  S  M  I  C  R  E  T  I  C  U  L  U  M
```

Phase of mitosis in which the chromosomes begin to segregate

Events encompassing the "life time" of one cell between the events of cell division, including cytokinesis, DNA synthesis and mitosis.

Cell structures comprising a hollow cylinder of microtubules. During cell division they produce spindle fibres that pull apart the chromosomes.

The division of the cytoplasm of parent eukaryotic cell into two daughter cells during the late stages of mitosis.

Cell with a membrane-bound nucleus, enclosing DNA in distinct chromosomes.

Organelle that resembles a series of flattened stacks. It modifies and packages proteins and also performs a secretory function by budding off vesicles.

The phase of the cell cycle in which the cell spends most of its time and carries out its functions.

Membrane-bound vacuolar organelle which contain enzymes that form part of the intracellular digestive system.

Phase of mitosis at which time the chromosomes line up along the centre of the cell and spindle fibres form.

These organelles convert chemical energy into ATP within a cell.

The phase of a cell cycle resulting in nuclear division.

Structure found within the nucleus composed of nucleic acids and proteins.

A conspicuous organelle containing most of the cell's DNA.

Microscopy that uses lenses to focus visible light waves passing through an object into an image.

A collection of tissues that form a functional unit to serve a particular purpose.

A structural and functional part of the cell usually bound within its own membrane. Examples include the mitochondria and chloroplasts.

Unicellular organisms which lack a membrane-bounded nucleus.

Phase of mitosis in which the DNA begins to condense into chromosomes and the nuclear membrane disintegrates.

Small particles found in large numbers in all cells and involved in protein synthesis.

A complex system of membranous sacs continuous with the nuclear envelope. Covered with ribosomes, giving it a rough appearance.

An electron microscopy technique that produces images of the surface features of objects.

Part of the endoplasmic reticulum that has no ribosomes, giving it a smooth appearance.

The phase of mitosis in plant cells at which time two nuclei and the cell plate form.

A collection of cells generated from the same origin that carry out a specific role within an organ.

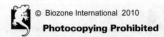

R 2

Variation and Heredity

KEY CONCEPTS

▶ Meiosis contributes to genetic variation.

▶ Sexual reproduction introduces genetic variation into a population.

▶ Stem cells can develop into any cell type. The type of cell they develop into is determined by differential gene expression.

▶ A phenotype is the product of the genotype and environmental interactions.

▶ Some phenotypes exhibit continuous variation.

KEY TERMS

acrosome reaction
allele
cancer
continuous variation
cortical reaction
crossing over
differentiation
egg (ova)
embryonic stem cell
epistasis
fertilisation
gametes
gametogenesis
gene
gene expression
genetic variation
genotype
independent assortment
meiosis
Mendel's laws of inheritance
micropropagation
oogenesis
phenotype
plant tissue culture
pluripotent
pollen tube
polygenes (polygenic
    inheritance)
polytene chromosome
specialisation
sperm
spermatogenesis
stem cell
totipotent
zygote

OBJECTIVES

☐ 1. Use the **KEY TERMS** to help you understand and complete these objectives.

Sexual Reproduction and Meiosis
page 148-160

☐ 2. Explain the role of **meiosis** in the production of **gametes**.

☐ 3. Explain how genetic variation arises through **independent assortment**, **crossing over**, and recombination of alleles.

☐ 4. Explain how mammalian **gametes**, i.e. the male **sperm** and the female **egg** (ova), are specialised for their function. Include reference to the size of the gametes and the number produced.

☐ 5. Describe the process of **fertilisation** in mammals beginning with the **acrosome reaction** and ending with the fusion of nuclei. Explain the role of the **cortical reaction** as a block to sperm entry after the egg is penetrated.

☐ 6. Describe the process of fertilisation in flowering plants, beginning with the growth of the **pollen tube** and ending with the fusion of nuclei. Appreciate the role of pollination in the plant life cycle.

☐ 7. Explain the importance of fertilisation in **sexual reproduction**.

Stem Cells and Specialisation
page 157-164

☐ 8. Explain the terms **stem cell**, **pluripotency**, and **totipotency**. Discuss how the unique properties of stem cells mean they could be used in medical therapies. Discuss the sources of stem cells, and the **ethical issues** surrounding their use.

☐ 9. Describe how the totipotent nature of plant cells is exploited in **micropropagation** of desirable phenotypes.[†]

☐ 10. Explain how cells become **specialised** through **differential gene expression**.

Genetic Variation
page 165-176

☐ 11. Explain that a **phenotype** is the result of an interaction between a **genotype** and **environmental factors**.

☐ 12. Describe the role of environmental factors in the development of some types of **cancer**. and provide examples.

☐ 13. Explain how phenotypes (e.g. height or skin colour) can be affected by **polygenes** as well as the environment. Show how this can give rise to **continuous variation**. Comment on the how the distribution of phenotypes is affected by the number of contributing genes. Understand the role of gene interactions and X inactivation can also contribute to variations in phenotype.

Periodicals:
listings for this
chapter are on page 229

Weblinks:
www.biozone.co.uk/
weblink/Edx-AS-2542.html

**Teacher Resource
CD-ROM:**
Pollination

Meiosis

Gametic meiosis produces gametes (eggs and sperm) for the purpose of sexual reproduction. Four **haploid** 'daughter cells', (cells with only half the full number of chromosomes), are formed from the parent cell in two cycles of division. In the first division of meiosis, the parent cell divides into two intermediate cells. Normally, only one chromosome from each homologous pair is passed to each of the intermediate cells. The second division simply pulls the two chromatids apart. The special events in meiosis (crossing over and recombination, and independent assortment), produce genetic variation in the gametes. In flowering plants, sporic meiosis produces haploid spores and the gametes are produced from mitosis from haploid gametophytes.

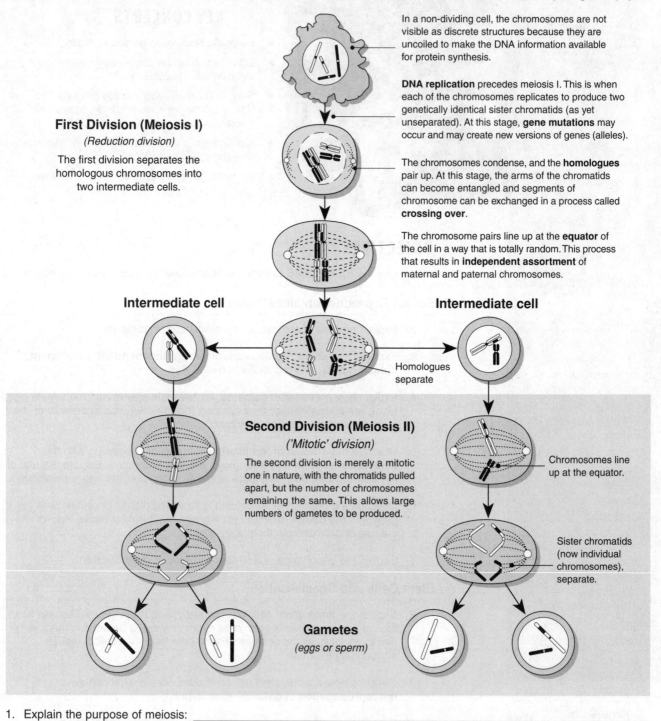

In a non-dividing cell, the chromosomes are not visible as discrete structures because they are uncoiled to make the DNA information available for protein synthesis.

First Division (Meiosis I)
(Reduction division)

The first division separates the homologous chromosomes into two intermediate cells.

DNA replication precedes meiosis I. This is when each of the chromosomes replicates to produce two genetically identical sister chromatids (as yet unseparated). At this stage, **gene mutations** may occur and may create new versions of genes (alleles).

The chromosomes condense, and the **homologues** pair up. At this stage, the arms of the chromatids can become entangled and segments of chromosome can be exchanged in a process called **crossing over**.

The chromosome pairs line up at the **equator** of the cell in a way that is totally random. This process that results in **independent assortment** of maternal and paternal chromosomes.

Intermediate cell

Intermediate cell

Homologues separate

Second Division (Meiosis II)
('Mitotic' division)

The second division is merely a mitotic one in nature, with the chromatids pulled apart, but the number of chromosomes remaining the same. This allows large numbers of gametes to be produced.

Chromosomes line up at the equator.

Sister chromatids (now individual chromosomes), separate.

Gametes
(eggs or sperm)

1. Explain the purpose of meiosis: _____

_____

2. (a) Name the two organs in a mammal that would carry out meiosis: _____

   (b) Name the two organs in a flower that would carry out meiosis: _____

3. (a) Name one way in which mitosis and meiosis are alike: _____

   (b) Describe the chromosome number of the cells produced by meiosis: _____

Related activities: Sources of Genetic Variation, Crossing Over
Web links: Meiosis Tutorial, Meiosis Animation

Periodicals:
Mechanisms of meiosis

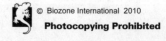
© Biozone International 2010
Photocopying Prohibited

Mitosis vs Meiosis

Cell division is fundamental to all life, as cells arise only by the division of existing cells. All types of cell division begin with replication of the cell's DNA. In eukaryotes, this is followed by division of the nucleus. There are two forms of nuclear division: **mitosis** and **meiosis**, and they have quite different purposes and outcomes. Mitosis is the simpler of the two and produces two identical daughter cells from each parent cell. Mitosis is

responsible for growth and repair processes in multicellular organisms and reproduction in single-celled and asexual eukaryotes. Gametic meiosis involves a **reduction division** in which haploid gametes are produced for the purposes of sexual reproduction. Fusion of haploid gametes in fertilisation restores the diploid cell number in the **zygote**. These two fundamentally different types of cell division are compared below.

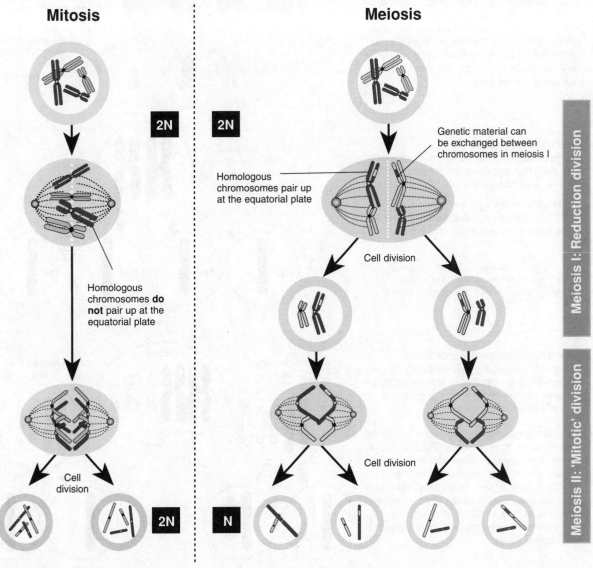

1. Describe the fundamental differences between mitosis and meiosis: _____

_____

_____

_____

_____

2. Explain why meiosis introduces genetic variability whereas (barring mutation) mitosis does not: _____

_____

_____

3. Explain why meiosis II is sometimes referred to as a 'mitotic division': _____

_____

_____

Related activities: Mitosis and the Cell Cycle, Meiosis
Web links: Comparing Mitosis and Meiosis

A 1

Mendel's Laws of Inheritance

From his work on the inheritance of phenotypic traits in peas, Mendel formulated a number of ideas about the inheritance of characters. These were later given formal recognition as Mendel's Laws of Inheritance. These are outlined below.

The Theory of Particulate Inheritance
Characteristics of both parents are passed on to the next generation as discrete entities (genes).

This model explained many observations that could not be explained by the idea of blending inheritance, which was universally accepted prior to this theory. The trait for flower colour (right) appears to take on the appearance of only one parent plant in the first generation, but reappears in later generations.

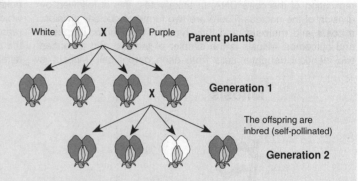

Law of Segregation
During gametic meiosis, the two members of any pair of alleles segregate unchanged and are passed into different gametes.

These gametes are eggs and sperm in animals, and pollen grains and ova in plants. The allele in the gamete will be passed on to the offspring.

> NOTE: This diagram has been simplified, omitting the stage where the second chromatid is produced for each chromosome.

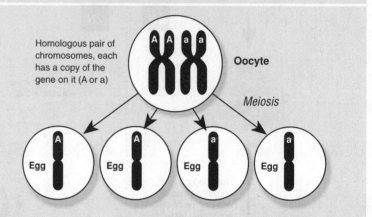

Law of Independent Assortment
Allele pairs separate independently during gamete formation, and traits are passed on to offspring independently of one another (this is only true for unlinked genes).

This diagram shows are two genes (A and B) that code for different traits. Each of these genes is represented twice, one copy (allele) on each of two homologous chromosomes. The genes A and B are located on different chromosomes and, because of this, they will be inherited independently of each other i.e. the gametes may contain any combination of the parental alleles.

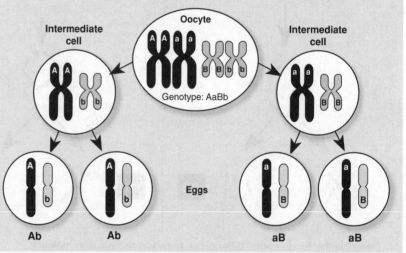

1. Briefly state what **property of genetic inheritance** allows parent pea plants that differ in flower colour to give rise to flowers of a single colour in the first generation, with both parental flower colours reappearing in the following generation:

   _____

2. The oocyte is the egg producing cell in the ovary of an animal. In the diagram illustrating the **law of segregation** above:

   (a) State the genotype for the oocyte (adult organism): _____

   (b) State the genotype of each of the **four** gametes: _____

   (c) State how many different kinds of gamete can be produced by this oocyte: _____

3. The diagram illustrating the **law of independent assortment** (above) shows only one possible result of the random sorting of the chromosomes to produce: Ab and aB in the gametes.
   (a) List another possible combination of genes (on the chromosomes) ending up in gametes from the same oocyte:

   _____

   (b) State how many different gene combinations are possible for the oocyte: _____

Related activities: Alleles, Mendel's Pea Plant Experiments
Web links: Independent Assortment

Periodicals:
Mendel's legacy

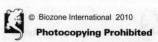
© Biozone International 2010
Photocopying Prohibited

Crossing Over

Crossing over refers to the mutual exchange of pieces of chromosome and involves the swapping of whole groups of genes between the **homologous** chromosomes. This process can occur only during the first division of **meiosis**. Errors in crossing over may cause **chromosome mutations**, which can be very damaging to development. Crossing over can also upset expected frequencies of offspring in genetic crosses. The frequency of crossing over (COV) for different genes (as followed by inherited, observable traits) can be used to determine the relative positions of genes on a chromosome (a **genetic map**). There has been a recent suggestion that crossing over may be necessary to ensure accurate cell division.

Pairing of Homologous Chromosomes

Every somatic cell contains a pair of each type of chromosome, one from each parent. These are called **homologous pairs** or **homologues**. Early in the first division of **meiosis**, the homologues pair up to form **bivalents**. This process is called **synapsis** and it brings the chromatids of the homologues into close contact.

Chiasma Formation and Crossing Over

Synapsis allows the homologous, non-sister chromatids to become entangled and the chromosomes exchange segments. This exchange occurs at regions called **chiasmata** (sing. chiasma). In the diagram (centre), a chiasma is forming and the exchange of pieces of chromosome has not yet taken place. Numerous chiasmata may develop between homologues.

Separation

Crossing over produces new allele combinations, a phenomenon known as **recombination**. When the homologues separate towards the end of meiosis I, each of the chromosomes pictured will have new mix of alleles that will be passed into the gametes soon to be formed. Recombination is an important source of variation in population gene pools.

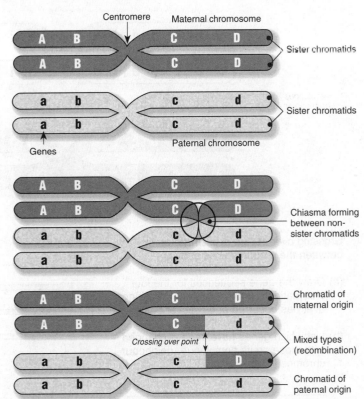

Centromere · Maternal chromosome · Sister chromatids · Genes · Paternal chromosome · Sister chromatids · Chiasma forming between non-sister chromatids · Chromatid of maternal origin · Crossing over point · Mixed types (recombination) · Chromatid of paternal origin

Gamete Formation

Once the final division of meiosis is complete, the two chromatids that made up each replicated chromosome become separated and are now referred to as chromosomes. As a result of the crossing over, **four** genetically different chromosomes are produced. If no crossing over had occurred, there would have been only two parental types (two copies of each).

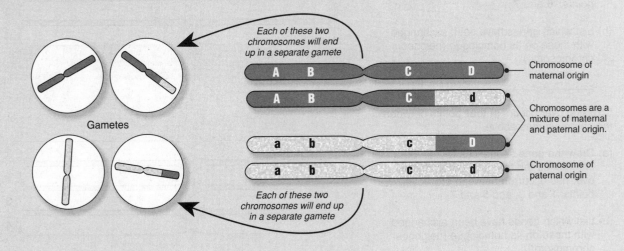

Each of these two chromosomes will end up in a separate gamete

Gametes

Chromosome of maternal origin

Chromosomes are a mixture of maternal and paternal origin.

Chromosome of paternal origin

Each of these two chromosomes will end up in a separate gamete

1. Briefly explain how the process of crossing over is going to alter the genotype of gametes: _____

_____

2. Describe the importance of crossing over in the process of evolution: _____

_____

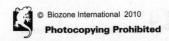

Crossing Over Problems

The diagram below shows a pair of homologous chromosomes about to form chiasma during the first division of meiosis. There are known crossover points along the length of the chromatids (the same on all four chromatids shown in the diagram). In the prepared spaces below, draw the gene sequences after crossing over has occurred on three unrelated and separate occasions (it would be useful to use different coloured pens to represent the genes from the two different chromosomes). See the diagrams on the previous page as a guide.

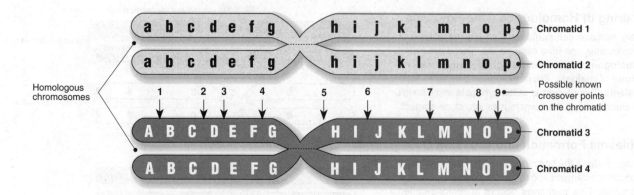

1. Crossing over occurs at a **single** point between the chromosomes above.

   (a) Draw the gene sequences for the four chromatids (on the right), after crossing over has occurred at crossover point: **2**.

   (b) List which genes have been exchanged with those on its homologue (neighbour chromosome):

   _____

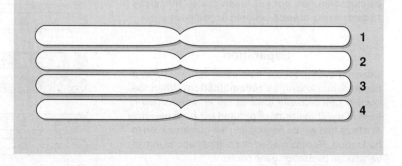

2. Crossing over occurs at **two** points between the chromosomes above.

   (a) Draw the gene sequences for the four chromatids (on the right), after crossing over has occurred between crossover points: **6** and **7**.

   (b) List which genes have been exchanged with those on its homologue (neighbour chromosome):

   _____

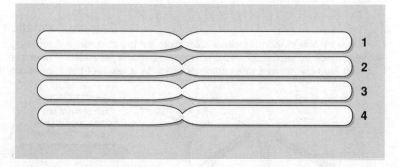

3. Crossing over occurs at **four** points between the chromosomes above.

   (a) Draw the gene sequences for the four chromatids (on the right), after crossing over has occurred between crossover points: **1** and **3**, and **5** and **7**.

   (b) List which genes have been exchanged with those on its homologue (neighbour chromosome):

   _____

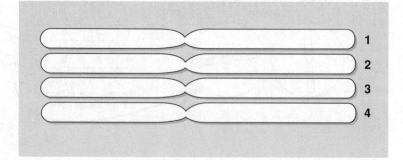

4. Explain the genetic significance of **crossing over**: _____

_____

_____

_____

Gametes

Gametes are the sex cells of organisms. The gametes of male and female mammals vary greatly in their size, shape, and number. These differences reflect their very different roles in fertilisation and reproduction. Male gametes are called **sperm** and female gametes are called eggs or **ovum** (plural ova). Mammalian sperm are highly motile and produced in large numbers. Eggs are large, few in number, and immobile in themselves. They move as a result of the wave like motion produced by the ciliated cells lining the reproductive system. Egg cells contain some food sources to nourish the developing embryo. In mammals this food source is small because once implantation into the uterus takes place, the foetus derives its nutrient supply from the mother's blood supply.

Egg Structure and Function

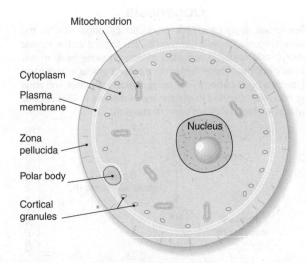

Mitochondrion
Cytoplasm
Plasma membrane
Zona pellucida
Polar body
Cortical granules
Nucleus

The ovum has no propulsion mechanism and is a simpler structure than the sperm cell. It is required to survive for a much longer time than a sperm, so it contains many more nutrients and metabolites and, as a result, it is much larger than a sperm cell (up to 100 μm).

The contents of the ovum are similar to that of a typical mammalian cell, although it is externally surrounded by a jelly-like glycoprotein called the zona pellucida. A small polar body (the remnants of a sister cell) lies between the plasma membrane and zona pellucida. Cortical granules around the inner edge of the plasma membrane contain enzymes that are released once a sperm has penetrated the egg, forming a block to prevent further sperm entry (the cortical reaction).

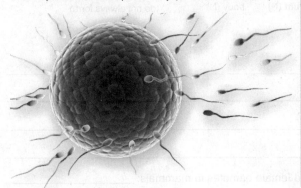

Sperm Structure and Function

Enzyme-filled acrosome
Nucleus
Mitochondria aligned in a helix
5 μm

The **midpiece** has many mitochondria to generate the energy for swimming.

The **headpiece** contains the nucleus and the acrosome, which contains the enzymes that help penetrate the egg.

The **tail** is a long flagellum that propels the sperm in its swim to the egg.

Mature spermatozoa (sperm) are produced by **spermatogenesis** in the testes. Meiotic division of spermatocytes produces spermatids, which then differentiate into mature sperm.

The sperm's structure reflect its purpose, which is to swim along the fluid environment of the female reproductive tract to the ovum, penetrate the ovum's protective barrier, and donate its genetic material. A sperm cell comprises three regions: a headpiece, containing the nucleus and pentetrative enzymes, an energy-producing midpiece, and a tail for propulsion.

Human sperm live only about 48 hours), but they swim quickly and there are so many of them (millions per ejaculation) that some are able to reach the egg to fertilise it.

A Long Way to Swim

Sperm face a tough challenge to make it to the egg. The lining of the female reproductive tract is slightly acidic, whereas sperm prefer a slightly alkaline environment. Even though the alkaline seminal fluid helps to neutralise these acids, only a few tens of thousands out of the hundreds of millions ejaculated ever make it to the uterus. Here, muscular contractions help sweep the sperm toward to the oviduct where a few thousand will encounter the egg. By this stage the sperm will have used up most of their limited energy supply, but must still make it through the outer zona of the egg. The enzymes in the acrosome help break down the zona pellucida, but only one sperm will make it through to fertilise the egg.

1. Explain why sperm need to be motile: _____

_____

_____

2. Explain why a mature ovum needs to be so many times larger than a sperm: _____

_____

_____

3. Explain why the sperm cell has a large number of mitochondria: _____

_____

Variation & Heredity

Gametogenesis

Gametogenesis involves meiotic division to produce male and female gametes for the purpose of sexual reproduction. Male mammals produce sperm in the testis by a process called **spermatogenesis**. In humans, sperm production begins at puberty and continues throughout a male's lifetime, but does decline with age. Thousands of sperm are produced every second, and take approximately two months to fully mature. Egg production in females occurs by **oogenesis**. Unlike spermatogenesis, no new eggs are produced after birth. Instead a human female is born with her entire complement of immature eggs which ripen and are released from the ovaries at regular monthly intervals (the menstrual cycle). Most commonly, one egg is released with every cycle. This takes place from the onset of puberty until menopause when menstruation ceases.

Spermatogenesis

Mature spermatozoa (sperm) are produced in the testis by spermatogenesis. In humans, they are produced at the rate of about 120 million per day. Spermatogenesis is regulated by pituitary hormones and testosterone. Spermatogonia (male germ cells) are produced throughout reproductive life, dividing by meiosis into spermatids, which mature into spermatozoa in the tubules of the testis.

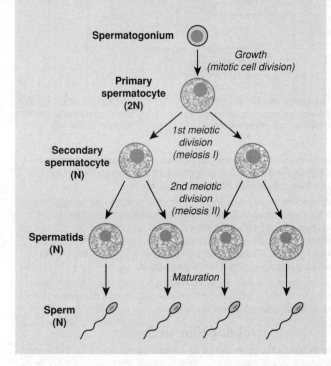

Oogenesis

Mature ova (egg cells) are produced by oogenesis in the ovary. Oogonia (female germ cells) are formed in the female embryo and undergo repeated mitotic divisions to form the primary oocytes. These remain in a resting phase in prophase of meiosis I until puberty. At puberty, meiosis resumes and eggs are released, arrested in metaphase of meiosis II. This second division is only completed upon fertilisation.

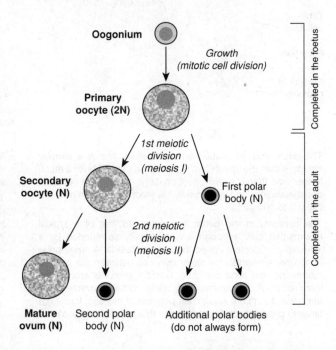

1. (a) Name the process by which mature sperm are formed: _____

   (b) Name where this process takes place: _____

   (c) Name the process by which mature ova are formed: _____

   (d) Name where this process takes place: _____

2. Discuss the main differences between the production of male and female gametes in mammals: _____

   _____

   _____

   _____

   _____

3. Explain why males can potentially be fertile all their life, but female fertility decreases and eventually ceases with age:

   _____

   _____

   _____

Fertilisation in Mammals

When an egg cell is released from the ovary it is arrested in metaphase of meiosis II and is termed a secondary oocyte. In mammals, sperm is transferred directly from the male to inside the female to fertilise the egg. This is called **internal fertilisation**, and increases the chances of the gametes meeting successfully. **Fertilisation** occurs when a sperm penetrates an egg cell at this stage and the sperm and egg nuclei unite to form the zygote. Fertilisation is always regarded as time 0 in a period of gestation (pregnancy) and has five distinct stages (below). After fertilisation, the zygote begins its **development,** i.e. its growth and differentiation into a multicellular organism.

Fertilisation (Time 0)

The stages in fertilisation are represented below in a numbered sequence (1-5)

1. Capacitation

The surface of the sperm cell undergoes changes that are essential to enabling the acrosome reaction and sperm entry.

2. The Acrosome Reaction

Enzymes from the acrosome (an enzyme-filled bag at the tip of the sperm) are released and digest a pathway through the follicle cells (not shown) and the jelly-like zona pellucida surrounding the egg cell (secondary oocyte).

3. Fusion of Sperm Head

The plasma membranes of the sperm and egg fuse, and the nucleus of the sperm enters the egg cytoplasm. Fusion causes a sudden membrane depolarisation that acts as a "fast block" to further sperm entry. The fusion of the two plasma membranes also triggers the completion of meiosis II in the egg cell and induces the cortical reaction (below).

4. The Cortical Reaction

The fusion of the two plasma membranes induces a permanent change in the egg surface that prevents further sperm entry. Cortical granules in the egg cytoplasm release their contents into the space between the plasma membrane and the vitelline layer. Substances released from the granules raise and harden the vitelline layer to form a slow (permanent) block to further sperm entry.

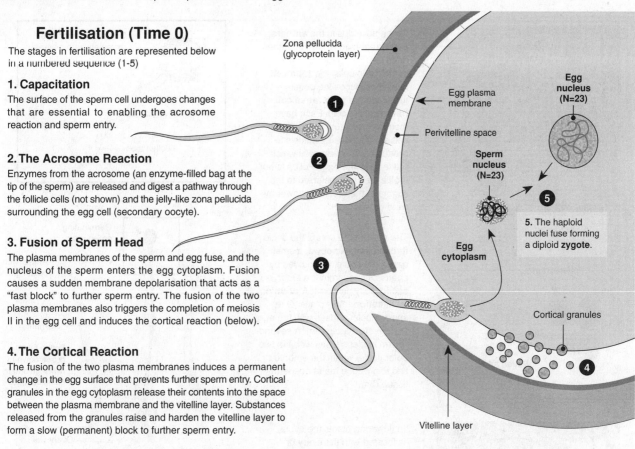

Zona pellucida (glycoprotein layer)

Egg plasma membrane

Perivitelline space

Egg nucleus (N=23)

Sperm nucleus (N=23)

5. The haploid nuclei fuse forming a diploid **zygote.**

Egg cytoplasm

Cortical granules

Vitelline layer

Variation & Heredity

---

1. Briefly describe the significant events (and their importance) occurring at each of the following stages of fertilisation:

   (a) Capacitation: _____
   _____

   (b) The acrosome reaction: _____
   _____

   (c) Fusion of egg and sperm plasma membranes: _____
   _____

   (d) The cortical reaction: _____
   _____

   (e) Fusion of egg and sperm nuclei: _____
   _____

2. Explain the significance of the blocks that prevent entry of more than one sperm into the egg (polyspermy):
   _____
   _____

3. When the egg cell is released from the ovary it is arrested in metaphase of meiosis II, and it is termed a secondary oocyte. State at which stage its mitotic division is completed:
   _____

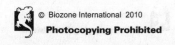
© Biozone International 2010
Photocopying Prohibited

Periodicals:
The great escape

Related activities: Meiosis
Web links: Fertilisation

A 2
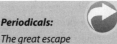

Fertilisation in Flowering Plants

As in animals, fertilisation in flowering plants (angiosperms) involves the production of the zygote by fusion of egg and sperm nuclei. In angiosperms, the sperm is contained within the pollen grain (the male gametophyte), and the egg is within the embryo sac (the female gametophyte) in the ovary of the flower.

Fertilisation occurs when a pollen tube from the pollen grain successfully penetrates the ovule and male and female nuclei fuse (below). During growth of the pollen tube, the nucleus of the sperm divides to form two male nuclei. This results in **double fertilisation** of the angiosperm egg cell.

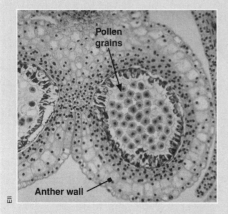

Pollen develops in the **anthers**, and is released when the anther dries out. The pollen grains contain two cells, the **tube cell** (which produces the pollen tube) and the **generative cell** (reproductive cell). Each have their own nucleus. The generative cell divides once by mitosis to produce two sperm cells which will fertilise the egg. Pollen is not motile so it is transferred to the stigma of the receiving flower by wind or animals.

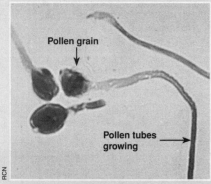

The pollen tube enters the ovule through the micropyle, a small gap in the ovule. It is guided by chemical cues from the embryo sac (usually calcium). **A double fertilisation** takes place. One sperm nucleus fuses with the egg to form the zygote (2n). A second sperm nucleus fuses with the two polar nuclei within the embryo sac to produce the endosperm tissue (3n).

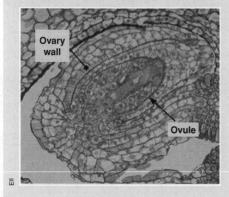

In flowering plants the ovule is located with the ovary of the flower. Some plant ovaries contain a single ovule, while others may have several, so many fertilisations may be required before the entire ovary can develop.

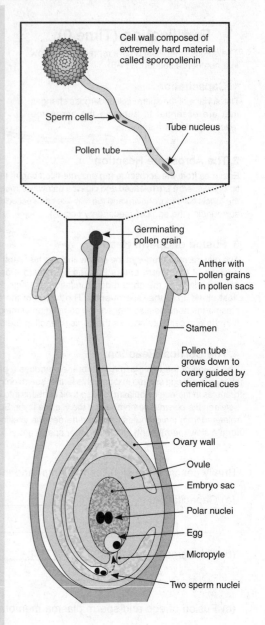

1. Briefly describe how the egg cell forms: _____

_____

_____

2. Explain the roles of the tube cell and generative cell of pollen: _____

_____

_____

_____

3. Name the main chemical responsible for pollen tube growth: _____

4. Describe the role of the double fertilisation in angiosperm reproduction: _____

_____

Web links: *Angiosperm Life cycle Animation, Pollination in Angiosperms*

Periodicals: *Flower power*

Stem Cells

Stem cells are undifferentiated cells found in multicellular organisms. They are characterised by two features. The first, **self renewal**, is the ability to undergo numerous cycles of cell division while maintaining an unspecialised state. The second, **potency**, is the ability to differentiate into specialised cells. **Totipotent** cells, produced in the first few divisions of a fertilised egg, can differentiate into any cell type, embryonic or extra-embryonic. **Pluripotent cells** are descended from totipotent cells and can give rise to any of the cells derived from the three germ layers (endoderm, mesoderm, and ectoderm). Embryonic stem cells at the blastocyst stage and foetal stem cells are pluripotent. Adult (somatic) stem cells are termed **multipotent**. They are undifferentiated cells found among differentiated cells in a tissue or organ. These cells can give rise to several other cell types, but those types are limited mainly to the cells of the blood, heart, muscle and nerves. The primary roles of adult stem cells are to maintain and repair the tissue in which they are found. A potential use of stem cells is making cells and tissues for medical therapies, such as **cell replacement therapy** and **tissue engineering** (for example, for bone and skin grafts).

Stem Cells and Blood Cell Production

New blood cells are produced in the red bone marrow, which becomes the main site of blood production after birth, taking over from the foetal liver. All types of blood cells develop from a single cell type: called a **multipotent stem cell** or haemocytoblast. These cells are capable of mitosis and of differentiation into 'committed' precursors of each of the main types of blood cell.

Each of the different cell lines is controlled by a specific **growth factor**. When a stem cell divides, one of its daughters remains a stem cell, while the other becomes a precursor cell, either a **lymphoid cell** or **myeloid cell**. These cells continue to mature into the various type of blood cells, developing their specialised features and characteristic roles as they do so.

Variation & Heredity

1. Describe the two defining features of stem cells:

    (a) _____

    (b) _____

2. Distinguish between embryonic stem cells and adult stem cells with respect to their **potency** and their potential applications in medical technologies:

    _____

    _____

    _____

3. Using an example, explain the purpose of stem cells in an adult: _____

    _____

    _____

4. Describe one potential advantage of using embryonic stem cells for tissue engineering technology: _____

    _____

    _____

    _____

Periodicals:
What is a stem cell?

Related activities: Stem Cell Therapy
Web links: Stem Cells in the Spotlight, Stem Cell Resources

A 2

Human Cell Specialisation

Animal cells are often specialised to perform particular functions. The eight specialised cell types shown below are representative of some 230 different cell types in humans. Each has specialised features that suit it to performing a specific role.

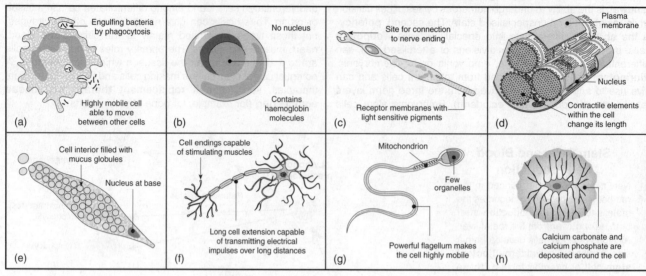

1. Identify each of the cells (b) to (h) pictured above, and describe their **specialised features** and **role** in the body:

(a) Type of cell: *Phagocytic white blood cell (neutrophil)*

Specialised features: *Engulfs bacteria and other foreign material by phagocytosis*

Role of cell within body: *Destroys pathogens and other foreign material as well as cellular debris*

(b) Type of cell: _____

Specialised features: _____

Role of cell within body: _____

(c) Type of cell: _____

Specialised features: _____

Role of cell within body: _____

(d) Type of cell: _____

Specialised features: _____

Role of cell within body: _____

(e) Type of cell: _____

Specialised features: _____

Role of cell within body: _____

(f) Type of cell: _____

Specialised features: _____

Role of cell within body: _____

(g) Type of cell: _____

Specialised features: _____

Role of cell within body: _____

(h) Type of cell: _____

Specialised features: _____

Role of cell within body: _____

RA 2

Related activities: Animal Cells, Human Cell Differentiation

Differentiation of Human Cells

A zygote commences development by dividing into a small ball of a few dozen identical cells called **embryonic stem cells**. These cells start to take different developmental paths to become specialised cells such as nerve stem cells which means they can no longer produce any other type of cell. **Differentiation** is cell specialisation that occurs at the end of a developmental pathway.

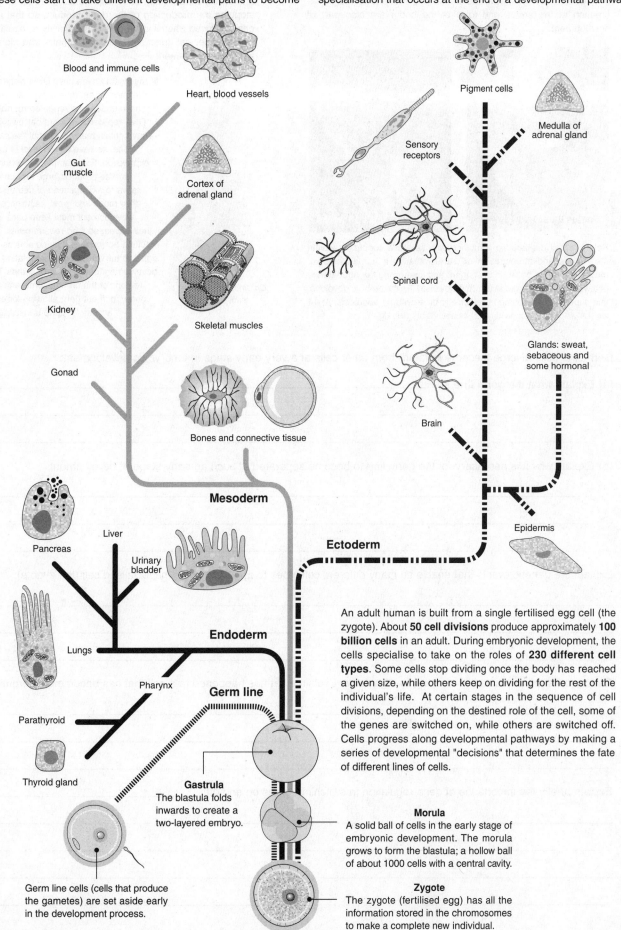

Blood and immune cells

Heart, blood vessels

Pigment cells

Medulla of adrenal gland

Gut muscle

Cortex of adrenal gland

Sensory receptors

Kidney

Skeletal muscles

Spinal cord

Gonad

Glands: sweat, sebaceous and some hormonal

Bones and connective tissue

Brain

Mesoderm

Pancreas

Liver

Urinary bladder

Ectoderm

Epidermis

Lungs

Endoderm

Pharynx

Parathyroid

Germ line

Thyroid gland

Gastrula
The blastula folds inwards to create a two-layered embryo.

Germ line cells (cells that produce the gametes) are set aside early in the development process.

An adult human is built from a single fertilised egg cell (the zygote). About **50 cell divisions** produce approximately **100 billion cells** in an adult. During embryonic development, the cells specialise to take on the roles of **230 different cell types**. Some cells stop dividing once the body has reached a given size, while others keep on dividing for the rest of the individual's life. At certain stages in the sequence of cell divisions, depending on the destined role of the cell, some of the genes are switched on, while others are switched off. Cells progress along developmental pathways by making a series of developmental "decisions" that determines the fate of different lines of cells.

Morula
A solid ball of cells in the early stage of embryonic development. The morula grows to form the blastula; a hollow ball of about 1000 cells with a central cavity.

Zygote
The zygote (fertilised egg) has all the information stored in the chromosomes to make a complete new individual.

Periodicals:
Cell differentiation

Related activities: The Simplest Case: Genes to Proteins

RA 2

Development

Development is the process of progressive change through the lifetime of an organism. It involves growth (increase in size), cell division (to generate the multicellular body), cellular **differentiation** (the generation of specialised cells) and **morphogenesis** (the creation of the shape and form of the body) are also part of development.

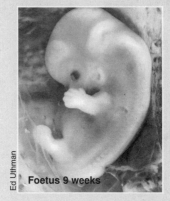 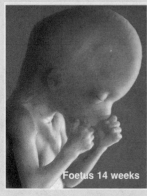

Foetus 9 weeks

Foetus 14 weeks

Selective cell proliferation combined with selective apoptosis sculpts the tissues and structures in all vertebrates. In a human embryo, mesoderm forms between the fingers and toes giving the appearance of a webbed, paddle like structure (above left). As the embryo develops, this superfluous webbing is selectively destroyed by apoptosis. At 14 weeks, each of the individual digits are visible (top right).

Control Over Genes

Gene expression is the process by which a cell regulates the production of gene products (DNA or RNA) to meet the cell's requirements. Regulation is achieved by switching on and off the transcription of specific genes, thus starting or stopping the production of the gene product, so that it is only produced when it is needed. Gene activity is regulated by cell type, cell function, chemical signals, and signals from the environment.

Many insect larvae have large **polytene chromosomes**, (chromosomes which have undergone repeated rounds of DNA replication without cell division). They contain many copies of the same gene, so have a high level of gene expression. For example, *Chironomus* larvae produce large amounts of saliva to aid digestion of detritus. As they moult and grow, salivary gland development must keep pace with their increased food requirements. The moulting hormone ecdysone acts as the signal to turn on the gene regulating the development of the salivary glands. The regions of the chromosome with this gene, puff out (left) allowing for easy access for gene transcription.

Chromosome puff

EII

1. Germ line cells diverge (become isolated) from other cells at a very early stage in embryonic development:

    (a) Explain what the **germ line** is: _____

    _____

    _____

    (b) Explain why it is necessary for the germ line to become separated at such an early stage of development:

    _____

    _____

    _____

2. Explain the genetic events that enable so many different cell types to arise from one unspecialised cell (the zygote):

    _____

    _____

    _____

3. Cancer cells are particularly damaging to organisms. Explain what has happened to a cell that has become cancerous:

    _____

    _____

    _____

4. Explain briefly the importance of gene regulation in switching genes on and off: _____

    _____

    _____

    _____

    _____

Stem Cell Therapy

The properties of **stem cells** (self renewal and potency) make them potentially very valuable in those aspects of medicine that require a disease-free and plentiful supply of cells of specific types. Stem cells have potential applications in studies of human development and gene regulation, in testing the safety and effectiveness of new drugs, in monoclonal antibody production, and as a renewable source of cells to treat diseased and damaged tissue. Therapeutic **stem cell cloning** (below right) is still in its very early stages but has potential for healing damaged tissue and organs, and treating diseases such as Parkinson's

and Alzheimer's. Stem cell cloning does require the production and destruction of human embryos (for the express purpose of medical research), which some people find morally wrong. The possibility of **reproductive cloning** (human cloning), also concerns many people. In the UK, the government controls stem cell cloning. Regulatory measures include the Human Reproductive Cloning Act (2001), (banning human reproductive cloning), and the Human Fertilisation Embryology Authority (HFEA), which regulates research on human embryos in the UK (including stem cell work).

Embryonic Stem Cells

Embryonic stem cells (ESC) are stem cells taken from **blastocysts** (below). Blastocysts are embryos which are about five days old and consist of a hollow ball containing 50-150 cells. In general, ESCs come from embryos which have been fertilised *in vitro* at fertilisation clinics and have then been donated for research.

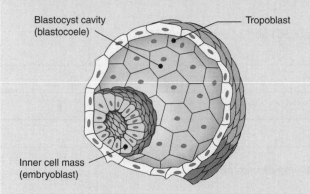

Blastocyst cavity (blastocoele) — Tropoblast
Inner cell mass (embryoblast)

Cells derived from the inner cell mass are **pluripotent**; they can become any cells of the body, with the exception of placental cells. When grown *in vitro*, without any stimulation for differentiation, these cells retain their pluripotency through multiple cell divisions. As a consequence of this, **ESC therapies** have potential use in regenerative medicine and tissue replacement. By manipulating the culture conditions, scientists are able to select and control the type of cells grown (e.g. heart cells). This ability could allow for specific cell types to be grown to treat specific diseases or replace damaged tissue. However, no ESC treatments have been approved to date due to ethical issues.

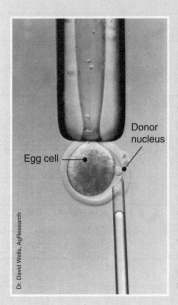

Egg cell — Donor nucleus

Poor **histocompatibility**, in which the recipient's immune system rejects foreign cells, is one of the major difficulties with transplantation. Stem cell cloning (also called **therapeutic cloning**) provides a way around this problem. Stem cell cloning produces cells that have been derived from the recipient and are therefore histocompatible. Such an approach would mean immunosuppressant drugs would no longer be required. Diseases such as leukaemia and Parkinson's could be treated using this technique.

Embryonic Stem Cell Cloning

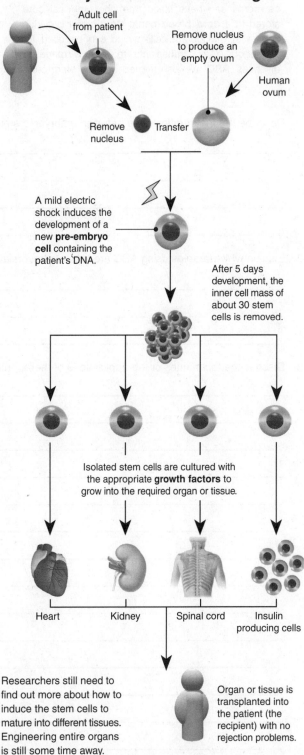

Adult cell from patient. Remove nucleus to produce an empty ovum. Human ovum. Remove nucleus. Transfer. A mild electric shock induces the development of a new **pre-embryo cell** containing the patient's DNA. After 5 days development, the inner cell mass of about 30 stem cells is removed. Isolated stem cells are cultured with the appropriate **growth factors** to grow into the required organ or tissue. Heart. Kidney. Spinal cord. Insulin producing cells.

Researchers still need to find out more about how to induce the stem cells to mature into different tissues. Engineering entire organs is still some time away.

Organ or tissue is transplanted into the patient (the recipient) with no rejection problems.

Periodicals: Grown to order, Embryonic stem cells

Related activities: Gene Therapy, Stem Cells
Web links: Stem Cells in the Spotlight

RA 3

Adult Stem Cells

Adult stem cells (ASC) are undifferentiated cells found in several types of tissues (e.g. brain, bone marrow, skin, and liver) in adults, children, and umbilical cord blood. Unlike ESCs, they are multipotent and can only differentiate into a limited number of cell types, usually related to the tissue of origin. In the body, the main role of ASC is to replace damaged or dying cells. There are fewer ethical issues associated with using ASC for therapeutic purposes, and for this reason ASC are already used to treat a number of diseases including leukaemia and other blood disorders.

In many countries, including the UK, the parents of newborns can have blood from the umbilical cord stored in a **cord blood bank**. Cord blood is a rich source of multipotent cells which can be used for autologous (self) transplants to treat a range of diseases. ASC are also termed somatic stem cells.

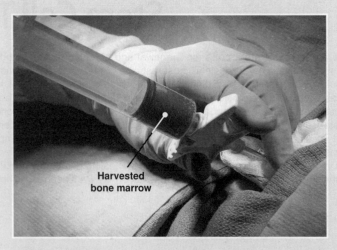

Harvested bone marrow

Cells obtained from bone marrow (above) or umbilical cord blood could be used to treat patients with a variety of diseases including leukemia, lymphomas, anemia, and a range of congenital diseases. Multipotent stem cells from marrow or cord blood give rise to the precursor cells for red blood cells, all white blood cell types, and platelets.

1. Describe the major differences between embryonic stem cells (ESC) and adult stem cells (ASC):

_____

_____

_____

_____

2. Explain why therapies using ASC are less controversial than therapies using ESC: _____

_____

_____

_____

3. Discuss the techniques or the applications of therapeutic stem cell cloning (including ethical issues where relevant):

_____

_____

_____

_____

_____

_____

_____

_____

_____

_____

_____

_____

_____

Plant Tissue Culture

Plants cells are **totipotent** throughout their life. This characteristic is exploited for plant tissue culture, or **micropropagation**, a method used for cloning plants. It is used widely for the rapid multiplication of commercially important plant species with superior genotypes, as well as for the recovery of endangered plant species. Cloning can rapidly improve plant productivity and quality, and increase resistance to disease, pollutants, and insect pests. However, continued culture of a limited number of cloned varieties can lead to a loss of genetic variation. New genetic stock may be introduced into cloned lines periodically to prevent this happening.

Micropropagation is possible because differentiated plant cells have the potential to give rise to all the cells of an adult plant. It has considerable advantages over traditional methods of plant propagation (see table below), but it is very labour intensive. In addition, the optimal conditions for growth and regeneration must be determined and plants propagated in this way may be genetically unstable or infertile, with chromosomes structurally altered or in unusual numbers. The success of tissue culture is affected by factors such as selection of **explant** material, the composition of the culturing media, plant hormone levels, lighting, and temperature.

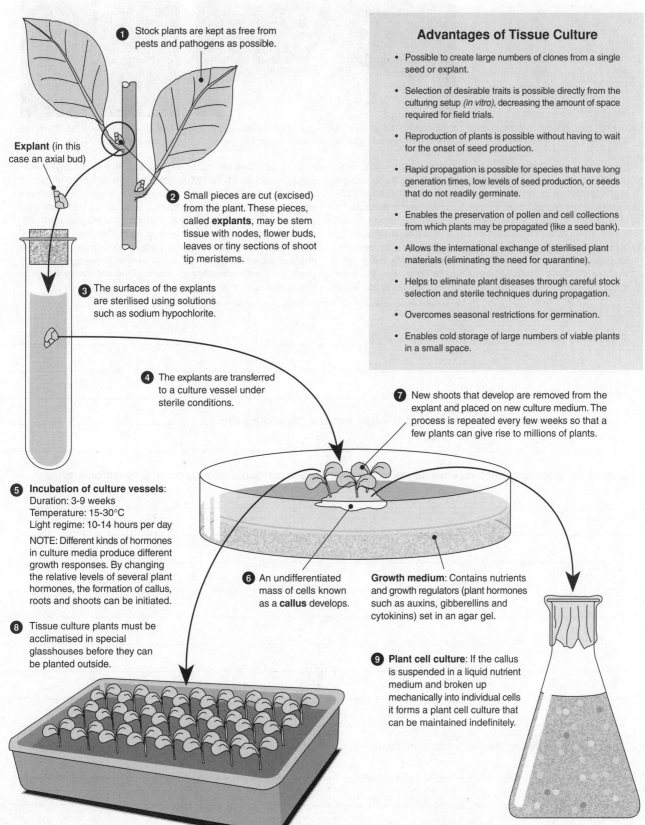

1 Stock plants are kept as free from pests and pathogens as possible.

Explant (in this case an axial bud)

2 Small pieces are cut (excised) from the plant. These pieces, called **explants**, may be stem tissue with nodes, flower buds, leaves or tiny sections of shoot tip meristems.

3 The surfaces of the explants are sterilised using solutions such as sodium hypochlorite.

4 The explants are transferred to a culture vessel under sterile conditions.

5 **Incubation of culture vessels**:
Duration: 3-9 weeks
Temperature: 15-30°C
Light regime: 10-14 hours per day

NOTE: Different kinds of hormones in culture media produce different growth responses. By changing the relative levels of several plant hormones, the formation of callus, roots and shoots can be initiated.

6 An undifferentiated mass of cells known as a **callus** develops.

7 New shoots that develop are removed from the explant and placed on new culture medium. The process is repeated every few weeks so that a few plants can give rise to millions of plants.

8 Tissue culture plants must be acclimatised in special glasshouses before they can be planted outside.

Growth medium: Contains nutrients and growth regulators (plant hormones such as auxins, gibberellins and cytokinins) set in an agar gel.

9 **Plant cell culture**: If the callus is suspended in a liquid nutrient medium and broken up mechanically into individual cells it forms a plant cell culture that can be maintained indefinitely.

Advantages of Tissue Culture

- Possible to create large numbers of clones from a single seed or explant.

- Selection of desirable traits is possible directly from the culturing setup *(in vitro)*, decreasing the amount of space required for field trials.

- Reproduction of plants is possible without having to wait for the onset of seed production.

- Rapid propagation is possible for species that have long generation times, low levels of seed production, or seeds that do not readily germinate.

- Enables the preservation of pollen and cell collections from which plants may be propagated (like a seed bank).

- Allows the international exchange of sterilised plant materials (eliminating the need for quarantine).

- Helps to eliminate plant diseases through careful stock selection and sterile techniques during propagation.

- Overcomes seasonal restrictions for germination.

- Enables cold storage of large numbers of viable plants in a small space.

Variation & Heredity

Related activities: Stem Cells
Web links: Artificial Vegetative Propagation

RA 2

Micropropagation of the Tasmanian blackwood tree (*Acacia melanoxylon*)

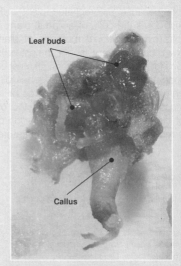
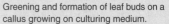

Greening and formation of leaf buds on a callus growing on culturing medium.

Normal shoots with juvenile leaves growing from a callus on media. They appear identical to those produced directly from seeds.

Seedling with juvenile foliage 6 months after transfer to greenhouse.

Micropropagation is increasingly used in conjunction with genetic engineering to propagate transgenic plants. Genetic engineering and micropropagation achieve similar results to conventional selective breeding but more precisely, quickly, and independently of growing season. The **Tasmanian blackwood** (above) is well suited to this type of manipulation. It is a versatile hardwood tree now being extensively trialled in some countries as a replacement for tropical hardwoods. The timber is of high quality, but genetic variations between individual trees lead to differences in timber quality and colour. Tissue culture allows the multiple propagation of trees with desirable traits (e.g. uniform timber colour). Tissue culture could also help to find solutions to problems that cannot be easily solved by forestry management. When combined with genetic engineering (introduction of new genes into the plant) problems of pest and herbicide susceptibility may be resolved. Genetic engineering may also be used to introduce a gene for male sterility, thereby stopping pollen production. This would improve the efficiency of conventional breeding programmes by preventing self-pollination of flowers (the manual removal of stamens is difficult and very labour intensive).

Information courtesy of Raewyn Poole, University of Waikato (Unpublished Msc. thesis).

1. Explain the general purpose of plant tissue culture: _____

_____

2. (a) Explain what a **callus** is: _____

   (b) Explain how a callus may be stimulated to initiate root and shoot formation: _____

   _____

3. Discuss the **advantages** and **disadvantages** of micropropagation compared with traditional propagation methods:

_____

_____

_____

_____

_____

_____

_____

_____

4. Describe a potential problem with micropropagation in terms of long term ability to adapt to environmental changes:

_____

_____

Sources of Variation

Variation refers to the diversity of genetic and phenotypic traits within and between species. Variation is a feature of sexually reproducing populations and the essential raw material for selection. Variation gives species greater opportunity to adapt to and survive in a dynamic environment. Populations of highly variable species will include individuals with different **fitness** in the prevailing environment. It is this variability that offers the chance of reproductive advantage if environments change. Species with low variability (such as populations of parthenogenetic clones)

can be highly successful in stable conditions, but reproduce sexually when environments become unfavourable for just this reason. Phenotype is influenced by both genotype and the environment. Even organisms of fixed genotype may differ in the degree to which they can alter their phenotype in response to environmental change; a phenomenon known as **phenotypic plasticity**. Phenotypic plasticity allows organisms a greater chance of survival in ever-changing surroundings and it is particularly well developed in immobile species, such as plants.

Mutations
gene mutations; chromosome mutations

Mutations are the source of all **new** alleles. Existing genes are modified by base substitutions and deletions, causing the formation of new alleles. Silent mutations may escape selection pressure until conditions change.

Mutation: Substitute **T** instead of **C**

Original DNA: A A A A T G C T T C   T C

Mutant DNA: A A A A T G T T T C   T C

Sexual Reproduction
independent assortment; crossing over and recombination; mate selection

Sexual reproduction rearranges and reshuffles the genetic material into new combinations.

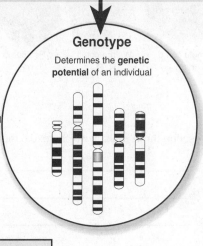

Phenotype

An individual's phenotype is the result of the ineraction of genetic and environmental factors during its lifetime (including during development). The genetic instructions for creating the individual may be modified by the internal and external environment both during and after development. Phenotypic plasticity in response to environment can be adaptive if it increases fitness.

Dominant, recessive, codominant and multiple allele systems, as well as interactions between genes and epigenetic inheritance, combine in their effects.

Genotype

Determines the **genetic potential** of an individual

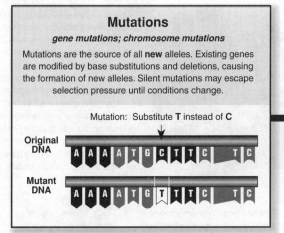

Some snail populations show phenotypic plasticity with respect to shell thickness; individuals can develop thicker shells when subjected to heavy predation.

Environmental Factors

The external and internal environments can influence the expression of the genotype. The external environment may include physical factors such as temperature or light intensity, or biotic factors such as competition. The internal environment, e.g presence or absence of hormones or growth factors during development, may also affect genotypic expression.

---

1. Using examples, explain how the environment of a particular genotype can affect the phenotype:

_____

_____

_____

2. Discuss the significance of variation in selection: _____

_____

_____

_____

_____

Periodicals:
What is variation?

Related activities: Meiosis, The Effect of Mutations

RA 2

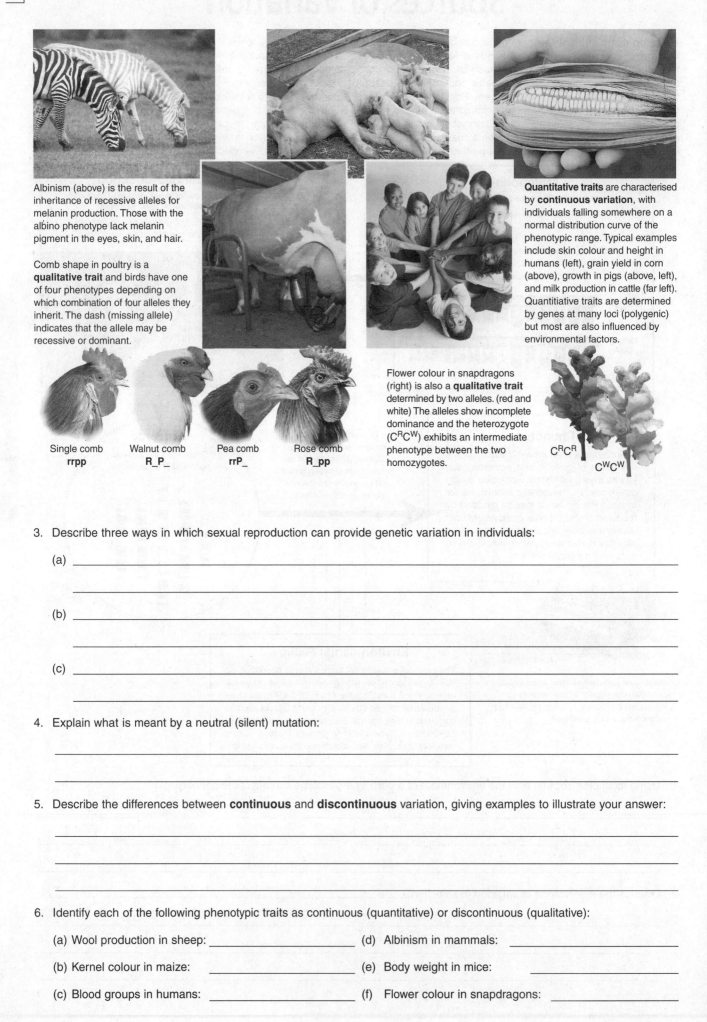

Albinism (above) is the result of the inheritance of recessive alleles for melanin production. Those with the albino phenotype lack melanin pigment in the eyes, skin, and hair.

Comb shape in poultry is a **qualitative trait** and birds have one of four phenotypes depending on which combination of four alleles they inherit. The dash (missing allele) indicates that the allele may be recessive or dominant.

Single comb	Walnut comb	Pea comb	Rose comb
rrpp	**R_P_**	**rrP_**	**R_pp**

Quantitative traits are characterised by **continuous variation**, with individuals falling somewhere on a normal distribution curve of the phenotypic range. Typical examples include skin colour and height in humans (left), grain yield in corn (above), growth in pigs (above, left), and milk production in cattle (far left). Quantitiative traits are determined by genes at many loci (polygenic) but most are also influenced by environmental factors.

Flower colour in snapdragons (right) is also a **qualitative trait** determined by two alleles. (red and white) The alleles show incomplete dominance and the heterozygote ($C^R C^W$) exhibits an intermediate phenotype between the two homozygotes.

$C^R C^R$

$C^W C^W$

3. Describe three ways in which sexual reproduction can provide genetic variation in individuals:

(a) _____

_____

(b) _____

_____

(c) _____

_____

4. Explain what is meant by a neutral (silent) mutation:

_____

_____

5. Describe the differences between **continuous** and **discontinuous** variation, giving examples to illustrate your answer:

_____

_____

_____

6. Identify each of the following phenotypic traits as continuous (quantitative) or discontinuous (qualitative):

(a) Wool production in sheep: _____

(d) Albinism in mammals: _____

(b) Kernel colour in maize: _____

(e) Body weight in mice: _____

(c) Blood groups in humans: _____

(f) Flower colour in snapdragons: _____

Gene-Environment Interactions

External environmental factors can modify the phenotype encoded by genes. This can occur both during development and later in life. Even identical twins have minor differences in their appearance due to environmental factors such as diet and intrauterine environment before birth. Environmental factors that affect the phenotype of plants and animals include nutrients or diet, temperature, altitude or latitude, and the presence of other organisms.

Sources of Variation in Organisms

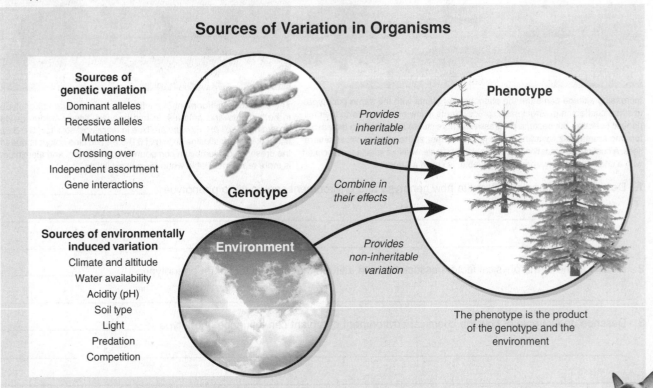

Sources of genetic variation
Dominant alleles
Recessive alleles
Mutations
Crossing over
Independent assortment
Gene interactions

Genotype

Provides inheritable variation

Combine in their effects

Phenotype

Sources of environmentally induced variation
Climate and altitude
Water availability
Acidity (pH)
Soil type
Light
Predation
Competition

Environment

Provides non-inheritable variation

The phenotype is the product of the genotype and the environment

Variation & Heredity

The Effect of Temperature

The sex of some animals is determined by the temperature at which they were incubated during their embryonic development. Examples include turtles, crocodiles, and the American alligator. In some species, high incubation temperatures produce males and low temperatures produce females. In other species, the opposite is true. Temperature regulated sex determination may be advantageous by preventing inbreeding (since all siblings will tend to be of the same sex).

Colour-pointing in breeds of cats and rabbits (e.g. Siamese, Himalyan) is a result of a temperature sensitive mutation in one of the enzymes in the metabolic pathway from tyrosine to melanin. The dark pigment is only produced in the cooler areas of the body (face, ears, feet, and tail), while the rest of the body is a paler version of the same colour, or white.

The Effect of Other Organisms

Female

Male

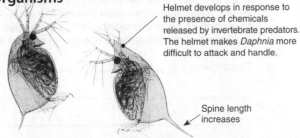

Helmet develops in response to the presence of chemicals released by invertebrate predators. The helmet makes *Daphnia* more difficult to attack and handle.

Spine length increases

Non-helmeted form   **Helmeted form with long tail spine**

The presence of other individuals of the same species may control sex determination for some animals. Some fish species, including some in the wrasse family (e.g. *Coris sandageri,* above), show this phenomenon. The fish live in groups consisting of a single male with attendant females and juveniles. In the presence of a male, all juvenile fish of this species grow into females. When the male dies, the dominant female will undergo physiological changes to become a male. The male has distinctive bands, whereas the female is pale in colour and has very faint markings.

Some organisms respond to the presence of other, potentially harmful, organisms by changing their morphology or body shape. Invertebrates such as *Daphnia* will grow a large helmet when a predatory midge larva is present. Such responses are usually mediated through chemicals produced by the predator (or competitor), and are common in plants as well as animals.

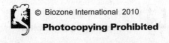
Periodicals:
What is variation?

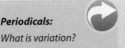

Related activities: Sources of Variation

RA 2

The Effect of Altitude

Severe stunting (krummholz)

Growth to genetic potential

Cline

Increasing altitude can stunt the phenotype of plants with the same genotype. In some conifers, e.g. Engelmann spruce, plants at low altitude grow to their full genetic potential, but become progressively more stunted as elevation increases, forming krummholz (gnarled bushy growth forms) at the highest, most severe sites. A continuous gradation in a phenotypic character within a species, associated with a change in an environmental variable, is called a **cline**.

The Effect of Chemical Environment

The chemical environment can influence the expressed phenotype in both plants and animals. In hydragenas, flower colour varies according to soil pH. Flowers are blue in more acid soils (pH 5.0-5.5), but pink in more alkaline soils (pH 6.0-6.5). The blue colour is due to the presence of aluminium compounds in the flowers and aluminium is more readily available when the soil pH is low.

1. Describe an example to illustrate how genotype and environment contribute to phenotype: _____

_____

_____

2. Identify some of the physical factors associated with altitude that could affect plant phenotype: _____

_____

3. Describe an example of how the chemical environment of a plant can influence phenotype: _____

_____

_____

_____

4. Explain why the darker patches of fur in colour-pointed cats and rabbits are found only on the face, paws and tail:

_____

_____

5. There has been much amusement over the size of record-breaking vegetables, such as enormous pumpkins, produced for competitions. Explain how you could improve the chance that a vegetable would reach its maximum genetic potential:

_____

_____

6. (a) Explain what is meant by a **cline**: _____

_____

(b) On a windswept portion of a coast, two different species of plant (species A and species B) were found growing together. Both had a low growing (prostrate) phenotype. One of each plant type was transferred to a greenhouse where "ideal" conditions were provided to allow maximum growth. In this controlled environment, species B continued to grow in its original prostrate form, but species A changed its growing pattern and became erect in form. Identify the **cause** of the prostrate phenotype in each of the coastal grown plant species and explain your answer:

Plant species A: _____

_____

_____

Plant species B: _____

_____

_____

(c) Identify which of these species (A or B) would be most likely to exhibit clinal variation: _____

Polygenes

When a phenotype is controlled by more than one gene it is said to be **polygenic**. Examples of phenotypes controlled by polygenes include height, weight, human skin colour, and kernel colour in maize. Polygenic traits produce a range of phenotypes, so a population will exhibit **continuous variation** for that characteristic. Generally, as the number of genes controlling the phenotype increases, the more continuous the distribution appears. The production of the skin pigment melanin in humans is controlled by at least three genes. The amount of melanin produced is directly proportional to the number of dominant alleles for either gene (from 0 to 6). Environmental factors also contribute to skin colour (see below).

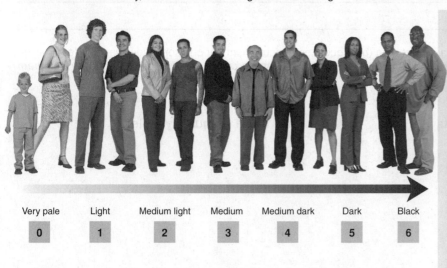

Light-skinned person

Medium-skinned person

Medium-dark skinned

Dark-skinned person

Very pale	Light	Medium light	Medium	Medium dark	Dark	Black
0	1	2	3	4	5	6

Number of dark alleles

aabbcc	aabbCc	aaBbCc	aaBBCc	aaBBCC	AaBBCC	AABBCC
	aaBbcc	aabbCC	aaBbCC	AaBBCc	AABBCc	
	Aabbcc	aaBBcc	AaBBCc	AaBbCC	AABbCC	
		AabbCc	AaBbCc	AABbCc		
		AaBbcc	AabbCC	AABBcc		
		AAbbcc	AABbcc	AAbbCC		
			AAbbCc			

There are seven shades skin colour ranging from very dark to very pale, with most individual being somewhat intermediate in skin colour. No dominant allele results in a lack of dark pigment (aabbcc). Full pigmentation (black) requires six dominant alleles (AABBCC).

1. State how many phenotypes are possible for this type of gene interaction: _____

2. State which alleles must be present/absent for the following phenotypes:

   Black: _____ Medium: _____ White: _____

3. Explain why in reality we see many more than seven shades of skin colour: _____

   _____

   _____

Environment and Skin Colour

Higher latitudes

Netherlands

Iraq

Equatorial latitudes

Burundi

Liberia

The distribution of skin colour globally is not random. People native to equatorial regions, where UV light levels are high, have darker skin tones than people from higher latitudes, where UV levels are low for most of the year. Pigmentation protects against penetration by UV, which destroys the body's stores of folate, a vitamin required for normal neural development in the foetus. However, some UV is needed to synthesise vitamin D, which is needed for bone development and strength. Skin colour reflects the delicate balance between these two opposing selection pressures.

Periodicals:
The colour code

Related activities: Sources of Genetic Variation, Gene Interactions
Web links: Summary of Gene Interactions

A 3

Variation & Heredity

4. Discuss the differences between **continuous** and **discontinuous** variation, giving examples to illustrate your answer:

_____

_____

_____

_____

_____

5. From a sample of no less than 30 adults, collect data (by request or measurement) for one continuous variable (e.g. height, weight, shoe size, or hand span). Record and tabulate your results in the space below, and then plot a frequency histogram of the data on the grid below:

Raw data	Tally Chart (frequency table)

Variable: _____

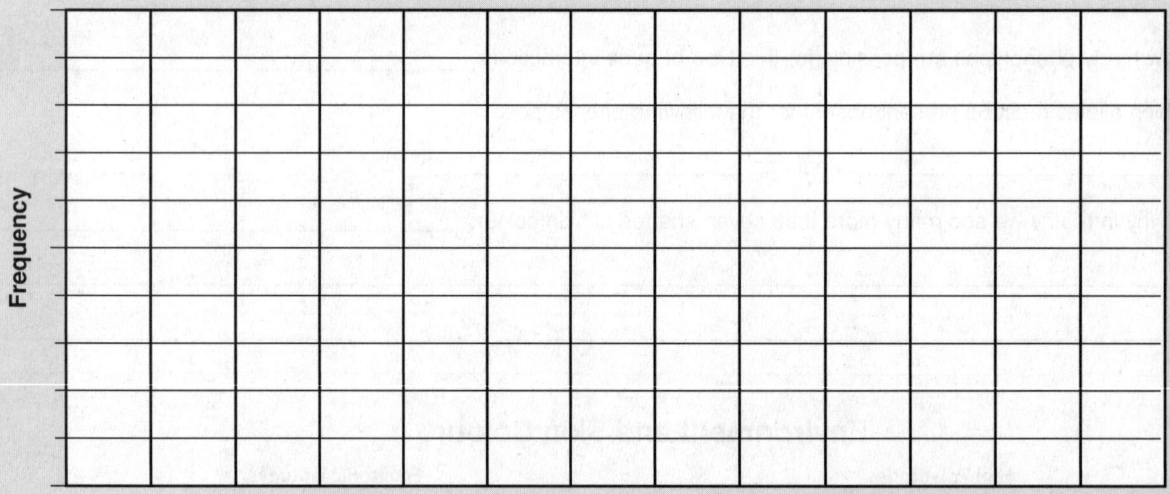

Frequency

(a) Calculate the mean, median, and mode of your data:

Mean: _____ **Mode:** _____ **Median:** _____

(b) Describe the pattern of distribution shown by the graph, giving a reason for your answer: _____

_____

_____

(c) Explain the genetic basis of this distribution: _____

_____

(d) Explain the importance of a large sample size when gathering data relating to a continuous variable:

_____

Gene Interactions and Gene Inactivation

It is not always the case the that one gene product produces a single phenotypic character without the influence of other genes. Instead, genes frequently interact to produce the phenotype we finally see. Several genes may control the expression of a single character, or one gene, because of its protein product may have wide ranging phenotypic effects. Some genes, called epistatic genes, can even override the expression of other genes,

regardless of their dominance status. For some of these gene interactions (e.g. epistasis in albinos and pleiotropy in sickle cell disease), the phenotypic effect is the end result of the absence or defectiveness of the functional gene product. In female mammals, even the normal random inactivation of X chromosomes during development plays a part in the eventual phenotype of those with a certain genotype.

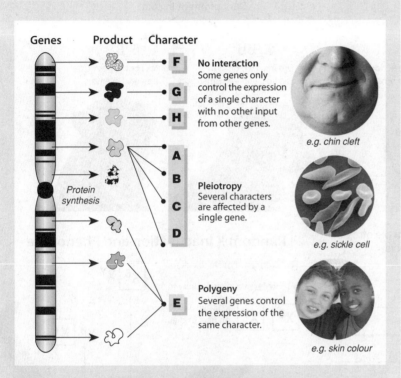

Genes   Product   Character

F No interaction
Some genes only control the expression of a single character with no other input from other genes.

e.g. chin cleft

Protein synthesis

A B C D Pleiotropy
Several characters are affected by a single gene.

e.g. sickle cell

E Polygeny
Several genes control the expression of the same character.

e.g. skin colour

The homozygous condition for the albino allele (cc) overrides any other genes present for hair colour.

A widow's peak in the hairline, which is a dominant trait (as seen on Jude Law, left) is not evident if the epistatic gene for male pattern baldness is present (right).

Polygeny, Pleitropy, and Epistasis

▶ Polygeny describes cases where a single phenotypic characteristic (such as skin colour) is controlled by more than one gene. This phenomenon is known as polygeny (polygenic inheritance) and leads to continuous phenotypic variation in the population.

▶ Pleiotropy describes the genetic effect of a single gene on multiple traits. As a consequence of pleiotropy, a mutation in a gene may have a phenotypic effect on some or all traits simultaneously. The human disease, PKU (phenylketonuria) is one example. This disease causes mental retardation and reduced hair and skin pigmentation, and results from any of a number of mutations in the gene coding for the conversion of phenylalanine to tyrosine, an intermediate in the metabolic pathway leading to melanin production.

Epistasis: One Gene Overrides Others

Epistasis involves two non-allelic genes (at different loci), where the action of one gene masks or otherwise alters the expression of other genes. A common example is **albinism**, which appears in rodents that are homozygous recessive for the 'albino' allele (a mutation) even if they have the alleles for agouti or black fur.

Albinism is widespread in animal phyla. In every case, the presence two albino alleles (cc) overrides pigment expression. Another epistatic gene is that encoding male pattern baldness.

1. Explain the differences between **polygeny** and **pleiotropy**, giving examples to illustrate your answer:

2. Explain how epistasis differs from pleiotropy:

3. Explain why the genes present for hair colour are irrelevant if an animal is homozygous recessive for the albino gene:

Allele Combinations and Coat Colour

Epistatic interactions also regulate coat colour in Labrador retrievers. The basic coat colours (yellow, chocolate, and black) are controlled by two the interaction of two genes at different loci, each with two alleles. The epistatic E gene determines if pigment will be present in the coat, and the B gene determines the pigment density (depth of colour). Dogs with genotype ee will always be yellow, but the presence of bb modifies the effect of any dark pigment. In labradors, coat colour variation also depends on the influence of the C gene, which always produces phaeomelanin no matter what the E or B genes are producing (in dark coats, it is not visible). As a result, four main coat colour variations are possible in Labradors, with a further three variations possible in the yellow coated dog from dark yellow to very pale (CC, Cc, and cc).

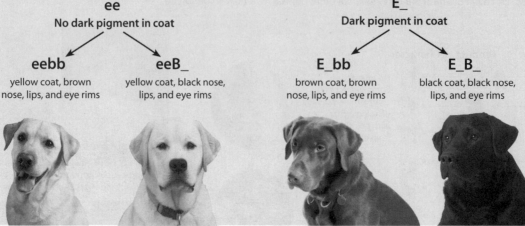

ee
No dark pigment in coat

eebb
yellow coat, brown nose, lips, and eye rims

eeB_
yellow coat, black nose, lips, and eye rims

E_
Dark pigment in coat

E_bb
brown coat, brown nose, lips, and eye rims

E_B_
black coat, black nose, lips, and eye rims

4. Using an example (e.g. PKU), suggest how a pleiotropic gene could exert its multiple phenotypic effects:

_____

_____

_____

_____

_____

_____

_____

5. For coat colour in Labradors, identify the alleles that must be present and absent for the following phenotypes:

Black: _____

Brown: _____

Yellow with brown nose: _____

Yellow with black nose: _____

Random X Inactivation and Phenotype

B Dominant allele (black fur)

Y Recessive allele (yellow fur)

X^B X^Y Inactive alleles on Barr bodies

$X^B X^Y$

$X^B X^Y$

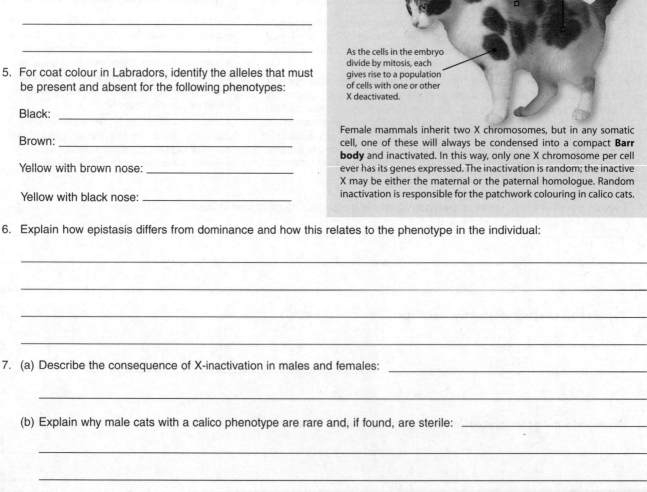

As the cells in the embryo divide by mitosis, each gives rise to a population of cells with one or other X deactivated.

Female mammals inherit two X chromosomes, but in any somatic cell, one of these will always be condensed into a compact **Barr body** and inactivated. In this way, only one X chromosome per cell ever has its genes expressed. The inactivation is random; the inactive X may be either the maternal or the paternal homologue. Random inactivation is responsible for the patchwork colouring in calico cats.

6. Explain how epistasis differs from dominance and how this relates to the phenotype in the individual:

_____

_____

_____

_____

7. (a) Describe the consequence of X-inactivation in males and females: _____

_____

(b) Explain why male cats with a calico phenotype are rare and, if found, are sterile: _____

_____

_____

Inheritance in Domestic Cats

Cats have been domesticated for thousands of years. During this time, certain traits or characteristics have been considered fashionable or desirable in a cat by people in different parts of the world. In the domestic cat, the 'wild type' is the short-haired tabby. All the other coat colors found in cats are modifications of this ancestral tabby pattern. Inheritance of coat characteristics and a few other features in cats is interesting because they exhibit the most common genetic phenomena. Some selected traits for domestic cats are identified below, together with a list of the kinds of genetic phenomena easily demonstrated in cats.

Inheritance Patterns in Domestic Cats

Dominance	The polydactylism gene with the dominant allele (Pd) produces a paw with extra digits.
Recessiveness	The dilution gene with the recessive allele (d) produces a diluted black to produce gray, or orange to cream.
Epistasis	The dominant agouti gene (A) must be present for the tabby gene (T) to be expressed.
Multiple alleles	The albino series (C) produces a range of phenotypes from full pigment intensity to true albino.
Incomplete dominance	The spotting gene (S) has three phenotypes ranging from extensive spotting to no spotting at all.
Lethal genes	The Manx gene (M) that produces a stubby or no tail is lethal when in the homozygous dominant condition (MM causes death in the womb).
Pleiotropy	The white gene (W) also affects eye color and can cause congenital deafness (one gene with three effects).
Sex linkage	The orange gene is sex (X) linked and can convert black pigment to orange. Since female cats have two X chromosomes they have three possible phenotypes (black, orange and tortoiseshell) whereas males can normally only exhibit two phenotypes (black and orange).
Environmental effects	The dark color pointing in Siamese and Burmese cats where the gene (cs) is only active in the cooler extremities such as the paws, tail and face.

(NOTE: Some of these genetic phenomena are covered elsewhere)

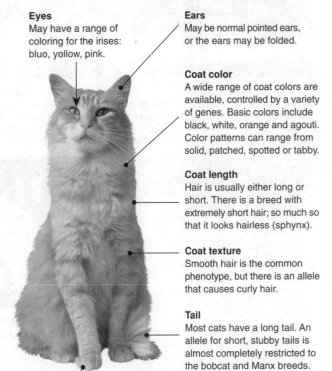

Eyes
May have a range of coloring for the irises: blue, yellow, pink.

Ears
May be normal pointed ears, or the ears may be folded.

Coat color
A wide range of coat colors are available, controlled by a variety of genes. Basic colors include black, white, orange and agouti. Color patterns can range from solid, patched, spotted or tabby.

Coat length
Hair is usually either long or short. There is a breed with extremely short hair; so much so that it looks hairless (sphynx).

Coat texture
Smooth hair is the common phenotype, but there is an allele that causes curly hair.

Tail
Most cats have a long tail. An allele for short, stubby tails is almost completely restricted to the bobcat and Manx breeds.

Paws
Most cats have five digits on the front paw and four on the rear. The occurrence of polydactyly with as many as six or seven digits affects as many as one out of five cats (in some parts of the world it is even higher than this).

Genes controlling inherited traits in domestic cats

Wild forms		Mutant forms		Wild forms		Mutant forms	
Allele	*Phenotype*	*Allele*	*Phenotype*	*Allele*	*Phenotype*	*Allele*	*Phenotype*
A	Agouti	**a**	Black (non-agouti)	**m**	Normal tail	**M**	Manx tail, shorter than normal (stubby)
B	Black pigment	**b**	Brown pigment	**o**	Normal colors (no red, usually black)	**O**	Orange (sex linked)
C	Unicolored	**cch**	Silver				
		cs	Siamese (pointing: dark at extremities)	**pd**	Normal number of toes	**Pd**	Polydactylism; has extra toes
		ca	Albino with blue eyes				
		c	Albino with pink eyes	**R**	Normal, smooth hair	**R**	Rex hair, curly
D	Dense pigment	**d**	Dilute pigment	**s**	Normal coat color without white spots	**S**	Color interspersed with white patches or spots (piebald white spotting)
fd	Normal, pointed ears	**Fd**	Folded ears				
Hr	Normal, full coat	**h**	Hairlessness	**T**	Tabby pattern (mackerel striped)	**Ta**	Abyssinian tabby
						tb	Blotched tabby, classic pattern of patches or stripes
i	Fur colored all over	**I**	Inhibitor: part of the hair is not colored (silver)				
				w	Normal coat color, not all white	**W**	All white coat color (dominant white)
L	Short hair	**l**	Long hair, longer than normal			**Wh**	Wirehair

Variation in Coat Color in Domestic Cats

Non-agouti
A completely jet black cat has no markings on it whatsoever. It would have the genotype: **aaB–D–** since no dominant agouti allele must be present, and the black pigment is not diluted.

Siamese
The color pointing of Siamese cats is caused by warm temperature deactivation of a gene that produces melanin pigment. Cooler parts of the body are not affected and appear dark.

Tortoiseshell
Because this is a sex linked trait, it is normally found only in female cats (**XO**, **Xo**). The coat is a mixture of orange and black fur irregularly blended together.

Agouti hair
Enlarged view of agouti hair. Note that the number of darkly pigmented stripes can vary on the same animal.

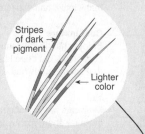

Stripes of dark pigment

Lighter color

Sex linked orange
The orange (**XO**, **XO**) cat has an orange coat with little or no patterns such as tabby showing.

Blotched tabby
Lacks stripes but has broad, irregular bands arranged in whorls (**tb**).

Wild type
Mackerel (striped) tabby (**A–B–T–**) with evenly spaced, well-defined, vertical stripes. The background color is **agouti** with the stripes being areas of completely black hairs.

Orange

Black

White

Calico
Similar to a tortoiseshell, but with substantial amounts of white fur present as well. Black, orange and white fur.

Golden yellow coat

Deeper colour stripes

Marmalade
The orange color (**XO**, **XO**) is expressed, along with the alleles for the tabby pattern. The allele for orange color shows epistatic dominance and overrides the expression of the agouti color so that the tabby pattern appears dark orange.

Other Inherited Features in Domestic Cats

No tail

Manx tail (Mm)
The Manx breed of cat has little or no tail. This dominant allele is lethal if it occurs in the homozygous condition.

6 digits on the paw

Polydactylism (Pd–)
This is a dominant mutation. The number of digits on the front paw should be five, with four digits on the rear paw.

Ears folded forwards

Ear fold (Fd–)
Most cats have normal pointed ears. A dominant mutation exists where the ear is permanently folded forwards.

What Genotype Has That Cat?

Consult the table of genes listed on the previous pages and enter the allele symbols associated with each of the phenotypes in the column headed 'Allele'. For this exercise, study the appearance of real cats around your home or look at color photographs of different cats. For each cat, complete the checklist of traits listed below by simply placing a tick in the appropriate spaces. These traits are listed in the same order as the genes for **wild forms** and **mutant forms** on page 173. On a piece of paper, write each of the cat's genotypes. Use a dash (-) for the second allele for characteristics that could be either heterozygous or homozygous dominant (see the sample at the bottom of the page).

NOTES:
1. *Agouti* fur coloring is used to describe black **hairs** with a light band of pigment close to its tip.
2. Patches of silver fur (called chinchilla) produces the silver tabby phenotype in agouti cats. Can also produce "smoke" phenotype in Persian long-haired cats, causing reduced intensity of the black.
3. Describes the dark extremities (face, tail and paws) with lighter body (e.g. Siamese).
4. The recessive allele makes black cats blue-gray in color and yellow cats cream.
5. Spottiness involving less than half the surface area is likely to be heterozygous.

Gene	Phenotype	Allele	Sample	Cat 1	Cat 2	Cat 3	Cat 4
Phenotype Record Sheet for Domestic Cats							
Agouti color	Agouti[1]						
	Non-agouti		✔				
Pigment color	Black		✔				
	Brown						
Color present	Uncolored		✔				
	Silver patches[2]						
	Pointed[3]						
	Albino with blue eyes						
	Albino with pink eyes						
Pigment density	Dense pigment						
	Dilute pigment[4]		✔				
Ear shape	Pointed ears		✔				
	Folded ears						
Hairiness	Normal, full coat		✔				
	Hairlessness						
Hair length	Short hair		✔				
	Long hair						
Tail length	Normal tail (long)		✔				
	Stubby tail or no tail at all						
Orange color	Normal colors (non-orange)		✔				
	Orange						
Number of digits	Normal number of toes		✔				
	Polydactylism (extra toes)						
Hair curliness	Normal, smooth hair		✔				
	Curly hair (rex)						
Spottiness	No white spots						
	White spots (less than half)[5]		✔				
	White spots (more than half)						
Stripes	Mackerel striped (tabby)						
	Blotched stripes						
White coat	Not all white		✔				
	All white coat color						

Sample cat: (see ticks in chart above)

To give you an idea of how to read the chart you have created, here is an example genotype of the author's cat with the following features: *A smoky gray uniform-coloured cat, with short smooth hair, normal tail and ears, with 5 digits on the front paws and 4 on the rear paws, small patches of white on the feet and chest. (Note that the stripe genotype is completely unknown since there is no agouti allele present).*

GENOTYPE: aa B– C– dd fdfd Hr– ii L– mm oo pdpd R– Ss ww

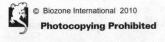

Cancer and the Environment

Cancer is a term describing a large group of diseases characterised by the progressive and uncontrolled growth of abnormal cells. Cancer cells become immortal, continuing to divide regardless of any damage incurred. Agents that cause cancer are termed **carcinogens**. There are many known carcinogens, some agents are genetic, some are biological (e.g. viral infection), and others are environmental. Chronic exposure to carcinogens accelerates the rate at which dividing cells make errors. Certain risk factors increase a person's chance of getting cancer. Some risk factors, such as exposure to tobacco smoke, are controllable, while others, such as gender, are not. Because cancers arise as a result of damage to DNA, those factors that cause cellular damage, e.g. exposure to UV in sunlight, increase the risk of cancers developing.

Features of Cancer Cells

The diagram below shows a single lung cell that has become cancerous. It no longer carries out the role of a lung cell, and instead takes on a parasitic lifestyle, taking from the body what it needs in the way of nutrients and contributing nothing in return. The rate of cell division is greater than in normal cells in the same tissue because there is no resting phase between divisions.

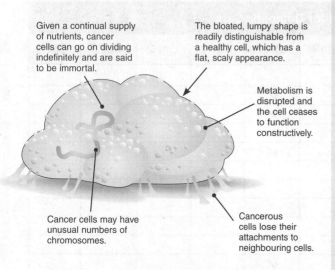

Given a continual supply of nutrients, cancer cells can go on dividing indefinitely and are said to be immortal.

The bloated, lumpy shape is readily distinguishable from a healthy cell, which has a flat, scaly appearance.

Metabolism is disrupted and the cell ceases to function constructively.

Cancer cells may have unusual numbers of chromosomes.

Cancerous cells lose their attachments to neighbouring cells.

Risk Factors for Cancer

Lifestyle factors (diet, alcohol intake, weight, exercise) are all significant contributors to cancer. Smoking (left) causes one third of all cancer deaths. Excessive alcohol has been linked to oral cancers. A highly processed, high fat diet is associated with colon cancer.

Unprotected exposure to ultraviolet light (e.g sunbathing, left) causes early ageing of the skin and damage that can lead to the development of skin cancers. Hazardous substances (e.g. asbestos) cause cell damage that can lead to cancer.

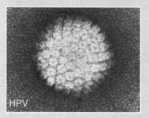
HPV

Some viruses and bacteria can cause genetic changes in cells making them more likely to become cancerous. Examples include hepatitis B and C viruses (liver cancer), and the *Human Papillomavirus* (HPV), which is linked to cervical cancer.

CDC

Skin cancers, such as melanoma (above), are the most common type of cancer in the UK. Increased rates over the last 30 years (left) are directly correlated with more people travelling abroad for holidays and exposing themselves to UV radiation in strong sunlight. Foreign holidays have increased three fold since 1985.

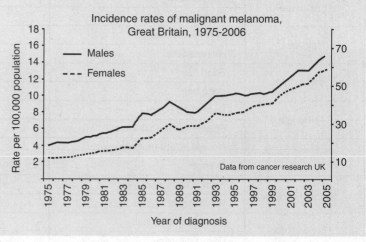

Incidence rates of malignant melanoma, Great Britain, 1975-2006

— Males
-- Females

Rate per 100,000 population

Year of diagnosis

Data from cancer research UK

1. Explain how cancerous cells differ from normal cells:

_____

_____

2. Explain the mechanism by which the risk factors described above increase the chance of developing cancer:

_____

_____

3. Explain why it can be difficult to determine the causative role of a single risk factor in the development of a cancer:

_____

_____

Related activities: Causes of mutations
Web links: The Cancer Cell, Inside Cancer

Periodicals:
Bring me sunshine

© Biozone International 2010
Photocopying Prohibited

KEY TERMS: What Am I?

THE OBJECT OF THIS GAME is to guess the unknown term from clues given to you by your team. Teams can be two or more, or you can play against individuals.
1) Cut out the cards below. You will need one set per team.
2) Shuffle the cards and deal them, face down, to each person in your team.

3) Affix tape to the back of the card so it can be stuck to your forehead. At no stage look at the word on the card!
4) One team starts. The members of your team give you a clue, one at a time, up to a maximum of **three** clues about what your term is. Do not use the word(s) on your card!

5) The clue should be a single point e.g. *"You feed on grass."*
6) If you guess correctly, your team receives another turn and the score is recorded. If you cannot guess, then the turn passes to the other team.
7) The game can be ended after one round or many.

Allele	Cancer	Crossing over
Epistasis	Gametes	Gene
Genotype	Independent assortment	Meiosis
Oogenesis	Phenotype	Pluripotent
Polygenes	Polytene chromosome	Spermatogenesis
Stem cell	Totipotent	Zygote

R 2

These cards have been deliberately left blank

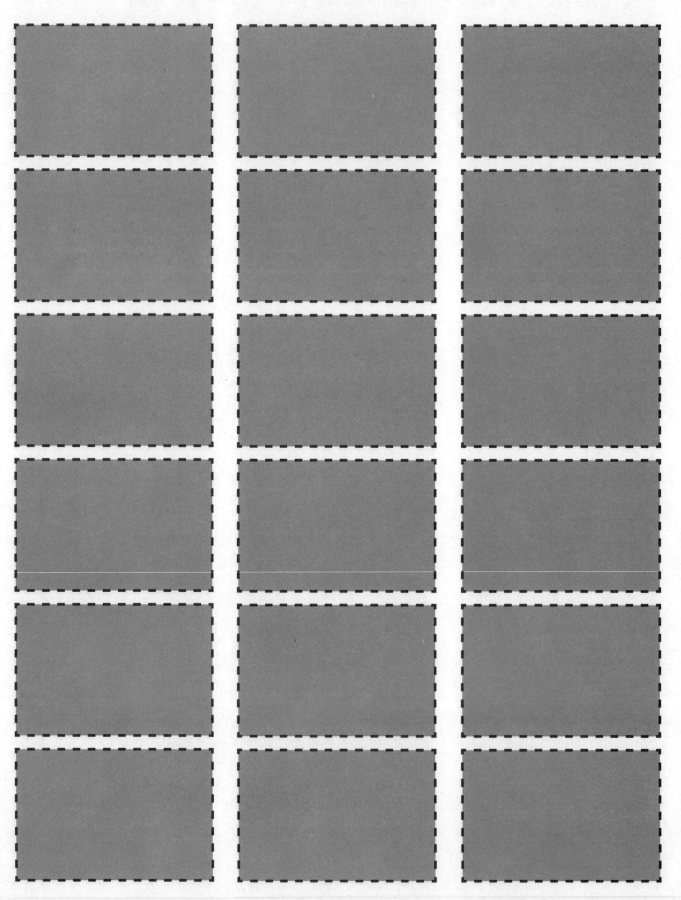

Plants as Resources

KEY CONCEPTS

▶ Plant cells are bound by a cellulose cell wall.

▶ Plant cells share many organelles in common with plant cells, but they also have organelles that are unique to them.

▶ Cellulose and starch have important structural and functional roles.

▶ Humans use plants in many applications, including the production of new pharmaceuticals.

OBJECTIVES

☐ 1. Use the **KEY TERMS** to help you understand and complete these objectives.

Plant Structure and Function   page 129-130, 180-182, 185-178

☐ 2. Compare the ultrastructure of a **plant cell** with that of an **animal cell**.

☐ 3. Identify the **organelles** of plant cells in slides and electron micrographs. Include reference to: **cell wall**, **chloroplast**, **amyloplast**, **vacuole**, **tonoplast**, **plasmodesmata**, **pits**, and **middle lamella**.

☐ 4. Compare and contrast the structure and location of **sclerenchyma fibres** (support) and **xylem vessels** (water and mineral ion transport). Identify these structures using a light microscope.

☐ 5. Compare the structure of the polysaccharides **starch** and **cellulose**. Describe the formation of **cellulose microfibrils** and discuss their importance to the plant.

Putting Plants to Use   page 183-186

☐ 6. Explain how the arrangement of cellulose microfibrils and secondary thickening in plants together produce strong fibres and discuss how these are exploited for human use.

☐ 7. Describe how **tensile strength** is measured. †

☐ 8. Describe the contribution of plant fibres to sustainable consumables, such as in the production of bioplastics.

Plant Transport and Nutrition   page 187-193

☐ 9. Explain the importance of **water** and **inorganic ions** (e.g.) to plants.

☐ 10. Explain how water and mineral ions are taken up by plant. Explain **transpiration** as an inevitable consequence of gas exchange in plants and appreciate its role in facilitating adequate mineral uptake.

☐ 11. Describe the effect of specific **mineral deficiencies** on a plant.

☐ 12. Describe how plant mineral deficiencies can be investigated. †

Drug Development and Testing   page 183, 194-196

☐ 13. Describe the importance of plants in developing new drugs and medicines.

☐ 14. Describe how plants can be tested for **antimicrobial properties**. †

☐ 15. Discuss the role of **William Withering**'s digitalis soup in the development of drug testing protocols. In particular, compare historic drug testing with modern drug testing. Include reference to **double blind trials**, use of **placebos**, and **three-phased clinical testing**.

Periodicals:
listings for this chapter are on page 230

Weblinks:
www.biozone.co.uk/
weblink/Edx-AS-2542.html

Teacher Resource CD-ROM:
Review of Eukaryotic Cells

Plant Cells

Plant cells share many structures and organelles in common with animal cells, but also have several unique features. Plant cells are enclosed in a cellulose cell wall, which gives them a more regular and uniform appearance than animal cells. The cell wall protects the cell, maintains its shape, and prevents excessive water uptake. It provides rigidity to plant structures but permits the free the passage of materials into and out of the cell. The diagram below illustrates structures of a typical plant cell.

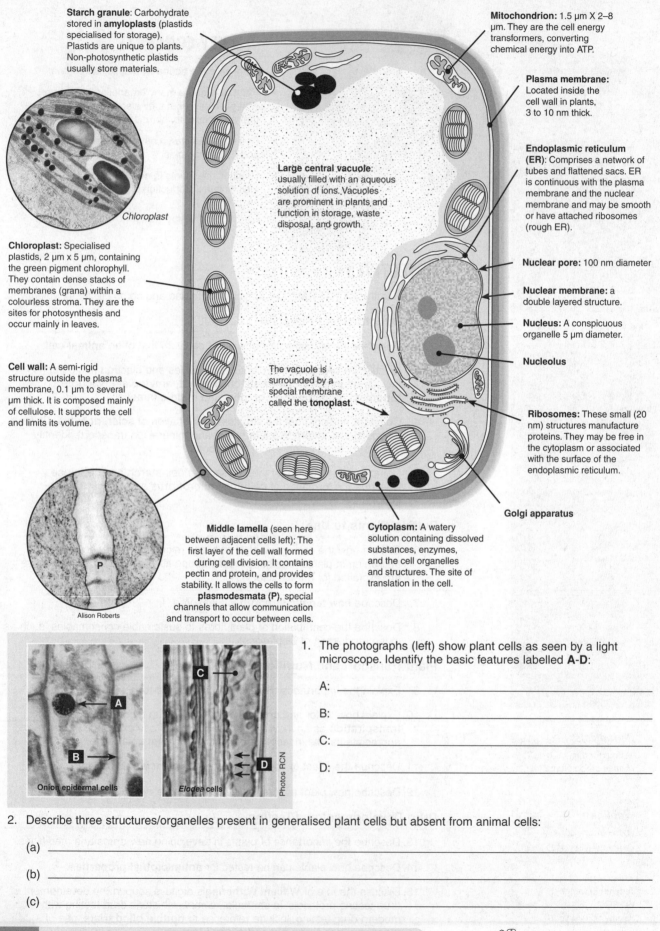

Starch granule: Carbohydrate stored in **amyloplasts** (plastids specialised for storage). Plastids are unique to plants. Non-photosynthetic plastids usually store materials.

Chloroplast

Chloroplast: Specialised plastids, 2 µm x 5 µm, containing the green pigment chlorophyll. They contain dense stacks of membranes (grana) within a colourless stroma. They are the sites for photosynthesis and occur mainly in leaves.

Cell wall: A semi-rigid structure outside the plasma membrane, 0.1 µm to several µm thick. It is composed mainly of cellulose. It supports the cell and limits its volume.

Mitochondrion: 1.5 µm X 2–8 µm. They are the cell energy transformers, converting chemical energy into ATP.

Plasma membrane: Located inside the cell wall in plants, 3 to 10 nm thick.

Endoplasmic reticulum (ER): Comprises a network of tubes and flattened sacs. ER is continuous with the plasma membrane and the nuclear membrane and may be smooth or have attached ribosomes (rough ER).

Nuclear pore: 100 nm diameter

Nuclear membrane: a double layered structure.

Nucleus: A conspicuous organelle 5 µm diameter.

Nucleolus

Ribosomes: These small (20 nm) structures manufacture proteins. They may be free in the cytoplasm or associated with the surface of the endoplasmic reticulum.

Large central vacuole: usually filled with an aqueous solution of ions. Vacuoles are prominent in plants and function in storage, waste disposal, and growth.

The vacuole is surrounded by a special membrane called the **tonoplast**.

Golgi apparatus

Middle lamella (seen here between adjacent cells left): The first layer of the cell wall formed during cell division. It contains pectin and protein, and provides stability. It allows the cells to form **plasmodesmata (P)**, special channels that allow communication and transport to occur between cells.

Alison Roberts

Cytoplasm: A watery solution containing dissolved substances, enzymes, and the cell organelles and structures. The site of translation in the cell.

Onion epidermal cells

Elodea cells

Photos RCN

1. The photographs (left) show plant cells as seen by a light microscope. Identify the basic features labelled **A-D**:

    A: _____

    B: _____

    C: _____

    D: _____

2. Describe three structures/organelles present in generalised plant cells but absent from animal cells:

    (a) _____

    (b) _____

    (c) _____

Related activities: Animal Cells
Web links: Eukaryotic Cells Interactive Animation, Review of Eukaryotic Cells

Identifying Plant Cell Structures

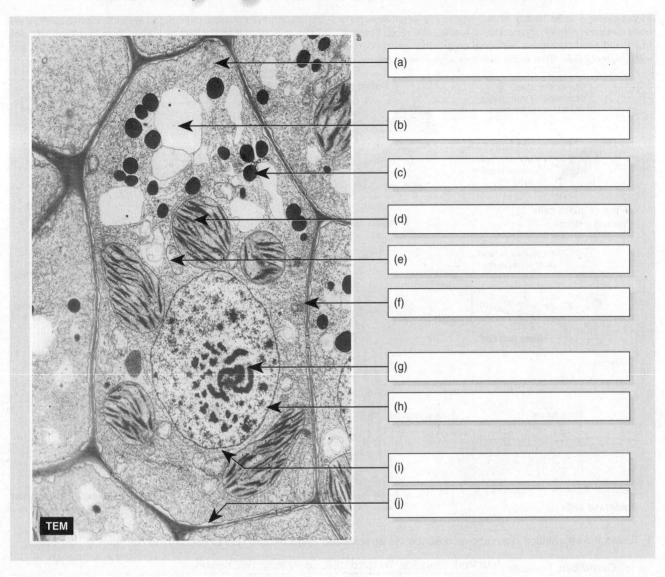

(a)

(b)

(c)

(d)

(e)

(f)

(g)

(h)

(i)

(j)

TEM

1. Identify and label the ten structures in the cell above using the following list of terms: *nuclear membrane, cytoplasm, endoplasmic reticulum, mitochondrion, starch granules, chromosome, vacuole, plasma membrane, cell wall, chloroplast*

2. State how many cells, or parts of cells, are visible in the electron micrograph above: _____

3. Name the features which identify this cell as a plant cell:

   _____

   _____

4. (a) Explain where cytoplasm is found in the cell: _____

   _____

   (b) Describe what cytoplasm is made up of: _____

   _____

5. Describe two structures, pictured in the cell above, that are associated with storage:

   (a) _____

   _____

   (b) _____

   _____

Plant Cell Specialisation

Plants show a wide variety of cell types. The vegetative plant body consists of three organs: stems, leaves, and roots. Flowers, fruits, and seeds comprise additional organs that are concerned with reproduction. The eight cell types illustrated below are representatives of these plant organ systems. Each has structural or physiological features that set it apart from the other cell types. The differentiation of cells enables each specialised type to fulfil a specific role in the plant.

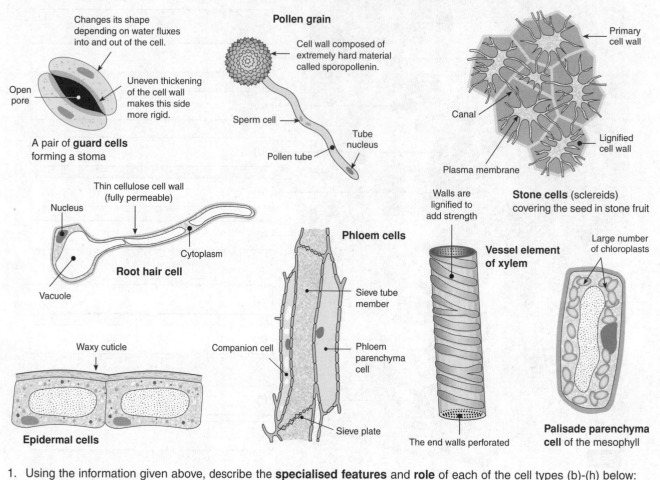

1. Using the information given above, describe the **specialised features** and **role** of each of the cell types (b)-(h) below:

(a) **Guard cell**: Features: _Curved, sausage shaped cell, unevenly thickened_____

　　　Role in plant: _____Turgor changes alter the cell shape to open or close the stoma_____

(b) **Pollen grain**: Features: _____

　　　Role in plant: _____

(c) **Palisade parenchyma cell**: Features: _____

　　　Role in plant: _____

(d) **Epidermal cell**: Features: _____

　　　Role in plant: _____

(e) **Vessel element**: Features: _____

　　　Role in plant: _____

(f) **Stone cell**: Features: _____

　　　Role in plant: _____

(g) **Sieve tube member**: Features: _____

　　　Role in plant: _____

(h) **Root hair cell**: Features: _____

　　　Role in plant:

Uses of Plants

Plants provide oxygen via photosynthesis and are the ultimate source of food and metabolic energy for nearly all consumers. Plants also provide the raw materials from which innumerable other products are made (see below). As technology has improved and more plant compounds are identified, the number of products derived from plants has also increased.

The Importance of Plants

Manufacturing Extracts

Plant extracts, including latex from rubber trees (above) is used in car tyres, latex gloves, rubber boots, and insulation. Gymnosperm resins are used to make varnish, adhesives and perfumes. Essential oils, and plant based pesticides are other examples of extracts.

Fuel

Coal, petroleum, and natural gas are fossil fuels formed over millions of years from the dead remains of plants and other organisms. With wood, they are important sources of fuel. However, the combustion of fossil fuels contributes to greenhouse gas emissions.

Clothing & Textiles

Many plants provide fibres used in a range of materials including linen (flax), coir (coconut husks) and cotton (above). The properties of the fabric (warmth, weight, water resistance) depend upon the type and number of the fibres used and diameter of the thread.

Plastics

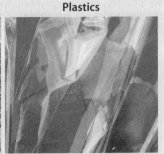

Plant fibres have been used to make plastics for many years. Cellophane, made from cellulose, was first produced in 1912 and is a common food packaging material. It is also used in the manufacture of Sellotape and dialysis tubing. Cellophane is fully biodegradable.

Shelter

Plant tissues can be utilised to provide shelter in the form of framing, cladding and roofing on both temporary and permanent structures. Most modern houses still have wooden frames and some use materials such as hay bales for insulation.

Medicines

Over 120 chemical substances extracted from higher plants are used in medicine. Aspirin (salicylic acid from willow bark) and digitalin (from foxglove, above) have been in medical use since antiquity. Others, are more recent discoveries.

Paper

Plants have been used to make paper for thousands of years. Papyrus scrolls (above) were made in Egypt 5500 years ago by overlapping strips of pith into layers then pounding and smoothing them. Paper today is made by digesting wood pulp with sulphates.

Food

Plant tissues provide energy for almost all heterotrophic life. Many plants produce delicious fruits in order to spread their seeds. Selective breeding has produced vegetables and fruits many times larger and more nutritious than the original wild types.

Plants as Resources

1. Using examples, describe how plant species are used by people for each of the following:

   (a) Food: _____

   _____

   (b) Fuel: _____

   _____

   (c) Clothing: _____

   _____

   (d) Shelter: _____

   _____

   (e) Paper: _____

   _____

   (f) Medicines: _____

   _____

Periodicals:
Designer starches

Related activities: *Cellulose and Starch, Antimicrobial Properties of Plants, Sustainable Plant Products*

A 2

Bioplastics

Plastic is used widely in packaging, food containers, plastic bags, CD's, glasses frames, and electronic cases to name a few products. Conventional plastics (petroleum plastics) are made from non-renewable fossil fuels, they are not biodegradable, their manufacture is unsustainable, and they contribute to carbon dioxide emissions. Some conventional plastics can not be recycled, but low recycling rates are observed even in types of plastics that are recyclable. As a result, plastic products rapidly fill landfills, and when they are thoughtlessly discarded they pollute the environment, causing substantial harm to wildlife. Bioplastics are plastics made from renewable biomass sources (e.g. plants or vegetable oil). Bioplastic technology has expanded rapidly since the 1980s, and bioplastics are beginning to replace conventional plastic products. The fact that bioplastics can be sustainably produced, and that some are biodegradable or can be composted makes them an attractive alternative to conventional plastic products. Starch based plastic (e.g. polylactic acid) are widely used (see below).

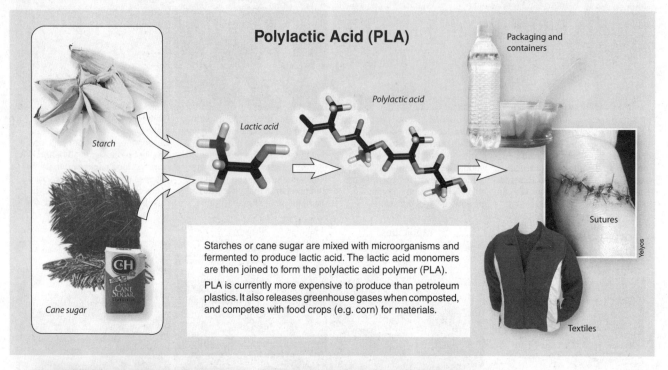

Polylactic Acid (PLA)

Starch

Cane sugar

Lactic acid

Polylactic acid

Packaging and containers

Sutures

Textiles

Starches or cane sugar are mixed with microorganisms and fermented to produce lactic acid. The lactic acid monomers are then joined to form the polylactic acid polymer (PLA).

PLA is currently more expensive to produce than petroleum plastics. It also releases greenhouse gases when composted, and competes with food crops (e.g. corn) for materials.

University of Hawaii

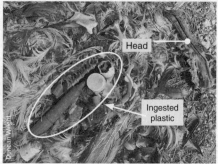

Head

Ingested plastic

Duncan Wright

Marine plastic debris, far left, in and around the oceans kill approximately 100,000 marine mammals each year. Reptiles, sea birds and fish species are also affected. The remains of this albatross chick (left), show that it died from ingesting large quantities of plastic rubbish. Biodegradable bioplastics would greatly reduce the loss of animal life from discarded plastic rubbish.

1. (a) Explain why plant-based plastics provide an environmentally friendly solution to the problems of plastic pollution:

_____

_____

_____

(b) Describe some of the disadvantages of manufacturing PLA products: _____

_____

_____

2. Explain why the manufacture of bioplastics from food crops (e.g. maize) is not considered as environmentally friendly as bioplastics produced from alternative crops (e.g. waste feedstock):

_____

_____

Cellulose and Starch

Cellulose is the most common molecule on Earth, making up one third of the volume of plants and one half of the volume of wood. Cellulose is made of thousands of β-glucose monomers arranged in a single, unbranched chain. In plants, cellulose makes up the bulk of the cell wall, providing both the strength required to keep the cell's shape and the support for the plant stem or trunk. As a material that can be exploited by humans, cellulose provides fibres which can be made into thread for textiles, and wood used for framing and cladding buildings. In contrast, **starch** is made of α-glucose monomers, and is a compact, branching molecule. It has no structural function, but can be hydrolysed to release soluble sugars for energy.

Cellulose Structure and Function

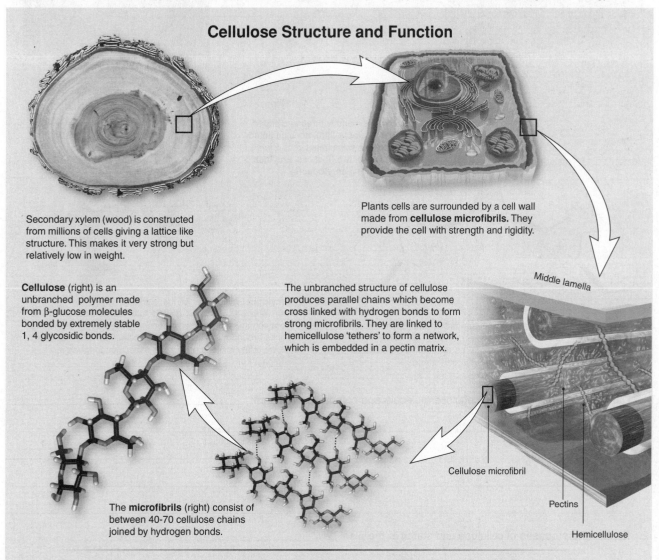

Secondary xylem (wood) is constructed from millions of cells giving a lattice like structure. This makes it very strong but relatively low in weight.

Plants cells are surrounded by a cell wall made from **cellulose microfibrils**. They provide the cell with strength and rigidity.

Cellulose (right) is an unbranched polymer made from β-glucose molecules bonded by extremely stable 1, 4 glycosidic bonds.

The unbranched structure of cellulose produces parallel chains which become cross linked with hydrogen bonds to form strong microfibrils. They are linked to hemicellulose 'tethers' to form a network, which is embedded in a pectin matrix.

Middle lamella

The **microfibrils** (right) consist of between 40-70 cellulose chains joined by hydrogen bonds.

Cellulose microfibril

Pectins

Hemicellulose

Cellulose and Tensile Strength

The **tensile strength** of a material refers to its ability to resist the strain of a pulling force. The simplest way to test tensile strength is to hang the material from a support and increase the tension by adding weight at the other end until the material separates. The tensile strength is then expressed as the force applied per the area of material in cross section. Typically this is written as pascals (Pa) or Newtons per square metre (Nm^{-2}), but is also often expressed as pounds per square inch (PSI). Cellulose fibrils show extremely high tensile strength (table below).

Material	Tensile strength (GPa)
Cellulose	17.8
Silk	1
Human hair	0.38
Rubber	0.15
Steel wire	2.2
Nylon (6/6)	0.75

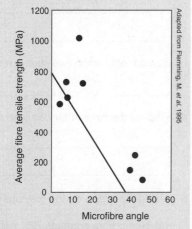

Adapted from Flemming, M. et al, 1995

Plants produce fibres of various strengths that can be used for textiles, ropes, and construction. In general, the greater the amount of cellulose in the fibre the greater its tensile strength (above left). The tensile strength of the plant fibre is also influenced by the orientation of the cellulose microfibrils. Microfibrils orientated parallel to the length of the fibre (0°) provide the greatest fibre strength (above right).

Periodicals:
Designer starches

Related activities: Carbohydrate Chemistry,
Carbohydrates for Energy

RA 3

Starch Structure and Function

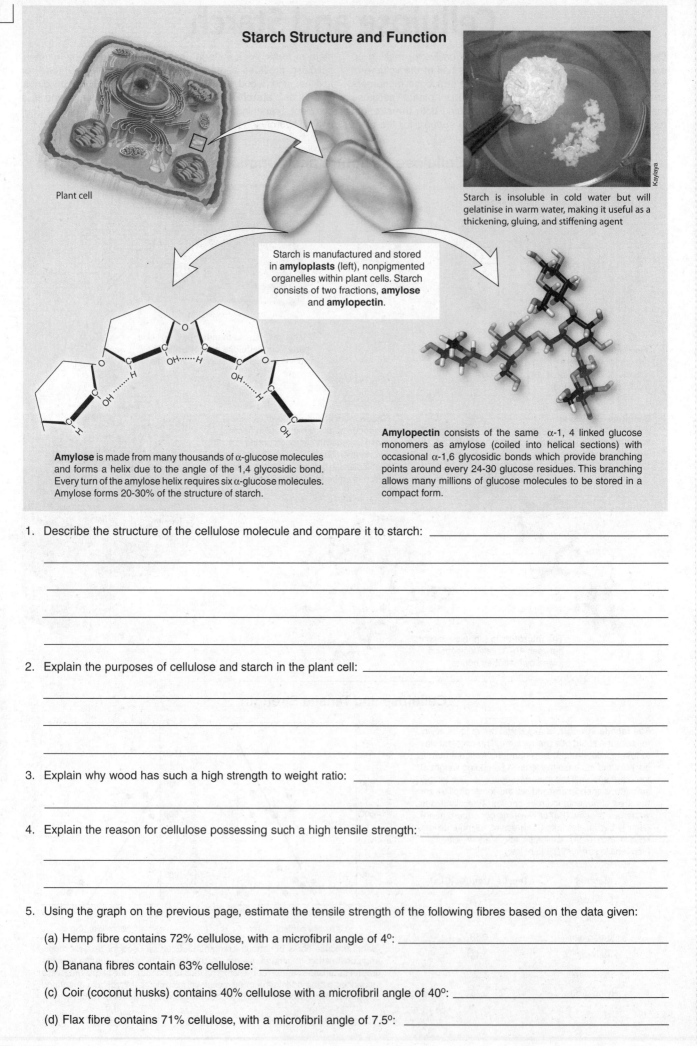

Plant cell

Starch is insoluble in cold water but will gelatinise in warm water, making it useful as a thickening, gluing, and stiffening agent

Starch is manufactured and stored in **amyloplasts** (left), nonpigmented organelles within plant cells. Starch consists of two fractions, **amylose** and **amylopectin**.

Amylose is made from many thousands of α-glucose molecules and forms a helix due to the angle of the 1,4 glycosidic bond. Every turn of the amylose helix requires six α-glucose molecules. Amylose forms 20-30% of the structure of starch.

Amylopectin consists of the same α-1, 4 linked glucose monomers as amylose (coiled into helical sections) with occasional α-1,6 glycosidic bonds which provide branching points around every 24-30 glucose residues. This branching allows many millions of glucose molecules to be stored in a compact form.

1. Describe the structure of the cellulose molecule and compare it to starch: _____

_____

_____

_____

2. Explain the purposes of cellulose and starch in the plant cell: _____

_____

_____

_____

3. Explain why wood has such a high strength to weight ratio: _____

_____

4. Explain the reason for cellulose possessing such a high tensile strength: _____

_____

_____

5. Using the graph on the previous page, estimate the tensile strength of the following fibres based on the data given:

   (a) Hemp fibre contains 72% cellulose, with a microfibril angle of 4°: _____

   (b) Banana fibres contain 63% cellulose: _____

   (c) Coir (coconut husks) contains 40% cellulose with a microfibril angle of 40°: _____

   (d) Flax fibre contains 71% cellulose, with a microfibril angle of 7.5°: _____

Xylem and Phloem

The two main kinds of supporting tissues in plants are **xylem** and **phloem**. As in animals, tissues in plants are groupings of different cell types that work together for a common function. Xylem and phloem are **complex tissues** composed of a number of cell types. They are specialised for the transport of water and dissolved sugars respectively. Most of xylem tissue is composed of large vessels, which have thickened and strengthened walls and conduct water. Xylem also contains packing cells and fibres, which provide support to the tissue. When mature, xylem is dead. Phloem comprises packing cells and supporting fibre cells, and two special cell types: **sieve tubes** and **companion cells**. Unlike xylem, phloem tissue is alive when mature.

Xylem Tissue

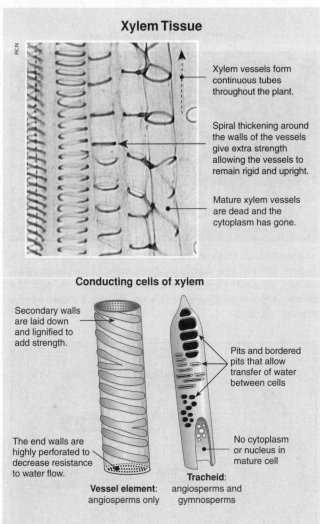

Xylem vessels form continuous tubes throughout the plant.

Spiral thickening around the walls of the vessels give extra strength allowing the vessels to remain rigid and upright.

Mature xylem vessels are dead and the cytoplasm has gone.

Conducting cells of xylem

Secondary walls are laid down and lignified to add strength.

Pits and bordered pits that allow transfer of water between cells

The end walls are highly perforated to decrease resistance to water flow.

No cytoplasm or nucleus in mature cell

Vessel element: angiosperms only

Tracheid: angiosperms and gymnosperms

Phloem Tissue

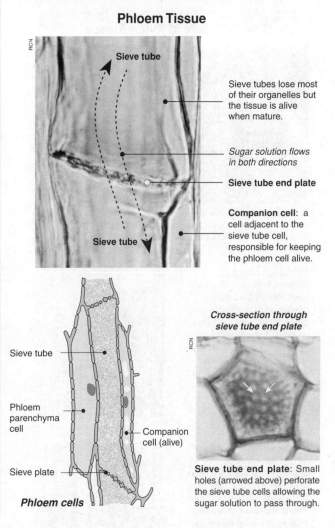

Sieve tube

Sieve tubes lose most of their organelles but the tissue is alive when mature.

Sugar solution flows in both directions

Sieve tube end plate

Sieve tube

Companion cell: a cell adjacent to the sieve tube cell, responsible for keeping the phloem cell alive.

Sieve tube

Phloem parenchyma cell

Companion cell (alive)

Sieve plate

Phloem cells

Cross-section through sieve tube end plate

Sieve tube end plate: Small holes (arrowed above) perforate the sieve tube cells allowing the sugar solution to pass through.

1. Discuss the structural and functional differences between xylem and phloem: _____

_____

_____

_____

2. Describe a way in which xylem is strengthened in a mature plant: _____

_____

3. Describe a difference between the xylem tissue of gymnosperms and angiosperms: _____

_____

4. Explain the purpose of the holes in the **sieve plate** at the ends of each sieve tube cell: _____

_____

5. (a) Name the cell type in the phloem that actually conducts the sugar solution: _____

   (b) Describe the purpose of the companion cell in phloem tissue: _____

_____

Related activities: Xylem
Web links: Photographic Atlas of Plant Anatomy

A 1

Plants as Resources

Identifying Xylem Tissue

Xylem is the principal **water conducting tissue** in vascular plants. It is also involved in conducting dissolved minerals, in food storage, and in supporting the plant body. In angiosperms, xylem is composed of several cell types: tracheids, vessels, xylem parenchyma, and **sclerenchyma**. Sclerenchyma is composed of two types of cell, stone shaped sclereids and long, thin fibres. Sclerenchyma cells are thick walled and lignified and give strength and support to the plant. The tracheids and vessel elements form the bulk of the tissue. They are heavily strengthened and are the conducting cells of the xylem. Parenchyma cells are involved in storage.

The Structure of Xylem Tissue

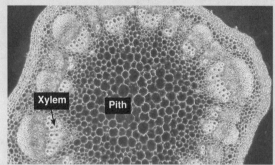

This cross section through the stem, *Helianthus* (sunflower) shows the central pith, surrounded by a peripheral ring of vascular bundles. Note the xylem vessels with their thick walls.

Fibres are a type of sclerenchyma cell. They are associated with vascular tissues and usually occur in groups. The cells are very elongated and taper to a point and the cell walls are heavily thickened. Fibres give mechanical support to tissues, providing both strength and elasticity.

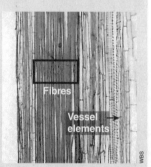

Vessel elements are found only in the xylem of angiosperms. They are large diameter cells that offer very low resistance to water flow. The possession of vessels (stacks of vessel elements) provides angiosperms with a major advantage over gymnosperms and ferns as they allow for very rapid water uptake and transport.

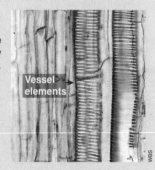

Identifying Xylem in Roots

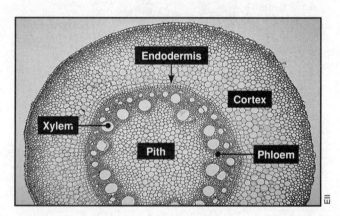

The appearance of xylem tissue in the roots of moncots and dicots is quite different. In a dicot root (above) the vascular tissue, xylem (X) and phloem (P) forms a central cylinder through the root. The primary xylem of dicot roots forms a star shape in the centre of the vascular cylinder.

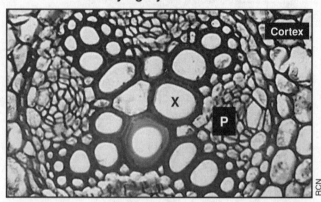

In monocot roots, such as corn root (above) the vascular tissue forms a ring like structure called the stele. The stele is very large compared to the size of the root, and there are many xylem points. Monocots have a central pith inside the vascular tissue that is absent in dicot roots.

1. Describe the function of **xylem:** _____

_____

2. Identify the four main cell types in xylem and explain their role in the tissue:

(a) _____

(b) _____

(c) _____

(d) _____

3. Describe one way in which xylem is strengthened in a mature plant: _____

4. Describe a feature of vessel elements that increases their efficiency of function: _____

_____

Related activities: Plant Tissues, Xylem and Phloem
Web links: Photographic Atlas of Plant Anatomy

Transpiration

Plants lose water all the time, despite the adaptations they have to help prevent it (e.g. waxy leaf cuticle). Approximately 99% of the water a plant absorbs from the soil is lost by evaporation from the leaves and stem. This loss, mostly through stomata, is called **transpiration** and the flow of water through the plant is called the **transpiration stream**. Plants rely on a gradient in solute concentration from the roots to the air to move water through their cells. Water flows passively from soil to air along a gradient of increasing solute (decreasing water) concentration. This gradient is the driving force in the ascent of water up a plant. A number of processes contribute to water movement up the plant: transpiration pull, cohesion, and root pressure. Transpiration may seem wasteful, but it has benefits; evaporative water loss cools the plant and the transpiration stream helps the plant to maintain an adequate mineral uptake, as many essential minerals occur in low concentrations in the soil.

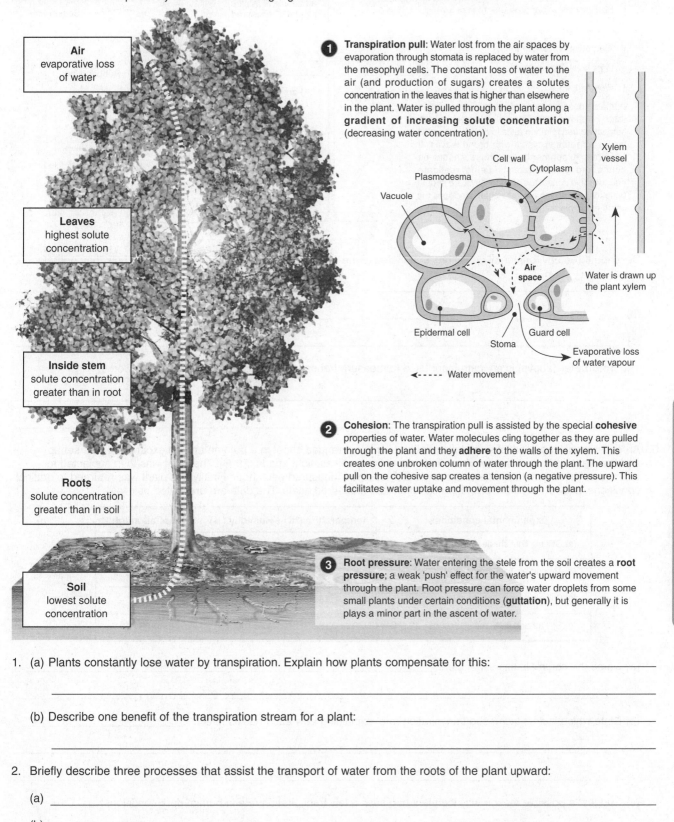

Air
evaporative loss of water

Leaves
highest solute concentration

Inside stem
solute concentration greater than in root

Roots
solute concentration greater than in soil

Soil
lowest solute concentration

1 **Transpiration pull**: Water lost from the air spaces by evaporation through stomata is replaced by water from the mesophyll cells. The constant loss of water to the air (and production of sugars) creates a solutes concentration in the leaves that is higher than elsewhere in the plant. Water is pulled through the plant along a **gradient of increasing solute concentration** (decreasing water concentration).

Cell wall
Cytoplasm
Plasmodesma
Vacuole
Xylem vessel
Water is drawn up the plant xylem
Air space
Epidermal cell
Guard cell
Stoma
Evaporative loss of water vapour
- - - - ◄ Water movement

2 **Cohesion**: The transpiration pull is assisted by the special **cohesive** properties of water. Water molecules cling together as they are pulled through the plant and they **adhere** to the walls of the xylem. This creates one unbroken column of water through the plant. The upward pull on the cohesive sap creates a tension (a negative pressure). This facilitates water uptake and movement through the plant.

3 **Root pressure**: Water entering the stele from the soil creates a **root pressure**; a weak 'push' effect for the water's upward movement through the plant. Root pressure can force water droplets from some small plants under certain conditions (**guttation**), but generally it is plays a minor part in the ascent of water.

Plants as Resources

1. (a) Plants constantly lose water by transpiration. Explain how plants compensate for this: _____

_____

(b) Describe one benefit of the transpiration stream for a plant: _____

_____

2. Briefly describe three processes that assist the transport of water from the roots of the plant upward:

(a) _____

(b) _____

(c) _____

Periodicals:
How trees lift water,
High tension

Related activities: Plant Mineral Nutrition
Web links: Transpiration Animation

The Potometer

A potometer is a simple instrument for investigating transpiration rate (water loss per unit time). The equipment is simple and easy to obtain. A basic potometer, such as the one shown right, can easily be moved around so that transpiration rate can be measured under different environmental conditions

Some of the physical conditions investigated are:

- Humidity or vapour pressure (high or low)

- Temperature (high or low)

- Air movement (still or windy)

- Light level (high or low)

- Water supply

It is also possible to compare the transpiration rates of plants with different adaptations e.g. comparing transpiration rates in plants with rolled leaves vs rates in plants with broad leaves. If possible, experiments like these should be conducted simultaneously using replicate equipment. If conducted sequentially, care should be taken to keep the environmental conditions the same for all plants used.

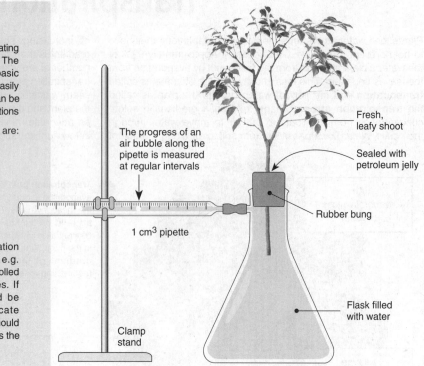

The progress of an air bubble along the pipette is measured at regular intervals

1 cm³ pipette

Clamp stand

Fresh, leafy shoot

Sealed with petroleum jelly

Rubber bung

Flask filled with water

3. Describe three environmental conditions that increase the rate of transpiration in plants, explaining how they operate:

(a) _____

(b) _____

(c) _____

4. The **potometer** (above) is an instrument used to measure transpiration rate. Briefly explain how it works:

_____

_____

5. An experiment was conducted on transpiration from a hydrangea shoot in a potometer. The experiment was set up and the plant left to stabilise (environmental conditions: still air, light shade, 20°C). The plant was then subjected to different environmental conditions and the water loss was measured each hour. Finally, the plant was returned to original conditions, allowed to stabilise and transpiration rate measured again. The data are presented below:

Experimental conditions	Temperature (°C)	Humidity (%)	Transpiration (gh⁻¹)
(a) Still air, light shade, 20°C	18	70	1.20
(b) Moving air, light shade, 20°C	18	70	1.60
(c) Still air, bright sunlight, 23°C	18	70	3.75
(d) Still air and dark, moist chamber, 19.5°C	18	100	0.05

(a) Name the control in this experiment: _____

_____

(b) State which factors increased transpiration rate, explaining how each has its effect: _____

_____

_____

(c) Suggest a possible reason why the plant had such a low transpiration rate in humid, dark conditions:

_____

_____

Uptake at the Root

Plants need to take up water and minerals constantly. They must compensate for the continuous loss of water from the leaves and provide the materials they need for the manufacture of food. The uptake of water and minerals is mostly restricted to the younger, most recently formed cells of the roots and the root hairs. Some water moves through the plant tissues via the **plasmodesmata** (cytoplasmic connections between cells), but most passes through the free spaces between cell walls. Water uptake is assisted by root pressure, the pressure created by the high concentration of the soil and root tissues. Two processes are involved in water and ion uptake: diffusion (osmosis in the case of water) and active transport.

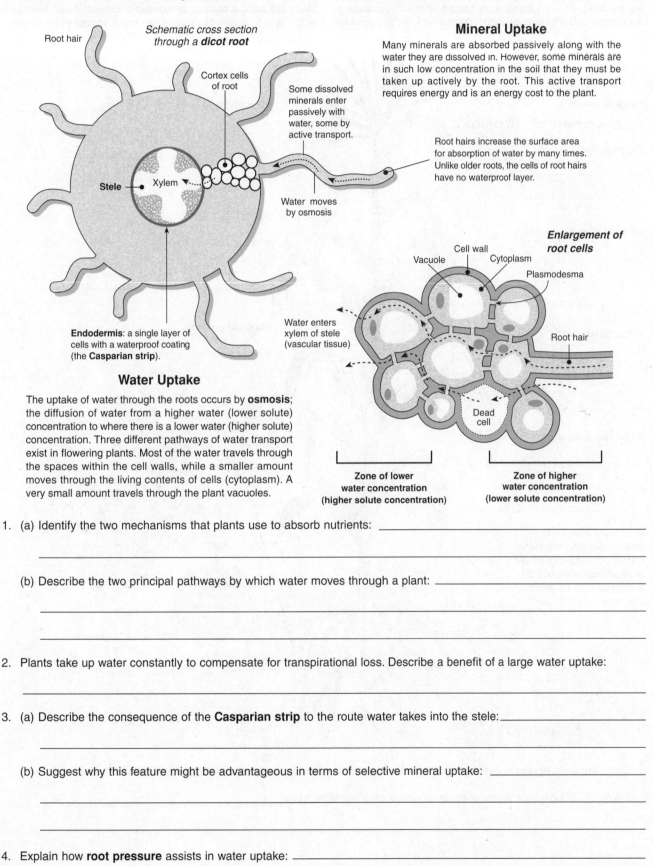

*Schematic cross section through a **dicot root***

Root hair

Cortex cells of root

Some dissolved minerals enter passively with water, some by active transport.

Stele

Xylem

Water moves by osmosis

Root hairs increase the surface area for absorption of water by many times. Unlike older roots, the cells of root hairs have no waterproof layer.

Endodermis: a single layer of cells with a waterproof coating (the **Casparian strip**).

Water enters xylem of stele (vascular tissue)

Enlargement of root cells

Cell wall
Vacuole
Cytoplasm
Plasmodesma

Root hair

Dead cell

Zone of lower water concentration (higher solute concentration)

Zone of higher water concentration (lower solute concentration)

Mineral Uptake

Many minerals are absorbed passively along with the water they are dissolved in. However, some minerals are in such low concentration in the soil that they must be taken up actively by the root. This active transport requires energy and is an energy cost to the plant.

Water Uptake

The uptake of water through the roots occurs by **osmosis**; the diffusion of water from a higher water (lower solute) concentration to where there is a lower water (higher solute) concentration. Three different pathways of water transport exist in flowering plants. Most of the water travels through the spaces within the cell walls, while a smaller amount moves through the living contents of cells (cytoplasm). A very small amount travels through the plant vacuoles.

1. (a) Identify the two mechanisms that plants use to absorb nutrients: _____

_____

(b) Describe the two principal pathways by which water moves through a plant: _____

_____

_____

2. Plants take up water constantly to compensate for transpirational loss. Describe a benefit of a large water uptake:

_____

3. (a) Describe the consequence of the **Casparian strip** to the route water takes into the stele:_____

_____

(b) Suggest why this feature might be advantageous in terms of selective mineral uptake: _____

_____

_____

4. Explain how **root pressure** assists in water uptake: _____

_____

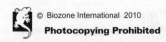
Related activities: Plant Mineral Nutrition

RA 3

Plant Mineral Nutrition

Plants are able to manufacture their own carbohydrates by photosynthesis, but must synthesise their own proteins and lipids by incorporating nutrients obtained from the soil. The availability of mineral ions to plant roots depends on soil texture, since this affects the permeability of the soil to air and water. Mineral ions may be available to the plant in the soil water, adsorbed on to clay particles, or via release from humus and soil weathering. Many higher plants increase their effective root surface area by the presence of mycorrhizal associations. This is an association between the plant root and a fungus, and increases their capacity to absorb minerals. **Macronutrients** (e.g. nitrogen, sulfur, phosphorus) are required in large amounts for building basic constituents such as proteins. **Trace elements** (e.g. manganese, copper, and zinc) are required in small amounts. Many are components of, or activators for, enzymes. Too much or too little of a specific nutrient can result in poor plant health.

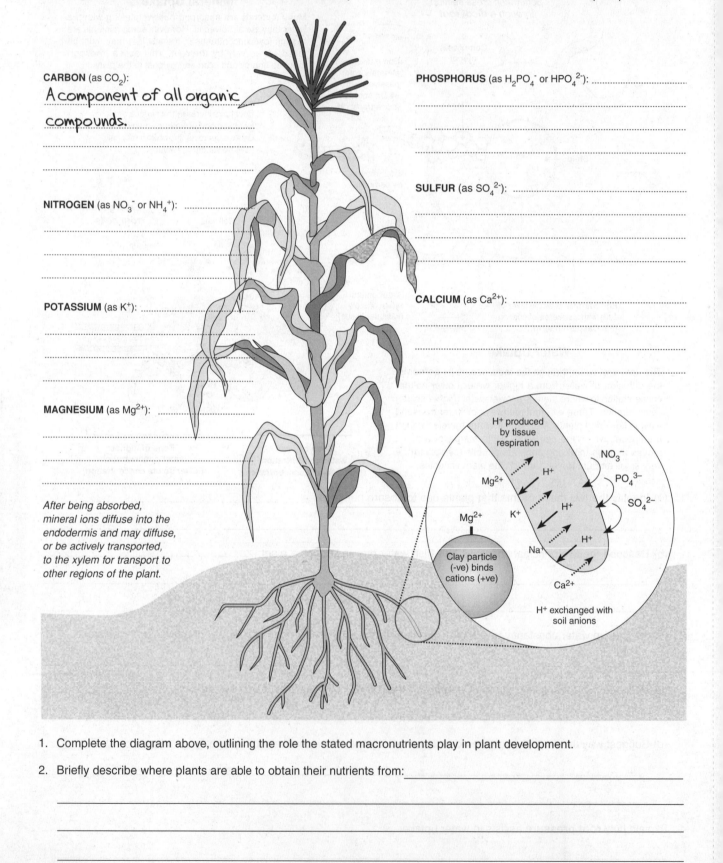

CARBON (as CO_2):

A component of all organic compounds.

NITROGEN (as NO_3^- or NH_4^+):

POTASSIUM (as K^+):

MAGNESIUM (as Mg^{2+}):

After being absorbed, mineral ions diffuse into the endodermis and may diffuse, or be actively transported, to the xylem for transport to other regions of the plant.

PHOSPHORUS (as $H_2PO_4^-$ or HPO_4^{2-}):

SULFUR (as SO_4^{2-}):

CALCIUM (as Ca^{2+}):

H^+ produced by tissue respiration

Mg^{2+}

Mg^{2+}

K^+

Na^+

H^+

H^+

H^+

H^+

NO_3^-

PO_4^{3-}

SO_4^{2-}

Ca^{2+}

Clay particle (-ve) binds cations (+ve)

H^+ exchanged with soil anions

1. Complete the diagram above, outlining the role the stated macronutrients play in plant development.

2. Briefly describe where plants are able to obtain their nutrients from:

Related activities: Uptake at the root, Plant Mineral Deficiencies

Plant Mineral Deficiencies

Plant mineral deficiencies can result in poor health and growth. For farmers growing these crops, deficiencies can result in low yields and poor economic returns. It is important to identify mineral deficiencies quickly so that the problem can be addressed, and plant health can be restored. Leaves and stems are often the first tissues analysed for mineral deficiencies because this is where the earliest symptoms usually appear. Symptoms may be similar for more than one nutrient deficiency (below). In these instances, chemical analysis of the soil or plant tissue enables an accurate assessment of a plants nutritional status.

Corn — Pale leaf

Symptoms of nitrogen deficiency include poor plant growth, and pale green or yellow leaves (above), which shows first in older foliage. Flowering or fruiting may be delayed.

Soybean — Curled leaf, Yellowed edges

The leaves on plants with potassium deficiency are curled at the tips, appear brown and scorched, and have yellow edges. They are prone to frost damage.

Coconut — Affected leaf areas

This coconut plant (above) is phosphorus deficient. Symptoms include poor growth and blueish-purple leaves. Flowering can be affected if phosphorus levels are too low.

Unhealthy leaf — Healthy leaf

Plants with magnesium deficiency develop pale yellowed and spotted foliage, which shows first in older leaves. Flower production is affected, and fruit is small.

Canola

Symptoms of sulfur deficiency are similar to nitrogen deficiency but the younger leaves are affected first. In general, the plants are small and have crop yield is low.

Apple

Calcium deficiency causes stunted growth. The leaves curl and terminal buds and root tips die. Apples (above), develop spots on the skin and flesh and have a bitter taste.

Investigating Mineral Deficiencies

The effects of mineral deficiencies on plants can be examined using comparative growth studies in the laboratory or field.

Plants of the same species and age are planted in the same growth medium and exposed to identical environmental conditions. They are divided into groups and given nutrient solutions lacking one specific mineral. The control group is given a complete nutrient solution (all minerals). Researchers can then observe the effect of the mineral deficiency on the plant.

A similar technique can be used to determine how different levels of a specific mineral affect plant health. Plants are provided with a complete nutrient solution, but one mineral is given at varying levels. This will show the level at which a deficiency begins to affect the plant. In the same way, it is also possible to test the effects of too much of a specific mineral on plant health.

Left: These rows of soybeans show the effects of potassium deficiency. The rows that produced large plants were given a complete nutrient solution. Rows that were potassium deprived produced smaller plants.

The effect of nitrogen deficiency was tested on growing lettuce plants (right). Plants given a nitrogen deprived solution were significantly smaller and lighter in colour than plants given a complete nutrient solution.

Gunter *et.al*

1. (a) Suggest why chemical analysis is sometimes required to identify which nutrient is deficient:

_____

_____

_____

_____

(b) Account for the similarity in symptoms between nitrogen deficiency and sulfur deficiency:

_____

_____

_____

2. (a) You have set up a laboratory experiment to demonstrate the effect of nitrogen deficiency on tomato plants. Explain why you need a control for the experiment:

_____

_____

_____

_____

(b) Describe an appropriate control: _____

_____

Related activities: Plant Mineral Nutrition, Designing Your Experiment

A 2

Plants as Resources

Antimicrobial Properties of Plants

Plants have been used for their medicinal properties for many centuries. Around half of the pharmaceuticals in use today are of plant origin. Although many plants have antimicrobial properties, very few are currently used commercially in this manner. This is generally because antibiotic production and extraction from fungi and bacteria is more straightforward. However, antimicrobial compounds from plants are being increasingly investigated. This is because there are concerns about the effectiveness of traditional fungal and bacterial-derived antibiotics, and there is also an increase in antibiotic resistance to these drugs.

1

A plant is chosen for testing. Plants produce a wide range of compounds, but those with aromatic rings often have medicinal properties.

2

Pulping

Solvent extraction

Extraction: Plant material is pulped, then soaked in a solvent (water, alcohol), or boiled to release and crudely fractionate the compounds.

3

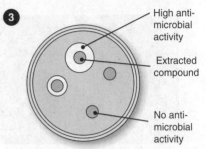

High anti-microbial activity

Extracted compound

No anti-microbial activity

Screening: Microbes are plated onto agar. Paper discs soaked with plant extract are placed on the agar, or the plant extract is pipetted into wells cut into the agar. The plates are then incubated. The effectiveness of the extract can be determined by measuring the clear zone around the plant extract after incubation.

4

Separation and analysis: Extracts that show antimicrobial properties undergo mass spectroscopy. This further separates and isolates the compounds and determines their chemical composition and structure.

5

Concentration: The compounds may be further separated, (e.g. by chromatography), before they are concentrated using techniques such as freeze drying or rotary evaporation (above).

6

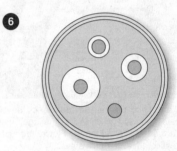

Testing: The isolated and concentrated compounds are tested for their antimicrobial properties. This identifies which specific component of the initial extract has the antimicrobial activity.

The antimicrobial compound in chili peppers, is capsaicin; an antibacterial agent of relatively high potency.

The essential oil produced from the herb rosemary (above), acts as an effective general antimicrobial agent.

Phloretin is a polyphenol extracted from the leaves of apple trees. It acts as a highly potent antimicrobial.

Saponins (compounds that foam in water) extracted from ginseng are effective against a range of bacteria.

1. Explain why plants are generally not used to produce antibiotics: _____

2. Explain how the effectiveness of a new antimicrobial compound can be tested: _____

3. Explain why research into antimicrobial plant extracts is increasing: _____

Testing New Drugs

The use of plants for their healing properties has been known to humans for several thousand years. The ancient Egyptians would chew on willow bark, which contains the compound salicylic acid, to relieve fever and minor pain. This compound is still widely used today in a modified form better know to us as aspirin. Plants form the basis of many commonly used drugs, but the drugs are often synthesised and modified in special manufacturing plants to make them safer and more effective. This also allows larger volumes to be produced. The development and testing of new drugs is highly regulated. This ensures that new drugs are safer and more effective, but the testing process can be very slow and costly. As a result, some drugs can be very expensive.

Digitalis Soup

A famous example of early drug testing is credited to an English doctor and scientist, **William Withering**. A patient came to Withering suffering from a serious heart condition called cardiac dropsy, which is fatal if left untreated. Withering had no treatment of his own for the man, so he visited a local "wise women" who prescribed her herbal tea for the patient. Amazingly, the patient recovered.

Withering's interest peaked, so he purchased the recipe, and discovered that the active ingredient was a compound from the foxglove plant called **digitalis**. He proceeded to test different concentrations on his patients. Initially, many of his patients become seriously ill from the side effects, so Withering introduced a standard procedure to ensure that his patients received the correct dosage. The amount of digitalis given to a patient was slowly increased until its side effects became apparent, and then the dose was decreased slightly. This ensured that the patients received a dose that was safe, but effective.

William Withering not only identified digitalis as a potent medicine, but his protocols to establish safe and effective doses form the basis of drug testing today.

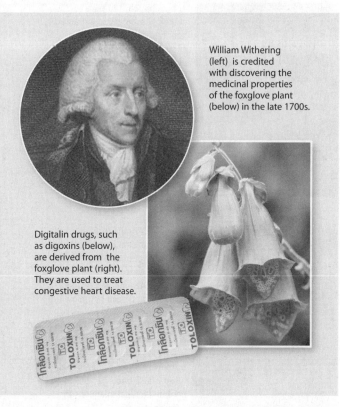

William Withering (left) is credited with discovering the medicinal properties of the foxglove plant (below) in the late 1700s.

Digitalin drugs, such as digoxins (below), are derived from the foxglove plant (right). They are used to treat congestive heart disease.

The Path to Drug Discovery and Release

Plants as Resources

Discovery

- Discovery or identification of a drug target.
- An ideal target is specific to a certain disease.
- Drug targets are typically proteins (e.g. receptors or kinases).
- Data is collected on how the target functions and its role in contributing to a specific disease.

Screening

- High throughput screening of potential chemicals against the drug target. This can involve computer modelling.
- Product characterisation; target specificity, mode of action and activity.
- Drug delivery mechanisms are developed.

Development

Preclinical
- Laboratory and animal studies.
- Pharmacology, toxicology, formulation, and dose are determined.
- Duration 3-4 years.

Clinical
- Human testing to develop safety and efficacy.
- Three phases of testing.
- Duration 3-5 years.

Compliance & Release

- Application for new drug approval filed.
- Approval given.
- Duration 2-3 years.
- Phase IV clinical testing may be required even after approval has been given.
- Release of new drug to market.

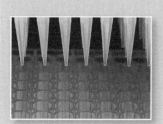

RDS Wellcom trust

Periodicals:
Clinical trials

Related activities: Antimicrobial Properties of Plants

A 2

Sources of New Medicines

Researchers are constantly trying to find and develop new medicines. Natural products (from plants, microbial metabolites, and marine invertebrates) play an important role in this search. Useful compounds extracted from these organisms often undergo modifications to improve their safety and efficacy. Often their chemical structures are reproduced synthetically so that large quantities can be produced.

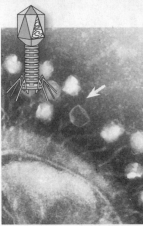

Approximately 120 pure chemical substances extracted from higher plants are used in medicine. Some, including **aspirin** (salicylic acid from willow bark), have been in medical use since antiquity. Others, such as the plant alkaloid **taxol** (an anticancer drug), are more recent discoveries.

*The bark and needles of the Pacific yew (right) provide the anti-cancer drug **taxol**.*

The use of drugs isolated from microorganisms began with the discovery of penicillin in 1928. The drug **botox**, derived from the toxin of *Clostridium botulinum*, is used to treat facial neuralgia as well as for cosmetic purposes.

New approaches to microbial medicines include using **bacteriophages** (viruses that infect attack bacteria, left) to control bacterial infections in different tissues.

The Testing Protocol

It is time consuming and expensive to develop new medicines. Only one in every thousand of new drugs identified ever makes it to the market. The total development time for a single drug can be up to 15 years and may cost between £500 million-1.1 billion. Testing, particularly at the preclinical and clinical stages (outlined below), is very expensive.

If preclinical animal trials are successful, then **clinical trials** are carried out on human volunteers. This involves three phases of testing:

Phase I: Testing is carried out on a small number (20-100) of healthy volunteers. It establishes the safety of the drug.

Phase II: Testing is carried out on a small group of a few hundred people who have the disease or condition the drug is targeting. The purpose is to determine effectiveness of the drug and discover if there are any side effects.

Phase III: Testing on thousands of people to gain additional information about safety, dosage, and effectiveness.

A **placebo** drug is given to a number of the patients within the study to act as a **control**. A placebo is an inactive compound with no pharmacological effect. It is included to eliminate any psychological effects patients feel simply from believing they are receiving a treatment.

Phase II and III trials are often **double blind**, meaning neither the doctor nor patient know whether a patient is receiving the drug or a placebo. It is designed to prevent biased responses or analysis.

1. Explain how William Withering ensured his patients received the correct dose of digitalis, and why his methods are considered to be pioneering work in the field of drug testing:

_____

_____

_____

_____

_____

2. (a) Describe the purpose of a placebo: _____

_____

_____

(b) Describe the purpose of a double blind trial: _____

_____

_____

3. Discuss reasons for the high cost of drug development: _____

_____

_____

_____

The cards below have a keyword or term printed on one side and its definition printed on the opposite side. The aim is to win as many cards as possible from the table. To play the game.....

1) Cut out the cards and lay them definition side down on the desk. You will need one set of cards between two students.

2) Taking turns, choose a card and, BEFORE you pick it up, state your own best definition of the keyword to your opponent.

3) Check the definition on the opposite side of the card. If both you and your opponent agree that your stated definition matches, then keep the card. If your definition does not match then return the card to the desk.

4) Once your turn is over, your opponent may choose a card.

Xylem	Cellulose	Amylose
Placebo	Amylopectin	Clinical trial
Middle lamella	Amyloplast	Bioplastic
Chloroplast	Cell wall	Microfibril
Starch	Fibres	Vessel
Antimicrobial	Vacuoles	Plasmodesmata

Plants as Resources

R 2

When you've finished the game keep these cutouts and use them as flash cards!

Helix shaped, unbranched macromolecule formed from the bonding of several hundred to several thousand a-glucose monomers.

Unbranched molecule that makes up the majority of a cell wall. It comprises several thousand repeating β-glucose monomers.

Vascular tissue that conducts water and mineral salts from the roots to the rest of the plant.

Trial in which the safety and efficacy (ability of a product to produce an effect) of a new product is tested for possible use in health care.

Molecule found in starch that contains around 4% α-glucose monomers that branch at carbon 6 to form a branched chain.

Simulated treatment with no medicinal component, used to evaluate the effect of both positive thought on the patient's behalf and the actual compared effect of the true medical treatment.

Plastic made from renewable plant material, normally cellulose or starch.

Nonpigmented organelle found in plant cells in which starch is manufactured and stored.

The outer layer of the cell wall formed during cell division. It contains pectin and protein, and helps bind cells together.

Fine, fibre-like strand consisting of several intermolecularly bonded chains of cellulose (in plants) or glycoproteins (in animals).

A structure, present in plants and bacteria, which is found outside the plasma membrane and gives rigidity to the cell.

An organelle found in plants that contains chlorophyll and in which the reactions of photosynthesis take place.

The primary water conducting cells of xylem in angiosperms.

Elongated sclerenchyma cells associated with vascular tissue.

Helical, unbranched macromolecule formed from the bonding of several hundred to several thousand α-glucose monomers.

Cytoplasmic connections through the middle lamella of plant cells.

These are usually large and prominent in plants and function in storage, waste disposal, and growth.

Ability to inhibit the growth of microorganisms

Biodiversity and Evolution

Unit 2
6BI02

4

KEY CONCEPTS

▶ Diversity is a broad term which can refer to species, genetic, ecosystem, or habitat diversity.

▶ Natural selection in the prevailing environment brings about evolutionary change and changes in diversity.

▶ Conservation plays a vital role in maintaining diversity.

KEY TERMS

adaptation
biodiversity
biological species
binomial nomenclature
classification scheme
conservation
competition
Darwin's theory
diversity
diversity indices
ecosystem
endemic
evolution
ex situ conservation
fitness
founder effect
Gause's principle
gene pool
genetic diversity
genetic drift
in situ conservation
interspecific competition
intraspecific competition
keystone species
molecular phylogeny
natural selection
niche
phylogeny
population bottleneck
seed bank
Simpson's index
species
species evenness
species richness
taxonomy

OBJECTIVES

☐ 1. Use the **KEY TERMS** to help you understand and complete these objectives.

Biodiversity and Conservation — page 200-206

☐ 2. Define the terms **biodiversity** and **endemism**. Recognise diversity as a broad term, which can refer to species, genetic, ecosystem, or habitat diversity.

☐ 3. Distinguish between the terms **species richness** and **species evenness**. Explain how biodiversity is measured, including sampling techniques and the use of diversity indices.

☐ 4. Evaluate the role of **conservation methods** in maintaining diversity. In particular, discuss the roles of zoos, seed banks, restoration programmes, and *in-situ* and *ex-situ* conservation methods. Discuss the role of local and regional programmes in maintaining biodiversity.

Species and Adaptation — page 207-211

☐ 5. Define the term **biological species** and recognise difficulties with the standard biological species definition.

☐ 6. Define the term **ecological niche**. Understand that within a niche an organism usually occupies a smaller area, the **realised niche**.

☐ 7. Discuss how organisms can become adapted to their environment. Include examples of behavioural, physiological, and anatomical adaptations.

Gene Pools and Evolution — page 212-221

☐ 8. Describe how changes to a species **gene pool** can bring about evolutionary change. Identify some of the factors which can alter a gene pool.

☐ 9. Explain **Darwin's theory of evolution by natural selection**. Describe how natural selection acts upon a phenotype to bring about change.

Classification — page 222-226

☐ 10. Describe why **classification** of organisms is useful for establishing relationships between organisms.

☐ 11. Discuss how advances in technology have lead to revising the classification of some taxa. Describe how these advances lead to the new **three domain system** for classifying life.

Periodicals:
listings for this chapter are on page 230

Weblinks:
www.biozone.co.uk/
weblink/Edx-AS-2542.html

Teacher Resource CD-ROM:
The Modern Theory of Evolution

Global Biodiversity

The species is the basic unit by which we measure biological diversity or **biodiversity**. Biodiversity is not distributed evenly on Earth, being consistently richer in the tropics and concentrated more in some areas than in others. The simplest definition of biodiversity is as the sum of all biotic variation from the level of genes to ecosystems, but often the components of total biodiversity are distinguished. **Species diversity** describes the number of different species in an area (**species richness**), **genetic diversity** is the diversity of genes within a species, and **ecosystem diversity** refers to the diversity at the higher ecosystem level of organisation. **Habitat diversity** is also sometimes described and is essentially a subset of ecosystem diversity expressed per given unit area. Total biological diversity is often threatened because of the loss of just one of these components. Conservation International recognises 25 **biodiversity hotspots**. These are biologically diverse and ecologically distinct regions under the greatest threat of destruction. They are identified on the basis of the number of species present, the amount of **endemism**, and the extent to which the species are threatened.

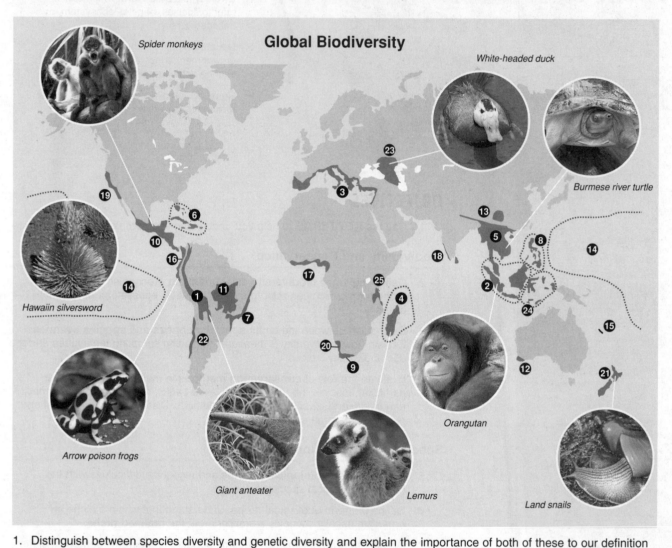

Global Biodiversity

Spider monkeys

White-headed duck

Burmese river turtle

Hawaiin silversword

Arrow poison frogs

Giant anteater

Lemurs

Orangutan

Land snails

1. Distinguish between species diversity and genetic diversity and explain the importance of both of these to our definition of total biological diversity:

_____

_____

_____

_____

2. Explain the importance of considering ecosystem (habitat) diversity when targeting regions for conservation purposes:

_____

_____

_____

3. Use your research tools (e.g. textbook, internet, or encyclopaedia) to identify each of the 25 biodiversity hotspots illustrated in the diagram above. For each region, summarise the characteristics that have resulted in it being identified as a biodiversity hotspot. Present your summary as a short report and attach it to this page of your workbook.

RA 2

Related activities: Britain's Biodiversity
Web links: Biodiversity Hotspots

Periodicals:
Biodiversity: taking stock,
Earth's nine lives

© Biozone International 2010
Photocopying Prohibited

Britain's Biodiversity

Biodiversity is measured by looking at species richness. For some taxa, e.g. bacteria, the true extent of species diversity remains unidentified. Some data on species richness for the UK are shown below. The biodiversity of the British Isles today is the result of a legacy of past climatic changes and a long history of human influence. Some of the most interesting, species-rich ecosystems, such as hedgerows, downland turf, and woodland, are maintained as a result of human activity. Many of the species characteristic of Britain's biodiversity are also found more widely in Europe. Other species (e.g. the Scottish crossbill), are **endemic** to Britain. Endemic simply means a species is found only within a particular geographical area. With increasing pressure on natural areas from urbanisation, roading, and other human encroachment, maintaining species diversity is paramount and should concern us all today.

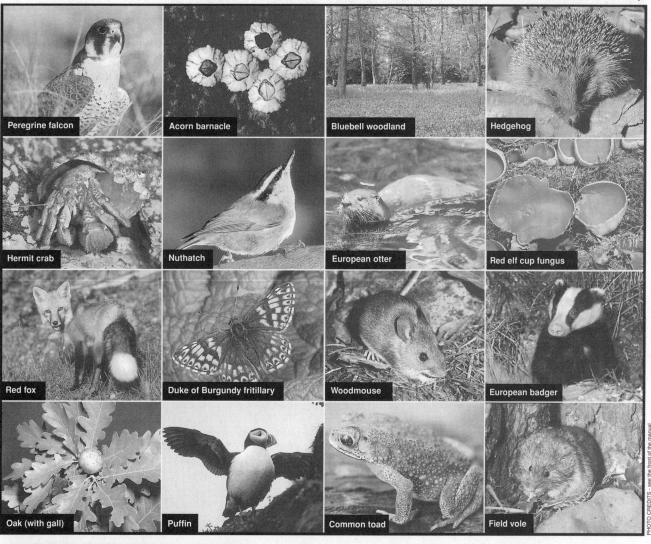

PHOTO CREDITS - see the front of the manual

Left: Fig. 1: British biodiversity, as numbers of terrestrial and freshwater species, compared with recent global estimates of described species in major taxonomic groups.

Major taxonomic group	Estimated no. of British species	Estimated no. of world species
Bacteria	*unknown*	> 4 000
Viruses	*unknown*	> 5 000
Protozoa	> 20 000	> 40 000
Algae	> 20 000	> 40 000
Fungi	> 15 000	> 70 000
Ferns and bryophytes	1 080	> 26 000
Lichens	1 500	> 17 000
Flowering plants	1 400	> 250 000
Invertebrate animals	> 28 500	> 1.28 million
Insects	22 500	> 1 million
Non-insect arthropods	> 3 000	> 190 000
All other invertebrates	> 3 000	> 90 000
Vertebrate animals	308	> 33 208
Fish (freshwater)	38	> 8 500
Amphibians	6	> 4 000
Reptiles	6	> 6 500
Birds (breeding residents)	210	9 881
Mammals	48	4 327

Source: Biodiversity: The UK Action Plan, 1994. HMSO

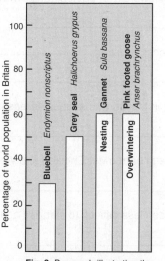

Fig. 2: Bar graph illustrating the percentage of world populations of various species permanently or temporarily resident in Britain.

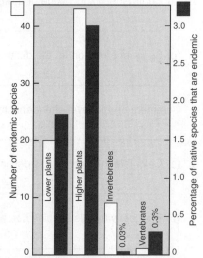

Fig 3: Bar graph illustrating the degree of endemism in Britain. Right axis indicates % endemism in relation to the number of described British native species.

Biodiversity & Evolution

Periodicals:
Biodiversity: taking stock

Related activities: Global Biodiversity
Web links: Space for Species

RA 3

Barn owls are predators of small mammals, birds, insects, and frogs. They are higher order consumers and, as such, have been badly affected by the bioaccumulation of pesticides in recent times. They require suitable nesting and bathing sites, and reliable sources of small prey.

Conservation of the barn owl *(Tyto alba)*

Status: The barn owl is one of the best known and widely distributed owl species in the world. It was once very common in Britain but has experienced severe declines in the last 50 years as a result of the combined impacts of habitat loss, changed farming practices, and increased sources of mortality.

Reasons for decline: Primarily, declines have been the result of changed farm management practices (e.g. increased land clearance and mechanisation) which have resulted in reduced prey abundance and fewer suitable breeding sites. Contributing factors include increases in road deaths as traffic speed and volume rises, and poorer breeding success and reduced chick survival as a result of pesticide bioaccumulation. In addition, more birds are drowned when attempting to bathe in the steep sided troughs which have increasingly replaced the more traditional shallow farm ponds.

Conservation management: A return to the population densities of 50 years ago is very unlikely, but current conservation measures have at least stabilised numbers. These involve habitat enhancement (e.g. provision of nest sites), reduction in pesticide use, and rearing of orphaned young followed by monitored release into suitable habitats.

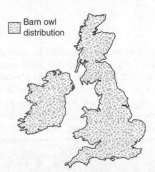

Barn owl distribution

Barn owls are widely distributed in Ireland and the UK, but numbers are not high.

Period of survey (England & Wales)	Breeding pairs (estimates)
1935	12 000
1968 – 1972	6000 – 9000
1983 – 1985	3800

1. Produce a pie graph below to show the proportions of British species in each taxonomic group (ignoring bacteria and viruses). Calculate the percentages from Fig. 1 (opposite) and tabulate the data (one has been completed for you). The chart has been marked in 5° divisions and each % point is equal to 3.6° on the pie chart. Provide a colour key in the space next to the tabulated figures. For the purposes of this exercise, use the values provided, ignoring the > sign:

Proportion of British species in different taxonomic groups

	Percentage of species in each taxon	Segment size	Key
Protozoa			
Algae			
Fungi			
Ferns and bryophytes			
Lichens			
Flowering plants			
Invertebrates	28 500 ÷ 87 788 X 100 = 32.5%	117°	■
Vertebrates			

2. Comment on the proportion of biodiversity within each taxonomic group: _____

_____

3. (a) Contrast our knowledge of the biodiversity of bacteria and invertebrates with that of vertebrates:

_____

(b) Suggest a reason for the difference: _____

_____

4. Comment on the level of endemism in the UK and suggest a reason for it: _____

_____

5. (a) Calculate the percentage decline in barn owls (England and Wales) over the 50 year period 1935 – 1985:

_____

(b) Suggest why this species has been less difficult to stabilise against decline than other (more endangered) species:

_____

_____

_____

Measuring Diversity in Ecosystems

Measurements of biodiversity have essentially two components: **species richness**, which describes the number of species, and **species evenness**, which quantifies how equally the community composition is distributed. Both are important, especially when rarity is a reflection of how threatened a species is in an environment. Information about the biodiversity of ecosystems is obtained through **sampling** the ecosystem in a manner that provides a fair (unbiased) representation of the organisms present and their distribution. This is usually achieved through **random sampling**, a technique in which every possible sample of a given size has the same chance of selection. Measures of biodiversity are commonly used as the basis for making conservation decisions and different measures of biodiversity may support different solutions. Often indicator species and species diversity indices are used as a way of quantifying biodiversity. Such indicators can be particularly useful when monitoring ecosystem change and looking for causative factors in species loss.

Quantifying the Diversity of Ecosystems

Reef community:
high density, clumped distribution

Measurements of biodiversity must be appropriate to the community being investigated. Communities in which the populations are at low density and have a random or clumped distribution will require a different sampling strategy to those where the populations are uniformly distributed and at higher density. There are many sampling options (below), each with advantages and drawbacks for particular communities. How would you estimate the biodiversity of this reef community?

Random point sampling	*Point sampling: systematic grid*	*Line and belt transects*	*Random quadrats*

Marine ecologists use quadrat sampling to estimate biodiversity prior to works such as dredging.

Line transects are appropriate to estimate biodiversity along an environmental gradient.

Photo: www.coastal-planning.net

Keystone Species in Ecosystems

The stability of an ecosystem refers to its apparently unchanging nature over time, something that depends partly on its ability to resist and recover from disturbance. Ecosystem stability is closely linked to biodiversity, and more biodiverse systems tend to be more stable, partly because the many species interactions that sustain them act as a buffer against change. Some species are more influential than others in the stability of an ecosystem because of their pivotal role in some ecosystem function such as nutrient recycling or productivity. Such species are called **keystone species** because of their disproportionate effect on ecosystem function.

The **European beaver**, *Castor fiber*, was originally distributed throughout most of Europe and northern Asia but populations have been decimated as a result of both hunting and habitat loss. The beaver is a keystone species; where they occur, beavers are critical to ecosystem function and a number of species depend partly or entirely on beaver ponds for survival. Their tree-felling activity is akin to a natural coppicing process and promotes vigorous regrowth, while historically they helped the spread of alder (a water-loving species) in Britain.

1. (a) Distinguish between the two measures of biodiversity: species richness and species evenness:

_____

_____

_____

(b) Explain why it is important to consider both these measures when considering species conservation:

_____

_____

_____

_____

Related activities: Global Biodiversity, Britain's Biodiversity

RA 2

Biodiversity & Evolution

Calculation and Use of Diversity Indices

One of the best ways to determine the health of an ecosystem is to measure the variety (rather than the absolute number) of organisms living in it. Certain species, called **indicator species**, are typical of ecosystems in a particular state (e.g. polluted or pristine). An objective evaluation of an ecosystem's biodiversity can provide valuable insight into its status, particularly if the species assemblages have changed as a result of disturbance.

Diversity can be quantified using a **diversity index (DI)**. Diversity indices attempt to quantify the degree of diversity and identify indicators for environmental stress or degradation. Most indices of diversity are easy to use and they are widely used in ecological work, particularly for monitoring ecosystem change or pollution. One example, which is a derivation of **Simpson's index**, is described below. Other indices produce values ranging between 0 and almost 1. These are more easily interpreted because of the more limited range of values, but no single index offers the "best" measure of diversity: they are chosen on their suitability to different situations.

Simpson's Index for finite populations

This diversity index (DI) is a commonly used inversion of Simpson's index, suitable for finite populations.

$$DI = \frac{N(N - 1)}{\Sigma n(n - 1)}$$

After Smith and Smith as per IOB.

Where:

- DI = Diversity index
- N = Total number of individuals (of all species) in the sample
- n = Number of individuals of each species in the sample

This index ranges between 1 (low diversity) and infinity. The higher the value, the greater the variety of living organisms. It can be difficult to evaluate objectively without reference to some standard ecosystem measure because the values calculated can, in theory, go to infinity.

Example of species diversity in a stream

The example describes the results from a survey of stream invertebrates. The species have been identified, but this is not necessary in order to calculate diversity as long as the different species can be distinguished. Calculation of the DI using Simpson's index for finite populations is:

Species	No. of individuals
A (Common backswimmer)	12
B (Stonefly larva)	7
C (Silver water beetle)	2
D (Caddis fly larva)	6
E (Water spider)	5
Total number of individuals = 32	

$$DI = \frac{32 \times 31}{(12 \times 11) + (7 \times 6) + (2 \times 1) + (6 \times 5) + (5 \times 4)} = \frac{992}{226} = 4.39$$

A stream community with a high macroinvertebrate diversity (above) in contrast to a low diversity stream community (below).

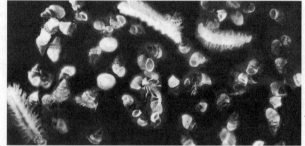

Photos: Stephen Moore

2. Describe two necessary considerations in attempting to make an unbiased measurement of biodiversity:

_____

_____

3. Explain why high biodiversity is generally associated with greater ecosystem stability: _____

_____

_____

4. Explain why the loss of a keystone species could be particularly disturbing for ecosystem diversity:

_____

_____

5. Describe a situation where a species diversity index may provide useful information: _____

_____

6. An area of forest floor was sampled and six invertebrate species were recorded, with counts of 7, 10, 11, 2, 4, and 3 individuals. Using Simpson's index for finite populations, calculate DI for this community:

(a) DI= _____   DI = _____

(b) Comment on the diversity of this community: _____

_____

Biodiversity and Conservation

One of the concerns facing conservationists today is the rapidly accelerating rate at which species are being lost. In 1992, the **Convention on Biological Diversity** was adopted in Rio de Janeiro. It is an international treaty and its aims are to conserve biodiversity, use biodiversity in a sustainable way, and ensure that the benefits of genetic resources are shared equitably. Various strategies are available to protect species already at risk, and help threatened species to return to sustainable population sizes. *Ex-situ* methods operate away from the natural environment and are particularly useful where species are critically endangered. *In-situ* methods use ecosystem management and legislation to protect and preserve diversity within the natural environment.

Ex-Situ Conservation Methods

Captive Breeding and Relocation

Individuals are captured and bred under protected conditions. If breeding programs are successful and there is suitable habitat available, captive individuals may be relocated to the wild where they can establish natural populations. Zoos now have an active role in captive breeding. *Photo left: A puppet 'mother' feeds a takahe chick.*

The Role of Zoos

Many zoos specialise in captive breeding programmes and have a major role in public education. Modern zoos tend to concentrate on particular species and are part of global programmes that work together to help retain genetic diversity in captive bred animals. *Photo right: Okapi a rare forest antelope.*

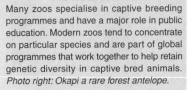

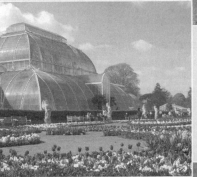

The Role of Botanic Gardens

Botanic gardens have years of collective expertise and resources and play a critical role in plant conservation. They maintain seed banks, nuture rare species, maintain a living collection of plants, and help to conserve indigenous plant knowledge. They also have an important role in both research and education. *Photo left: The palm house at Kew Botanic Gardens.*

Seed and Gene Banks

Seed and **gene banks** around the world have a role in preserving the genetic diversity of species. A seed bank (right) stores seeds as a source for future planting in case seed reserves elsewhere are lost. The seeds may be from rare species whose genetic diversity is at risk, or they may be the seeds of crop plants, in some cases of ancient varieties no longer used in commercial production.

In-Situ Conservation Methods

Woodland-pond restoration (UK)

Habitat Protection and Restoration

Most countries have a system of parks and reserves focussed on **whole ecosystem conservation**. These areas aim to preserve habitats with special importance and they may be intensively managed through pest and weed control programs, revegetation, and reintroduction of threatened species. A *"research by management"* approach is associated with careful population monitoring and management to return threatened species to viable levels.

Orangutan (endangered species)

Ban on Trade in Endangered Species

The Convention on International Trade in Endangered Species (CITES) is an international agreement between governments which aims to ensure that international trade in species of wild animals and plants does not threaten their survival. Unfortunately, even under CITES, species are not guaranteed safety from illegal trade.

In-Situ Conservation: Ecosystem Management

Stoat pest

Ecosystem management involves intensive management of a well defined area with a goal of ecosystem restoration and recovery of one or more at-risk species. In New Zealand, this strategy has been used successfully to restore populations of the endangered wattled crow, kokako (above).

An ecosystem management approach involves careful population monitoring and intensive pest control programmes. Areas must be large enough to sustain a viable population of the at-risk species, yet small enough to implement management strategies such as replanting and pest control.

In New Zealand, kokako (above) are at risk through forest clearance and predation by introduced mammals. Ecosystem management and intensive pest control has seen a reversal in the species decline. In seven years of management, the population of birds doubled. *Above: kokako chick.*

Biodiversity & Evolution

The UK's natural environment is highly modified as a result of a legacy of human exploitation. Few areas have escaped modification. The conservation problems faced by the UK are typical of many other developed nations: loss of biodiversity, natural habitat loss, pollution, waste disposal, and inadequate recycling. The main government agencies for conservation in the UK are **English Nature** (in England), **Scottish Nature Heritage** (Scotland), and **The Countryside Council for Wales** (Wales). These agencies are supported in their roles by a number of voluntary organisations that provide an additional source of expertise, labour, and finance for assisting conservation work. **Conservation** is a term describing the management of a resource so that it is maintained into the future. In contrast, **preservation** aims to keep untouched resources or habitats in their pristine state, without human interference.

The European Union Habitats Directive

In 1992, the Council of the European Communities adopted a directive for the conservation of natural habitats and wild flora and fauna, known as the **Habitats Directive**. The global objective of the Habitats Directive is "to contribute towards ensuring biodiversity through the conservation of natural habitats and of wild fauna and flora in the European territory of the Member States to which the Treaty applies". Within the Habitats Directive, is the ecological network of special areas of conservation called **Natura 2000**. Natura 2000 areas aim to conserve natural habitats and species of plants and animals that are rare, endangered, or vulnerable in the European Community. The Natura 2000 network will include two types of areas:

Special Areas of Conservation (SAC): areas with rare, endangered, or vulnerable natural habitats, and plant or animal species (other than birds).

Special Protection Areas (SPAs): areas with significant numbers of wild birds and their habitats.

Areas of very great importance on land and sea may become both SAC and SPA sites.

Environmental cleanup (pond, Glasgow)

High value habitat: woodland and lakes

Paper recycling

Recovery of birds affected by oil spills

Any conservation or restoration programme must be multifaceted: preserving or restoring valuable habitat, repairing damage and aiding species recovery, and educating people to consider environmentally friendly options (e.g. recycling and reuse) in their general lives.

1. Compare and contrast *in-situ* and *ex-situ* methods of conservation, including reference to the advantages and disadvantages of each approach:

_____

_____

_____

_____

_____

_____

_____

_____

_____

2. Distinguish between vulnerable, endangered , and extinct: _____

_____

_____

3. Explain the purpose of the following areas in the conservation of habitats and species diversity in the UK:

(a) The EU Habitats Directive: _____

_____

_____

(b) Natura 2000: _____

_____

_____

What is a Species?

The **species** is the basic unit used for classifying organisms and assigning taxonomic rank. Species can be defined in several ways. The most widely used definition is that of the **biological species**, which is defined as a group of organisms capable of interbreeding in nature and producing fertile offspring. However, defining a species is not as simple as it may first appear. The occurrence of cryptic species and closely related species that interbreed to produce fertile hybrids (e.g. species of *Canis*), indicate that the boundaries of a species gene pool can be unclear. Also, the biological species definition cannot be applied to organisms that reproduce asexually (e.g. bacteria). In these cases, the **phylogenetic species** definition is more useful. Phylogenetic species are defined on the basis of their shared evolutionary ancestry, which is determined by genetic analyses. Species definition was once based on morphological similarities, but this was an unreliable guide (e.g. males and females of the same species can have very different morphologies). Molecular analysis is now being widely used to define a species.

Geographical distribution of selected *Canis* species

The global distribution of most of the species of *Canis* (dogs and wolves) is shown on the map, right. The grey wolf inhabits the forests of North America, northern Europe, and Siberia. The red wolf and Mexican wolf (original distributions shown) were once distributed more widely, but are now extinct in the wild except for reintroduction efforts. In contrast, the coyote has expanded its original range and is now found throughout North and Central America. The range of the three jackal species overlap in the open savannah of Eastern Africa. The dingo is distributed throughout the Australian continent. Distribution of the domesticated dog is global as a result of the spread of human culture. The dog has been able to interbreed with all other members of the genus listed here to form fertile hybrids.

Interbreeding between *Canis* species

The *Canis* species illustrate problems with the traditional species concept. The domesticated dog is able to breed with other members of the same genus to produce fertile hybrids. Red wolves, grey wolves, Mexican wolves, and coyotes are all capable of interbreeding to produce fertile hybrids. Red wolves are very rare, and it is possible that hybridisation with coyotes has been a factor in their decline. By contrast, the ranges of the three distinct species of jackal overlap in the Serengeti of Eastern Africa. These animals are highly territorial, but simply ignore members of the other jackal species and no interbreeding takes place.

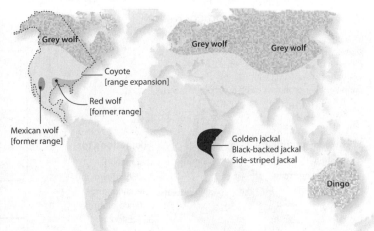

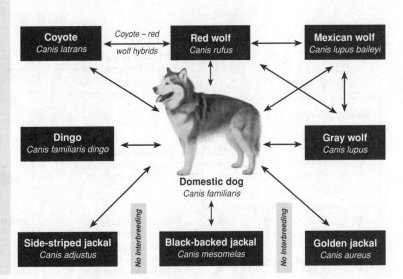

1. Explain what you understand by the term species: _____

_____

_____

2. Explain some of the limitations of using the biological species concept to assign species: _____

_____

_____

_____

3. Discuss the advantage of using genetic techniques to help assign species: _____

_____

_____

_____

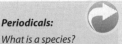

Periodicals:
What is a species?

Related activities: *Classification System*

Ecological Niche

The **ecological niche** describes the functional position of a species in its ecosystem; how it responds to the distribution of resources and how it, in turn, alters those resources for other species. The full range of environmental conditions (biological and physical) under which an organism can exist describes its **fundamental niche**. As a result of direct and indirect interactions with other organisms, species are usually forced to occupy a niche that is narrower than this and to which they are best adapted. This is termed the **realised niche**. From the concept of the niche arose the idea that two species with the same niche requirements could not coexist, because they would compete for the same resources, and one would exclude the other. This is known as **Gause's competitive exclusion principle**. If two species compete for some of the same resources (e.g. food items of a particular size), their resource use curves will overlap. Within the zone of overlap, competition will be intense.

The Ecological Niche

The physical conditions influence the habitat. The organism's tolerance to different factors in the abiotic environment will vary, presenting it with suitable conditions, or problems to be overcome.

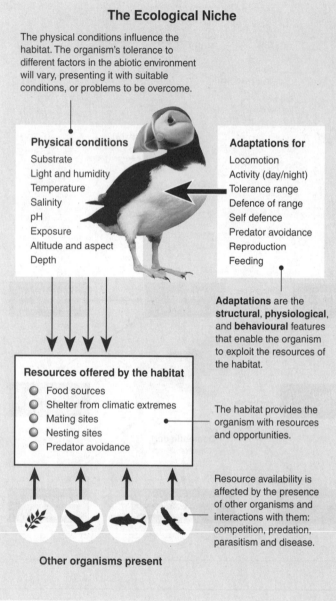

Physical conditions

Substrate
Light and humidity
Temperature
Salinity
pH
Exposure
Altitude and aspect
Depth

Adaptations for

Locomotion
Activity (day/night)
Tolerance range
Defence of range
Self defence
Predator avoidance
Reproduction
Feeding

Adaptations are the **structural**, **physiological**, and **behavioural** features that enable the organism to exploit the resources of the habitat.

Resources offered by the habitat

- Food sources
- Shelter from climatic extremes
- Mating sites
- Nesting sites
- Predator avoidance

The habitat provides the organism with resources and opportunities.

Resource availability is affected by the presence of other organisms and interactions with them: competition, predation, parasitism and disease.

Other organisms present

Competition and Niche Size

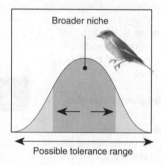

Realised niche of species A

Possible tolerance range

The realised niche

The tolerance range represents the potential (**fundamental**) niche a species could exploit. The actual or **realised** niche of a species is narrower than this because of competition with other species.

Broader niche

Possible tolerance range

Intraspecific competition

If two (or more) species compete for some of the same resources, their resource use curves will overlap. Within the zone of overlap, resource competition will be intense and selection will favour specialisation to occupy a narrower niche.

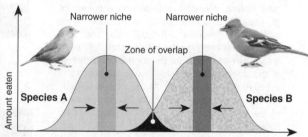

Narrower niche   Narrower niche

Zone of overlap

Species A   Species B

Amount eaten

Resource use as measured by food item size

Interspecific competition

If two (or more) species compete for some of the same resources, their resource use curves will overlap. Within the zone of overlap, resource competition will be intense and selection will favour niche specialisation so that one or both species occupy a narrower niche.

---

1. (a) Explain in what way the realised niche could be regarded as flexible: _____

_____

_____

(b) Describe factors that might constrain the extent of the realised niche: _____

_____

2. Explain the contrasting effects of interspecific competition and intraspecific competition on niche breadth:

_____

_____

_____

Related activities: Adaptations to Niche

Adaptations to Niche

The adaptive features that evolve in species are the result of selection pressures on them through the course of their evolution. These features enable an organism to function most effectively in its niche, enhancing its exploitation of its environment and therefore its survival. The examples below illustrate some of the adaptations of two species: a British placental mammal and a migratory Arctic bird. Note that adaptations may be associated with an animal's structure (morphology), its internal physiology, or its behaviour.

Northern or Common Mole
(Talpa europaea)

Head-body length: 113-159 mm, tail length: 25-40 mm, weight range: 70-130 g.

Moles (photos above) spend most of the time underground and are rarely seen at the surface. Mole hills are the piles of soil excavated from the tunnels and pushed to the surface. The cutaway view above shows a section of tunnels and a nest chamber. Nests are used for sleeping and raising young. They are dug out within the tunnel system and lined with dry plant material.

The northern (common) mole is a widespread insectivore found throughout most of Britain and Europe, apart from Ireland. They are found in most habitats but are less common in coniferous forest, moorland, and sand dunes, where their prey (earthworms and insect larvae) are rare. They are well adapted to life underground and burrow extensively, using enlarged forefeet for digging. Their small size, tubular body shape, and heavily buttressed head and neck are typical of burrowing species.

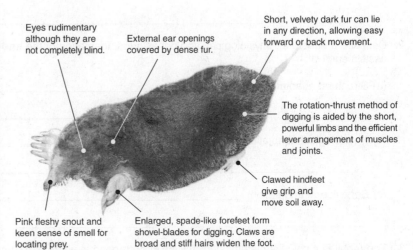

Eyes rudimentary although they are not completely blind.

External ear openings covered by dense fur.

Short, velvety dark fur can lie in any direction, allowing easy forward or back movement.

The rotation-thrust method of digging is aided by the short, powerful limbs and the efficient lever arrangement of muscles and joints.

Clawed hindfeet give grip and move soil away.

Pink fleshy snout and keen sense of smell for locating prey.

Enlarged, spade-like forefeet form shovel-blades for digging. Claws are broad and stiff hairs widen the foot.

Habitat and ecology: Moles spend most of their lives in underground tunnels. Surface tunnels occur where their prey is concentrated at the surface (e.g. land under cultivation). Deeper, permanent tunnels form a complex network used repeatedly for feeding and nesting, sometimes for several generations. **Senses and behaviour**: Keen sense of smell but almost blind. Both sexes are solitary and territorial except during breeding. Life span about 3 years. Moles are prey for owls, buzzards, stoats, cats, and dogs. Their activities aerate the soil and they control many soil pests. Despite this, they are regularly trapped and poisoned as pests.

Snow Bunting
(Plectrophenax nivalis)

The snow bunting is a small ground feeding bird that lives and breeds in the Arctic and sub-Arctic islands. Although migratory, snow buntings do not move to traditional winter homes but prefer winter habitats that resemble their Arctic breeding grounds, such as bleak shores or open fields of northern Britain and the eastern United States.

Snow buntings have the unique ability to moult very rapidly after breeding. During the warmer months, the buntings are a brown colour, changing to white in winter (right). They must complete this colour change quickly, so that they have a new set of feathers before the onset of winter and before migration. In order to achieve this, snow buntings lose as many as four or five of their main flight wing feathers at once, as opposed to most birds, which lose only one or two.

Very few small birds breed in the Arctic, because most small birds lose more heat than larger ones. In addition, birds that breed in the brief Arctic summer must migrate before the onset of winter, often travelling over large expanses of water. Large, long winged birds are better able to do this. However, the snow bunting is superbly adapted to survive in the extreme cold of the Arctic region.

White feathers are hollow and filled with air, which acts as an insulator. In the dark coloured feathers the internal spaces are filled with pigmented cells.

Less heat is lost from white plumage compared to dark plumage.

Snow buntings, on average, lay one or two more eggs than equivalent species further south. They are able to rear more young because the continuous daylight and the abundance of insects at high latitudes enables them to feed their chicks around the clock.

During snow storms or periods of high wind, snow buntings will burrow into snowdrifts for shelter.

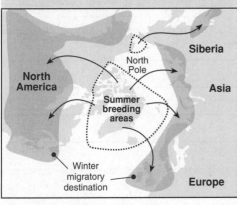

Habitat and ecology: Widespread throughout Arctic and sub-Arctic Islands. Active throughout the day and night, resting for only 2-3 hours in any 24 hour period. Snow buntings may migrate up to 6000 km but are always found at high latitudes. **Reproduction and behaviour**: The nest, which is concealed amongst stones, is made from dead grass, moss, and lichen. The male bird feeds his mate during the incubation period and helps to feed the young.

© Biozone International 2010
Photocopying Prohibited

Related activities: Ecological Niche
Web links: Some Adaptations to Habitats, Desert Plant Survival

RA 2

Biodiversity & Evolution

1. Describe a structural, physiological, and behavioural adaptation of the **common mole**, explaining how each adaptation assists survival:

   (a) Structural adaptation: _____

   _____

   (b) Physiological adaptation: _____

   _____

   (c) Behavioural adaptation: _____

   _____

2. Describe a structural, physiological, and behavioural adaptation of the **snow bunting**, explaining how each adaptation assists survival:

   (a) Structural adaptation: _____

   _____

   (b) Physiological adaptation: _____

   _____

   (c) Behavioural adaptation: _____

   _____

3. The rabbit is a colonial mammal which lives underground in warrens (burrow systems) and feeds on grasses, cereal crops, roots, and young trees. Rabbits are a hugely successful species worldwide and often reach plague proportions. Through discussion, or your own knowledge and research, describe **six adaptations** of rabbits, identifying them as structural (S), physiological (P), or behavioural (B). The examples below are typical:

   Structural: *Widely spaced eyes gives wide field of vision for surveillance and detection of danger.*

   Physiological: *High reproductive rate; short gestation and high fertility aids rapid population increases when food is available.*

   Behavioural: *Freeze behaviour when startled reduces the possibility of detection by wandering predators.*

   (a) _____

   (b) _____

   (c) _____

   (d) _____

   (e) _____

   (f) _____

4. Examples of adaptations are listed below. Identify them as predominantly structural, physiological, and/or behavioural:

   (a) Relationship of body size and shape to latitude (tropical or Arctic): _____

   (b) The production of concentrated urine in desert dwelling mammals: _____

   (c) The summer and winter migratory patterns in birds and mammals: _____

   (d) The C_4 photosynthetic pathway and CAM metabolism of plants: _____

   (e) The thick leaves and sunken stomata of desert plants: _____

   (f) Hibernation or torpor in small mammals over winter: _____

   (g) Basking in lizards and snakes: _____

Adaptations and Fitness

An **adaptation**, is any heritable trait that suits an organism to its natural function in the environment (its niche). These traits may be structural, physiological, or behavioural. The idea is important for evolutionary theory because adaptive features promote fitness. **Fitness** is a measure of an organism's ability to maximise the numbers of offspring surviving to reproductive age. Adaptations are distinct from properties which, although they

may be striking, cannot be described as adaptive unless they are shown to be functional in the organism's natural habitat. Genetic adaptation must not be confused with **physiological adjustment** (acclimatisation), which refers to an organism's ability to adapt during its lifetime to changing environmental conditions (e.g. a person's acclimatisation to altitude). Examples of adaptive features arising through evolution are illustrated below.

Ear Length in Rabbits and Hares

The external ears of many mammals are used as important organs to assist in thermoregulation (controlling loss and gain of body heat). The ears of rabbits and hares native to hot, dry climates, such as the jack rabbit of south-western USA and northern Mexico, are relatively very large. The Arctic hare lives in the tundra zone of Alaska, northern Canada and Greenland, and has ears that are relatively short. This reduction in the size of the extremities (ears, limbs, and noses) is typical of cold adapted species.

Arctic hare: *Lepus arcticus*

Black-tail jackrabbit: *Lepus californicus*

Body Size in Relation to Climate

Regulation of body temperature requires a large amount of energy and mammals exhibit a variety of structural and physiological adaptations to increase the effectiveness of this process. Heat production in any endotherm depends on body volume (heat generating metabolism), whereas the rate of heat loss depends on surface area. Increasing body size minimises heat loss to the environment by reducing the surface area to volume ratio. Animals in colder regions therefore tend to be larger overall than those living in hot climates. This relationship is know as **Bergman's rule** and it is well documented in many mammalian species. Cold adapted species also tend to have more compact bodies and shorter extremities than related species in hot climates.

Fennec fox

Arctic fox

The **fennec fox** of the Sahara illustrates the adaptations typical of mammals living in hot climates: a small body size and lightweight fur, and long ears, legs, and nose. These features facilitate heat dissipation and reduce heat gain.

The **Arctic fox** shows the physical characteristics typical of cold-adapted mammals: a stocky, compact body shape with small ears, short legs and nose, and dense fur. These features reduce heat loss to the environment.

Number of Horns in Rhinoceroses

Not all differences between species can be convincingly interpreted as adaptations to particular environments. Rhinoceroses charge rival males and predators, and the horn(s), when combined with the head-down posture, add effectiveness to this behaviour. Horns are obviously adaptive, but it is not clear that the possession of one (Indian rhino) or two (black rhino) horns is necessarily related directly to the environment in which those animals live.

Great Indian rhino

African black rhino

1. Distinguish between adaptive features (genetic) and acclimatisation: _____

2. Explain the nature of the relationship between the length of extremities (such as limbs and ears) and climate:

3. Explain the adaptive value of a larger body size at high latitude: _____

© Biozone International 2010
Photocopying Prohibited
Periodicals: Optimality
Related activities: Natural Selection
Web links: Adaptation
A 2
Biodiversity & Evolution

Gene Pools and Evolution

The diagram below illustrates the dynamic nature of **gene pools**. It portrays two imaginary populations of one beetle species. Each beetle is a 'carrier' of genetic information, represented here by the alleles (A and a) for a single **codominant gene** that controls the beetle's colour. Normally, there are three versions of the phenotype: black, dark, and pale. Mutations may create other versions of the phenotype. Some of the **microevolutionary processes** that can affect the genetic composition (**allele frequencies**) of the gene pool are illustrated. Some are discussed in more detail in subsequent activities.

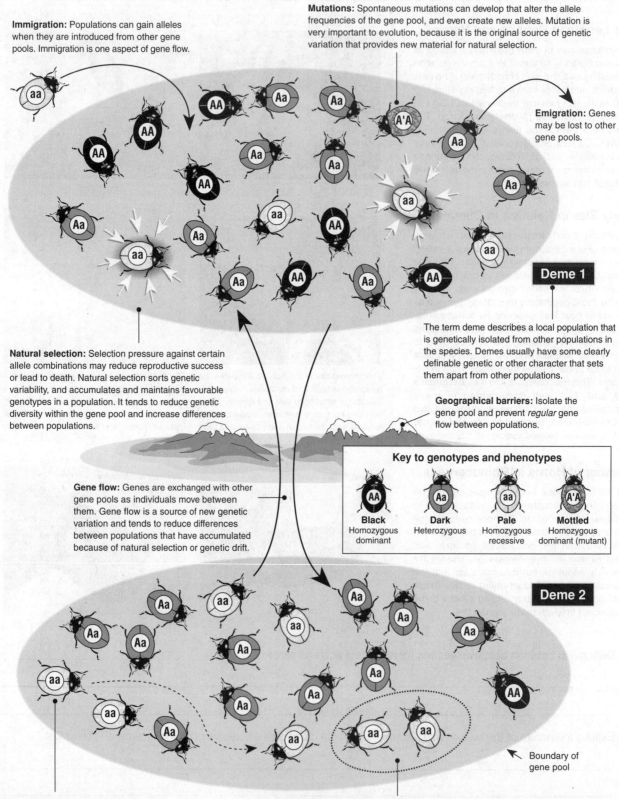

Immigration: Populations can gain alleles when they are introduced from other gene pools. Immigration is one aspect of gene flow.

Mutations: Spontaneous mutations can develop that alter the allele frequencies of the gene pool, and even create new alleles. Mutation is very important to evolution, because it is the original source of genetic variation that provides new material for natural selection.

Emigration: Genes may be lost to other gene pools.

Deme 1

Natural selection: Selection pressure against certain allele combinations may reduce reproductive success or lead to death. Natural selection sorts genetic variability, and accumulates and maintains favourable genotypes in a population. It tends to reduce genetic diversity within the gene pool and increase differences between populations.

The term deme describes a local population that is genetically isolated from other populations in the species. Demes usually have some clearly definable genetic or other character that sets them apart from other populations.

Geographical barriers: Isolate the gene pool and prevent *regular* gene flow between populations.

Gene flow: Genes are exchanged with other gene pools as individuals move between them. Gene flow is a source of new genetic variation and tends to reduce differences between populations that have accumulated because of natural selection or genetic drift.

Key to genotypes and phenotypes

Black Homozygous dominant	**Dark** Heterozygous	**Pale** Homozygous recessive	**Mottled** Homozygous dominant (mutant)
AA	Aa	aa	A'A

Deme 2

Boundary of gene pool

Mate selection (non-random mating): Individuals may not select their mate randomly and may seek out particular phenotypes, increasing the frequency of these "favoured" alleles in the population.

Genetic drift: Chance events can cause the allele frequencies of small populations to "drift" (change) randomly from generation to generation. Genetic drift can play a significant role in the microevolution of very small populations. The two situations most often leading to populations small enough for genetic drift to be significant are the **bottleneck effect** (where the population size is dramatically reduced by a catastrophic event) and the **founder effect** (where a small number of individuals colonise a new area).

Web links: *Natural Selection in Populations, Changes in a Gene Pool*

1. For each of the two demes shown on the previous page (treating the mutant in deme 1 as AA):

   (a) Count up the numbers of **allele types** (**A** and **a**).

   (b) Count up the numbers of **allele combinations** (**AA, Aa, aa**).

2. Calculate the frequencies as percentages (%) for the allele types and combinations:

Deme 1		Number counted	%	Deme 2		Number counted	%
Allele types	A			Allele types	A		
	a				a		
Allele combinations	AA			Allele combinations	AA		
	Aa				Aa		
	aa				aa		

3. One of the fundamental concepts for population genetics is that of **genetic equilibrium**, stated as: *"For a very large, randomly mating population, the proportion of dominant to recessive alleles remains constant from one generation to the next"*. If a gene pool is to remain unchanged, it must satisfy all of the criteria below that favour gene pool stability. Few populations meet all (or any) of these criteria and their genetic makeup must therefore be continually changing. For each of the five factors (a-e) below, state briefly **how** and **why** each would affect the allele frequency in a gene pool:

(a) Population size: _____

_____

_____

_____

(b) Mate selection: _____

_____

_____

_____

(c) Gene flow between populations: _____

_____

_____

_____

(d) Mutations: _____

_____

_____

_____

(e) Natural selection: _____

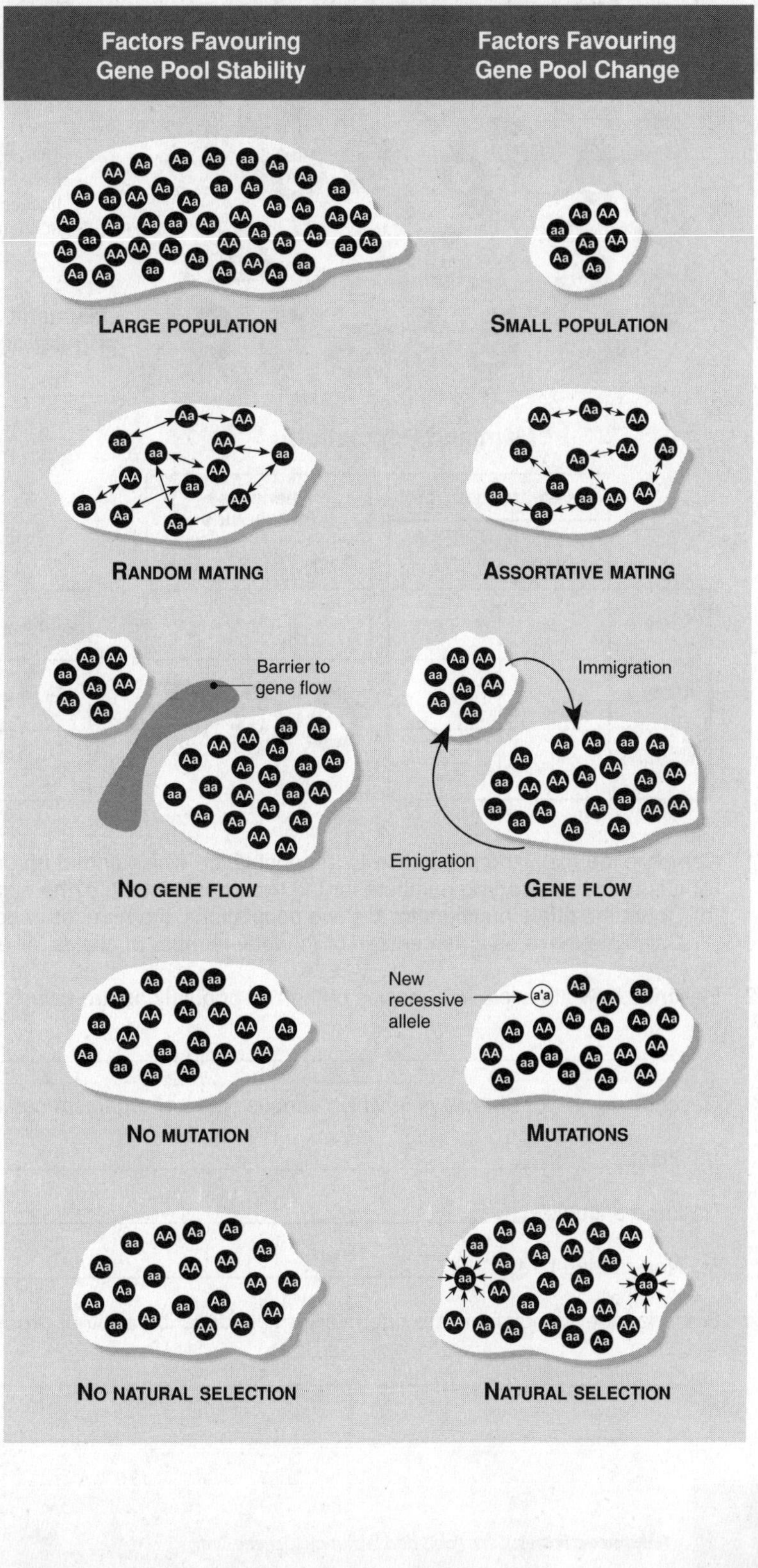

_____

_____

_____

4. Identify the factors that tend to:

(a) Increase genetic variation in populations:

_____

_____

(b) Decrease genetic variation in populations:

_____

_____

Factors Favouring Gene Pool Stability — Factors Favouring Gene Pool Change

LARGE POPULATION — SMALL POPULATION

RANDOM MATING — ASSORTATIVE MATING

NO GENE FLOW (Barrier to gene flow) — GENE FLOW (Immigration, Emigration)

NO MUTATION — MUTATIONS (New recessive allele)

NO NATURAL SELECTION — NATURAL SELECTION

Biodiversity & Evolution

The Founder Effect

Occasionally, a small number of individuals from a large population may migrate away, or become isolated from, their original population. If this colonising or 'founder' population is made up of only a few individuals, it will probably have a *non-representative sample* of alleles from the parent population's gene pool. As a consequence of this **founder effect**, the colonising population may evolve differently from that of the parent population, particularly since the environmental conditions for the isolated population may be different. In some cases, it may be possible for certain alleles to be missing altogether from the individuals in the isolated population. Future generations of this population will not have this allele.

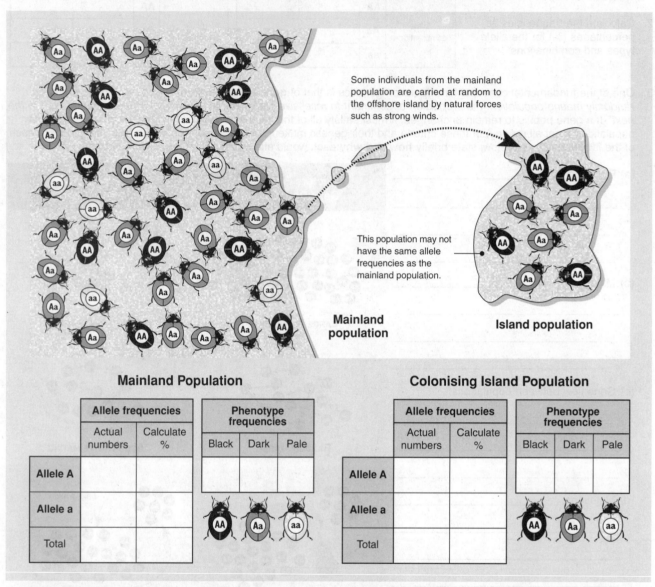

Some individuals from the mainland population are carried at random to the offshore island by natural forces such as strong winds.

This population may not have the same allele frequencies as the mainland population.

Mainland population

Island population

Mainland Population

	Allele frequencies		Phenotype frequencies		
	Actual numbers	Calculate %	Black	Dark	Pale
Allele A					
Allele a					
Total					

Colonising Island Population

	Allele frequencies		Phenotype frequencies		
	Actual numbers	Calculate %	Black	Dark	Pale
Allele A					
Allele a					
Total					

1. Compare the mainland population to the population which ended up on the island (use the spaces in the tables above):
   (a) Count the **phenotype** numbers for the two populations (i.e. the number of black, dark and pale beetles).
   (b) Count the **allele** numbers for the two populations: the number of dominant alleles (A) and recessive alleles (a). Calculate these as a percentage of the total number of alleles for each population.

2. Describe how the allele frequencies of the two populations are different: _____

_____

3. Describe some possible ways in which various types of organism can be carried to an offshore island:

   (a) Plants: _____

   (b) Land animals: _____

   (c) Non-marine birds: _____

4. Since founder populations are often very small, describe another process that may further alter the allele frequencies:

_____

_____

Related activities: Gene Pools and Evolution, Genetic Drift

Population Bottlenecks

Populations may sometimes be reduced to low numbers by predation, disease, or periods of climatic change. A population crash may not be 'selective': it may affect all phenotypes equally. Large scale catastrophic events, such as fire or volcanic eruption, are examples of such non-selective events. Humans may severely (and selectively) reduce the numbers of some species through hunting and/or habitat destruction. These populations may recover, having squeezed through a 'bottleneck' of low numbers.

The diagram below illustrates how population numbers may be reduced as a result of a catastrophic event. Following such an event, the small number of individuals contributing to the gene pool may not have a representative sample of the genes in the pre-catastrophe population, i.e. the allele frequencies in the remnant population may be severely altered. Genetic drift may cause further changes to allele frequencies. The small population may return to previous levels but with a reduced genetic diversity.

Population numbers

Low — High

Large population with plenty of genetic diversity.

Population crashes to a very low number and loses most of its genetic diversity.

Population grows to a large size again, but has lost much of its genetic diversity.

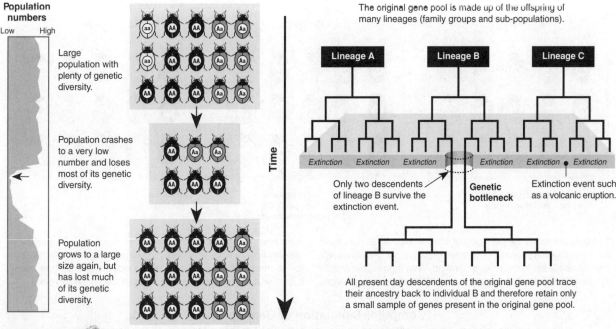

The original gene pool is made up of the offspring of many lineages (family groups and sub-populations).

Lineage A   **Lineage B**   **Lineage C**

Time

Extinction   Extinction   Extinction   Extinction   Extinction   Extinction

Only two descendents of lineage B survive the extinction event.

Genetic bottleneck

Extinction event such as a volcanic eruption.

All present day descendents of the original gene pool trace their ancestry back to individual B and therefore retain only a small sample of genes present in the original gene pool.

Modern Examples of Population Bottlenecks

Cheetahs: The world population of cheetahs currently stands at fewer than 20 000. Recent genetic analysis has found that the entire population exhibits very little genetic diversity. It appears that cheetahs may have narrowly escaped extinction at the end of the last ice age, about 10-20 000 years ago. If all modern cheetahs arose from a very limited genetic stock, this would explain their present lack of genetic diversity. The lack of genetic variation has resulted in a number of problems that threaten cheetah survival, including sperm abnormalities, decreased fecundity, high cub mortality, and sensitivity to disease.

Illinois prairie chicken: When Europeans first arrived in North America, there were millions of prairie chickens. As a result of hunting and habitat loss, the Illinois population of prairie chickens fell from about 100 million in 1900 to fewer than 50 in the 1990s. A comparison of the DNA from birds collected in the mid-twentieth century and DNA from the surviving population indicated that most of the genetic diversity has been lost.

Photo: Dept. of Natural Resources, Illinois

1. Endangered species are often subjected to population bottlenecks. Explain how population bottlenecks affect the ability of a population of an endangered species to recover from its plight:

_____

_____

_____

2. Explain why the lack of genetic diversity in cheetahs has increased their sensitivity to disease:

_____

_____

3. Describe the effect of a population bottleneck on the potential of a species to adapt to changes (i.e. its ability to evolve):

_____

_____

Biodiversity & Evolution

Genetic Drift

Not all individuals, for various reasons, will be able to contribute their genes to the next generation. **Genetic drift** (also known as the Sewall-Wright Effect) refers to the *random changes in allele frequency* that occur in all populations, but are much more pronounced in small populations. In a small population, the effect of a few individuals not contributing their alleles to the next generation can have a great effect on allele frequencies. Alleles may even become **lost** from the gene pool altogether (frequency becomes 0%) or **fixed** as the only allele for the gene present (frequency becomes 100%).

The genetic makeup (allele frequencies) of the population changes randomly over a period of time

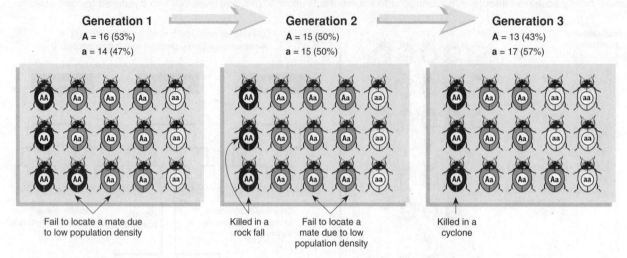

Generation 1
A = 16 (53%)
a = 14 (47%)

Generation 2
A = 15 (50%)
a = 15 (50%)

Generation 3
A = 13 (43%)
a = 17 (57%)

Fail to locate a mate due to low population density

Killed in a rock fall

Fail to locate a mate due to low population density

Killed in a cyclone

This diagram shows the gene pool of a hypothetical small population over three generations. For various reasons, not all individuals contribute alleles to the next generation. With the random loss of the alleles carried by these individuals, the allele frequency changes from one generation to the next. The change in frequency is directionless as there is no selecting force. The allele combinations for each successive generation are determined by how many alleles of each type are passed on from the preceding one.

Computer Simulation of Genetic Drift

Below are displayed the change in allele frequencies in a computer simulation showing random genetic drift. The breeding population progressively gets smaller from left to right. Each simulation was run for 140 generations.

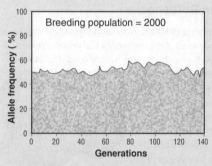

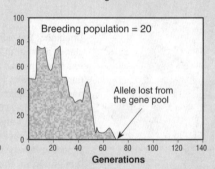

Large breeding population
Fluctuations are minimal in large breeding populations because the large numbers buffer the population against random loss of alleles. On average, losses for each allele type will be similar in frequency and little change occurs.

Small breeding population
Fluctuations are more severe in smaller breeding populations because random changes in a few alleles cause a greater percentage change in allele frequencies.

Very small breeding population
Fluctuations in very small breeding populations are so extreme that the allele can become fixed (frequency of 100%) or lost from the gene pool altogether (frequency of 0%).

1. Explain what is meant by **genetic drift**: _____

_____

2. Explain how genetic drift affects the amount of genetic variation within very small populations: _____

_____

_____

_____

3. Identify a known small breeding population of animals in Britain or Europe in which genetic drift could be occurring:

_____

RA 3

Related activities: *Gene Pools and Evolution, Founder Effect*
Web links: *Genetic Drift Simulation*

Darwin's Theory

In 1859, Darwin and Wallace jointly proposed that new species could develop by a process of natural selection. Natural selection is the term given to the mechanism by which better adapted organisms survive to produce a greater number of viable offspring. This has the effect of increasing their proportion in the population so that they become more common. It is Darwin who is best remembered for the theory of evolution by natural selection through his famous book: '**On the origin of species by means of natural selection**', written 23 years after returning from his voyage on the Beagle, from which much of the evidence for his theory was accumulated. Although Darwin

could not explain the origin of variation nor the mechanism of its transmission (this was provided later by Mendel's work), his basic theory of evolution by natural selection (outlined below) is widely accepted today. The study of population genetics has greatly improved our understanding of evolutionary processes, which are now seen largely as a (frequently gradual) change in allele frequencies within a population. Students should be aware that scientific debate on the subject of evolution centres around the relative merits of various alternative hypotheses about the nature of evolutionary processes. The debate is not about the existence of the phenomenon of evolution itself.

Darwin's Theory of Evolution by Natural Selection

Overproduction
Populations produce too many young: many must die

Populations tend to produce more offspring than are needed to replace the parents. Natural populations normally maintain constant numbers. There must therefore be a certain number dying.

Variation
Individuals show variation: some are more favourable than others

Individuals in a population vary in their phenotype and therefore, their genotype. Some variants are better suited in the prevailing environment and have greater survival and reproductive success.

Natural Selection
Natural selection favours the best suited at the time

The struggle for survival amongst individuals competing for limited resources will favour those with the most favourable variations. Relatively more of those without favourable variations will die.

Inherited
Variations are Inherited.
The best suited variants leave more offspring.

The variations (both favourable and unfavourable) are passed on to offspring. Each new generation will contain proportionally more descendents of individuals with favourable characters.

Andrew Dunn www.andrewdunnphoto.com

The banded or grove snail, *Cepaea nemoralis*, is famous for the highly variable colours and banding patterns of its shell. These **polymorphisms** are thought to have a role in differential survival in different regions, associated with both the risk of predation and maintenance of body temperature. Dark brown grove snails are more abundant in dark woodlands, whilst snails with light yellow shells and thin banding are more commonly found in grasslands.

1. In your own words, describe how Darwin's theory of evolution by natural selection provides an explanation for the change in the appearance of a species over time:

_____

_____

_____

_____

Biodiversity & Evolution

Periodicals:
Was Darwin wrong?

Related activities: Natural Selection
Web links: The Modern Theory of Evolution, Variation: Snails

Natural Selection

Natural selection operates on the phenotypes of individuals, produced by their particular combinations of alleles. The differential survival of some genotypes over others is called **natural selection** and, as a result of it, organisms with phenotypes most suited to the prevailing environment are more likely to survive and breed than those with less suited phenotypes. Favourable phenotypes become more numerous while unfavourable phenotypes become less common or may disappear altogether. Natural selection is not a static phenomenon; it is always linked to phenotypic suitability in the prevailing environment. It may favour existing phenotypes or shift the phenotypic median one way or another, as is shown in the diagrams below. The top row of diagrams represents the population phenotypic spread before selection, and the bottom row the spread afterwards. Note that balancing selection is similar to disruptive selection, but the polymorphism that results is not associated with phenotypic extremes.

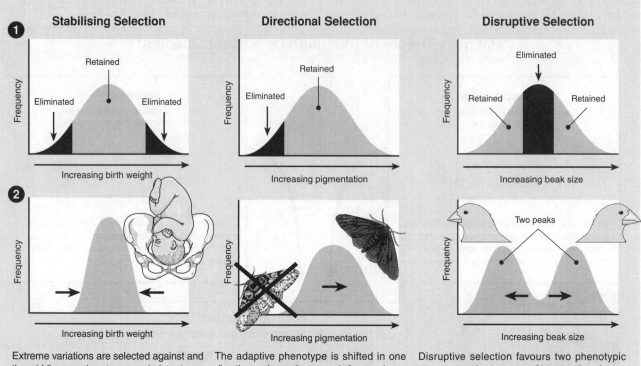

Extreme variations are selected against and the middle range (most common) phenotypes are retained in greater numbers. Stabilising selection results in decreased variation for the phenotypic character involved. This type of selection operates most of the time in most populations and acts to prevent divergence of form and function, e.g. birth weight of human infants.

The adaptive phenotype is shifted in one direction and one phenotype is favoured over another. Directional selection was observed in peppered moths in England during the Industrial Revolution. In England's current environment, the selection pressures on the moths are more typically balanced, and the proportions of each morph vary in different regions of the country.

Disruptive selection favours two phenotypic extremes at the expense of intermediate forms. During a prolonged drought on Santa Cruz Island in the Galapagos, it resulted in a population of ground finches that was bimodal for beak size. Competition for the usual seed sources was so intense that birds able to exploit either small or large seeds were favoured, although intermediate phenotypes remained in low numbers.

1. Explain why fluctuating (as opposed to stable) environments favour disruptive (diversifying) selection:

_____

_____

_____

2. Disruptive selection can be important in the formation of new species:

(a) Describe the evidence from the ground finches on Santa Cruz Island that lends support to this: _____

_____

_____

(b) The ground finches on Santa Cruz Island are one interbreeding population with a strongly bimodal distribution for the phenotypic character beak size. Suggest what conditions could lead to the two phenotypic extremes diverging further:

_____

_____

(c) Predict the consequences of the end of the drought and an increased abundance of medium size seeds as food:

_____

Related activities: Selection for Human Birth Weight
Web links: Natural Selection, Changes in a Gene Pool

Periodicals:
Skin deep

© Biozone International 2010
Photocopying Prohibited

The Evolution of Darwin's Finches

The Galápagos Islands, off the West coast of Ecuador, comprise 16 main islands and six smaller islands. They are home to a unique range of organisms, including 13 species of finches, each of which is thought to have evolved from a single species of grassquit. After colonising the islands, the grassquits underwent adaptive radiation in response to the availability of unexploited feeding niches on the islands. This adaptive radiation is most evident in the present beak shape of each species. The beaks are adapted for different purposes such as crushing seeds, pecking wood, or probing flowers for nectar. Current consensus groups the finches into ground finches, tree finches, warbler finches, and the Cocos Island finches. Between them, the 13 species of this endemic group fill the roles of seven different families of South American mainland birds. DNA analyses have confirmed Darwin's insight and have shown that all 13 species evolved from a flock of about 30 birds arriving a million years ago.

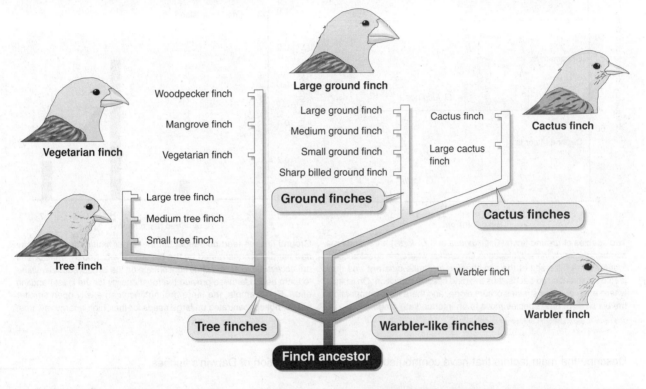

Vegetarian finch

Woodpecker finch
Mangrove finch
Vegetarian finch

Large ground finch

Large ground finch
Medium ground finch
Small ground finch
Sharp billed ground finch

Ground finches

Cactus finch
Large cactus finch

Cactus finch

Cactus finches

Tree finch

Large tree finch
Medium tree finch
Small tree finch

Tree finches

Warbler finch

Warbler finch

Warbler-like finches

Finch ancestor

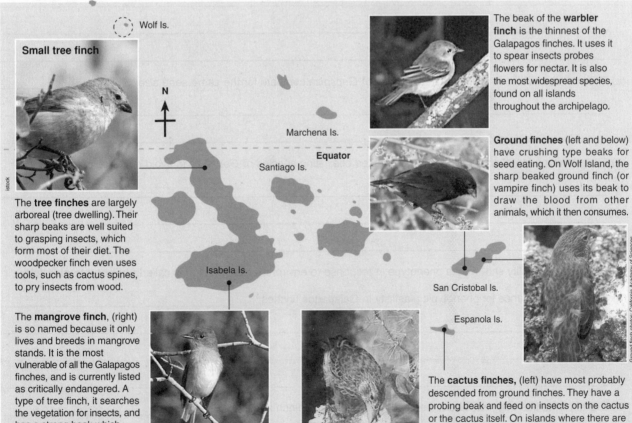

Wolf Is.

Small tree finch

N

Marchena Is.

Equator
Santiago Is.

Isabela Is.

San Cristobal Is.

Espanola Is.

The **tree finches** are largely arboreal (tree dwelling). Their sharp beaks are well suited to grasping insects, which form most of their diet. The woodpecker finch even uses tools, such as cactus spines, to pry insects from wood.

The **mangrove finch**, (right) is so named because it only lives and breeds in mangrove stands. It is the most vulnerable of all the Galapagos finches, and is currently listed as critically endangered. A type of tree finch, it searches the vegetation for insects, and has a strong beak which allows it to remove bark to access insects.

The beak of the **warbler finch** is the thinnest of the Galapagos finches. It uses it to spear insects probes flowers for nectar. It is also the most widespread species, found on all islands throughout the archipelago.

Ground finches (left and below) have crushing type beaks for seed eating. On Wolf Island, the sharp beaked ground finch (or vampire finch) uses its beak to draw the blood from other animals, which it then consumes.

The **cactus finches**, (left) have most probably descended from ground finches. They have a probing beak and feed on insects on the cactus or the cactus itself. On islands where there are no ground finches, there is more variation in beak size than on the islands where the species coexist.

Ground finch photos: California Academy of Sciences

Biodiversity & Evolution

Related activities: Darwin's Theory, Adaptation and Fitness
Web links: Darwin's Finches

A 2

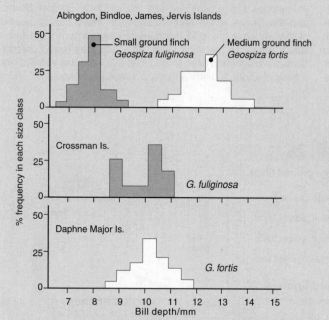

Adaptation in response to resource competition on bill size in small and medium ground finches

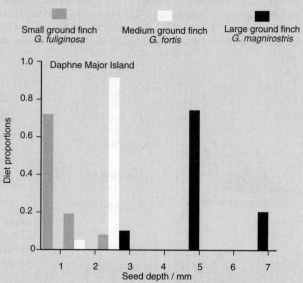

Proportions in the diet of various seed sizes in three species of ground finch

Two species of ground finch (*G. fuliginosa* and *G. fortis*) are found on a number of islands in the Galapagos. On islands where the species occur together, the bill sizes of the two species are quite different and they feed on different sized seeds, thus avoiding direct competition. On islands where each of these species occurs alone, and there is no competition, the bill sizes of both species move to an intermediate range.

Data based on an adaptation by Strickberger (2000)

Ground finches feed on seeds, but the upper limit of seed size they can handle is constrained by the bill size. Even though small seeds are accessible to all, the birds concentrate on the largest seeds availabe to them because these provide the most energy for the least handling effort. For example, the large ground finch can easily open smaller seeds, but concentrates on large seeds for their high energy rewards.

1. Describe the main factors that have contributed to the adaptive radiation of Darwin's finches: _____

_____

_____

_____

2. (a) Describe the evidence indicating that species of *Geospiza* compete for the same seed sizes: _____

_____

_____

(b) Explain how adaptations in bill size have enabled coexisting species of *Geospiza* to avoid resource competition:

_____

_____

_____

3. The range of variability shown by a phenotype in response to environmental variation is called **phenotypic plasticity**.

(a) Discuss the evidence for phenotypic plasticity in Galapagos finches: _____

_____

_____

_____

(b) Explain what this suggests about the biology of the original finch ancestor: _____

_____

_____

Selection for Human Birth Weight

Selection pressures operate on populations in such a way as to reduce mortality. For humans, giving birth is a special, but often traumatic, event. In a study of human birth weights it is possible to observe the effect of selection pressures operating to constrain human birth weight within certain limits. This is a good example of **stabilising selection**. This activity explores the selection pressures acting on the birth weight of human babies. Carry out the steps below:

Step 1: Collect the birth weights from 100 birth notices from your local newspaper (or 50 if you are having difficulty getting enough; this should involve looking back through the last 2-3 weeks of birth notices). If you cannot obtain birth weights in your local newspaper, a set of 100 sample birth weights is provided in the Model Answers booklet.

Step 2: Group the weights into each of the 12 weight classes (of 0.5 kg increments). Determine what percentage (of the total sample) fall into each weight class (e.g. 17 babies weigh 2.5-3.0 kg out of the 100 sampled = 17%).

Step 3: Graph these in the form of a histogram for the 12 weight classes (use the graphing grid provided right). Be sure to use the scale provided on the left vertical (y) axis.

Step 4: Create a second graph by plotting percentage mortality of newborn babies in relation to their birth weight. Use the scale on the right y axis and data provided (below).

Step 5: Draw a line of 'best fit' through these points.

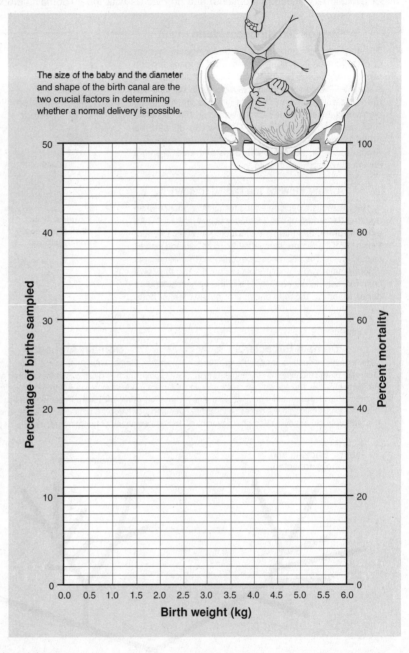

The size of the baby and the diameter and shape of the birth canal are the two crucial factors in determining whether a normal delivery is possible.

Mortality of newborn babies related to birth weight

Weight (kg)	Mortality (%)
1.0	80
1.5	30
2.0	12
2.5	4
3.0	3
3.5	2
4.0	3
4.5	7
5.0	15

Source: Biology: The Unity & Diversity of Life (4th ed), by Starr and Taggart

1. Describe the shape of the histogram for birth weights: _____

2. State the optimum birth weight in terms of the lowest newborn mortality: _____

3. Describe the relationship between the newborn mortality and the birth weights: _____

_____

4. Describe the selection pressures that are operating to control the range of birth weight: _____

_____

_____

5. Explain how medical intervention methods during pregnancy and childbirth may have altered these selection pressures:

_____

_____

© Biozone International 2010
Photocopying Prohibited

Related activities: Natural Selection, Drawing Histograms

A 3

Biodiversity & Evolution

The New Tree of Life

With the advent of more efficient genetic (DNA) sequencing technology, the genomes of many bacteria began to be sequenced. In 1996, the results of a scientific collaboration examining DNA evidence confirmed the proposal that life comprises three major evolutionary lineages (domains) and not two as was the convention. The recognised lineages were the **Eubacteria**, the **Eukarya** and the **Archaea** (formerly the Archaebacteria). The new classification reflects the fact that there are very large differences between the Archaea and the Eubacteria. All three domains probably had a distant common ancestor.

A Five (or Six) Kingdom World (right)

The diagram (right) represents the **five kingdom system** of classification commonly represented in many biology texts. It recognises two basic cell types: prokaryote and eukaryote. The domain Prokaryota includes all bacteria and cyanobacteria. Domain Eukaryota includes protists, fungi, plants, and animals. More recently, based on 16S ribosomal RNA sequence comparisons, Carl Woese divided the prokaryotes into two kingdoms, the Eubacteria and Archaebacteria. Such **six-kingdom systems** are also commonly recognised in texts.

A New View of the World (below)

In 1996, scientists deciphered the full DNA sequence of an unusual bacterium called *Methanococcus jannaschii*. An **extremophile**, this methane-producing archaebacterium lives at 85°C; a temperature lethal for most bacteria as well as eukaryotes. The DNA sequence confirmed that life consists of three major evolutionary lineages, not the two that have been routinely described. Only 44% of this archaebacterium's genes resemble those in bacteria or eukaryotes, or both.

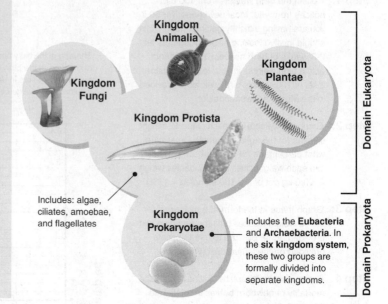

Includes: algae, ciliates, amoebae, and flagellates

Kingdom Prokaryotae Includes the **Eubacteria** and **Archaebacteria**. In the **six kingdom system**, these two groups are formally divided into separate kingdoms.

Domain Eubacteria

Lack a distinct nucleus and cell organelles. Generally prefer less extreme environments than Archaea. Includes well-known pathogens, many harmless and beneficial species, and the cyanobacteria (photosynthetic bacteria containing the pigments chlorophyll *a* and phycocyanin).

Domain Archaea

Closely resemble eubacteria in many ways but cell wall composition and aspects of metabolism are very different. Live in extreme environments similar to those on primeval Earth. They may utilise sulfur, methane, or halogens (chlorine, fluorine), and many tolerate extremes of temperature, salinity, or pH.

Domain Eukarya

Complex cell structure with organelles and nucleus. This group contains four of the kingdoms classified under the more traditional system. Note that Kingdom Protista is separated into distinct groups: e.g. amoebae, ciliates, flagellates.

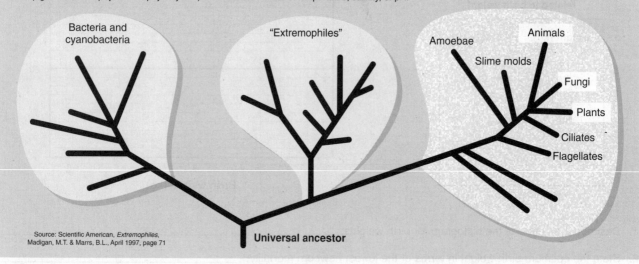

Source: Scientific American, *Extremophiles*, Madigan, M.T. & Marrs, B.L., April 1997, page 71

1. Explain why some scientists have recommended that the conventional classification of life be revised so that the Archaea, Eubacteria and Eukarya are three separate domains:

_____

_____

2. Describe one feature of the three domain system that is very different from the five kingdom classification:

_____

_____

3. Describe one way in which the three domain system and the six kingdom classification are alike:

_____

Related activities: *Classification System*
Web links: *Introduction to the Archaea, Types of Microbes*

Periodicals:
What is a species?

© Biozone International 2010
Photocopying Prohibited

Phylogenetic Systematics

The aim of classification is to organise species in a way that most accurately reflects their evolutionary history (**phylogeny**). Each successive group in the taxonomic hierarchy should represent finer and finer branching from a common ancestor. Traditional classification systems emphasise morphological similarities in order to group species into genera and other higher level taxa. In contrast, **cladistic analysis** relies on **shared derived characteristics** (**synapomorphies**), and emphasises features that are the result of shared ancestry (homologies), rather than convergent evolution. Technology has assisted taxonomy by providing biochemical evidence for the relatedness of species. Traditional and cladistic schemes do not necessarily conflict, but there have been reclassifications of some taxa (notably the primates, but also the reptiles, dinosaurs, and birds). Popular classifications will probably continue to reflect similarities and differences in appearance, rather than a strict evolutionary history. In this respect, they are a compromise between phylogeny and the need for a convenient filing system for species diversity.

Constructing a Simple Cladogram

A table listing the features for comparison allows us to identify where we should make branches in the **cladogram**. An outgroup (one which is known to have no or little relationship to the other organisms) is used as a basis for comparison.

	Jawless fish (outgroup)	Bony fish	Amphibians	Lizards	Birds	Mammals
Vertebral column	✔	✔	✔	✔	✔	✔
Jaws	✘	✔	✔	✔	✔	✔
Four supporting limbs	✘	✘	✔	✔	✔	✔
Amniotic egg	✘	✘	✘	✔	✔	✔
Diapsid skull	✘	✘	✘	✔	✔	✘
Feathers	✘	✘	✘	✘	✔	✘
Hair	✘	✘	✘	✘	✘	✔

The table above lists features shared by selected taxa. The outgroup (jawless fish) shares just one feature (vertebral column), so it gives a reference for comparison and the first branch of the cladogram (tree).

As the number of taxa in the table increases, the number of possible trees that could be drawn increases exponentially. To determine the most likely relationships, the rule of **parsimony** is used. This assumes that the tree with the least number of evolutionary events is most likely to show the correct evolutionary relationship.

Three possible cladograms are shown on the right. The top cladogram requires six events while the other two require seven events. Applying the rule of parsimony, the top cladogram must be taken as correct.

Parsimony can lead to some confusion. Some evolutionary events have occurred multiple times. An example is the evolution of the four chambered heart, which occurred separately in both birds and mammals. The use of fossil evidence and DNA analysis can help to solve problems like this.

Possible Cladograms

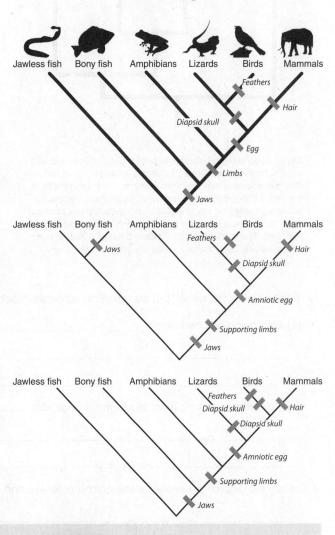

Using DNA Data

DNA analysis has allowed scientists to confirm many phylogenies and refute or redraw others. In a similar way to morphological differences, DNA sequences can be tabulated and analysed. The ancestry of whales has been in debate since Darwin. The radically different morphologies of whales and other mammals makes it difficult work out the correct phylogenetic tree. However recently discovered fossil ankle bones, as well as DNA studies, show whales are more closely related to hippopotami than to any other mammal. Coupled with molecular clocks, DNA data can also give the time between each split in the lineage.

The DNA sequences on the right show part of a the nucleotide subset 141-200 and some of the matching nucleotides used to draw the cladogram. Although whales were once thought most closely related to pigs, based on the DNA analysis the most parsimonious tree disputes this.

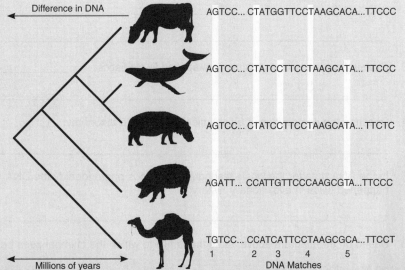

Periodicals: Uprooting the tree of life

Related activities: Classification System, The New tree of Life

Biodiversity & Evolution

A 2

224

A Classical Taxonomic View	A Cladistic View

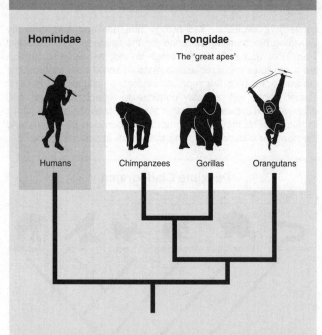

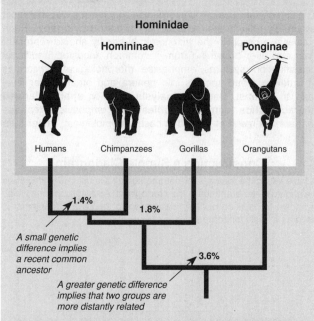

On the basis of overall anatomical similarity (e.g. bones and limb length, teeth, musculature), apes are grouped into a family (Pongidae) that is separate from humans and their immediate ancestors (Hominidae). The family Pongidae (the great apes) is not monophyletic (of one phylogeny), because it stems from an ancestor that also gave rise to a species in another family (i.e. humans). This traditional classification scheme is now at odds with schemes derived after considering genetic evidence.

Based on the evidence of genetic differences (% values above), chimpanzees and gorillas are more closely related to humans than to orangutans, and chimpanzees are more closely related to humans than they are to gorillas. Under this scheme there is no true family of great apes. The family Hominidae includes two subfamilies: Ponginae and Homininae (humans, chimpanzees, and gorillas). This classification is monophyletic: the Hominidae includes all the species that arise from a common ancestor.

1. Briefly explain the benefits of classification schemes based on:

(a) Morphological characters: _____

_____

_____

(b) Relatedness in time (from biochemical evidence): _____

_____

_____

2. Explain the difference between a shared characteristic and a shared derived characteristic: _____

_____

_____

_____

3. Explain how the rule of parsimony is applied to cladistics: _____

_____

4. Describe the contribution of biochemical evidence to taxonomy: _____

_____

5. In the DNA data for the whale cladogram (previous page) identify the DNA match that shows a mutation event must have happened twice in evolutionary history:

_____

6. Based on the diagram above, state the family to which the chimpanzees belong under:

(a) A traditional scheme: _____ (b) A cladistic scheme: _____

Classification System

The classification of organisms is designed to reflect how they are related to each other. The fundamental unit of classification of living things is the **species**. Its members are so alike genetically that they can interbreed. This genetic similarity also means that they are almost identical in their physical and other characteristics. Species are classified further into larger, more comprehensive categories (higher taxa). It must be emphasised that all such higher classifications are human inventions to suit a particular purpose.

1. The table below shows part of the classification for humans using the seven major levels of classification. For this question, use the example of the classification of the European hedgehog, on the next page, as a guide.

   (a) Complete the list of the classification levels on the left hand side of the table below:

	Classification level	Human classification
1.	_____	_____
2.	_____	_____
3.	_____	_____
4.	_____	_____
5.	Family	Hominidae
6.	_____	_____
7.	_____	_____

   (b) The name of the Family that humans belong to has already been entered into the space provided. Complete the classification for humans (*Homo sapiens*) on the table above.

2. Describe the two-part scientific naming system (called the **binomial system**) that is used to name organisms:

   _____

   _____

3. Give two reasons why the classification of organisms is important:

   (a) _____

   _____

   (b) _____

   _____

4. Traditionally, the classification of organisms has been based largely on similarities in physical appearance. More recently, new methods involving biochemical comparisons have been used to provide new insights into how species are related. Describe an example of a biochemical method for comparing how species are related:

   _____

   _____

   _____

5. As an example of physical features being used to classify organisms, mammals have been divided into three major sub-classes: monotremes, marsupials, and placentals. Describe the main physical feature distinguishing each of these taxa:

   (a) Monotreme: _____

   _____

   (b) Marsupial: _____

   _____

   (c) Placental: _____

   _____

© Biozone International 2010

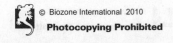

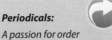

Periodicals:
A passion for order

Related activities: *New Classification Schemes*

RA 2

Biodiversity & Evolution

226

Classification of the European Hedgehog

Below is the classification for the **European hedgehog**. Only one of each group is subdivided in this chart showing the levels that can be used in classifying an organism. Not all possible subdivisions have been shown here. For example, it is possible to indicate such categories as **super-class** and **sub-family**. The only natural category is the **species**, often separated into geographical **races**, or **sub-species**, which generally differ in appearance.

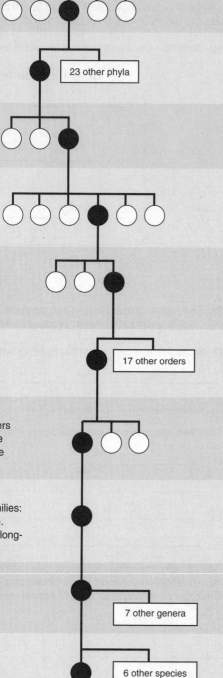

Kingdom: **Animalia**
Animals: one of five kingdoms

Phylum: **Chordata**
Animals with a notochord (supporting rod of cells along the upper surface).
tunicates, salps, lancelets, and vertebrates

23 other phyla

Sub-phylum: **Vertebrata**
Animals with backbones.
fish, amphibians, reptiles, birds, mammals

Class: **Mammalia**
Animals that suckle their young on milk from mammary glands.
placentals, marsupials, monotremes

Sub-class: **Eutheria**
Mammals whose young develop for some time in the female's reproductive tract gaining nourishment from a placenta.
placentals

Order: **Insectivora**
Insect eating mammals.
An order of over 300 species of primitive, small mammals that feed mainly on insects and other small invertebrates.

17 other orders

Sub-order: **Erinaceomorpha**
The hedgehog-type insectivores. One of the three suborders of insectivores. The other suborders include the tenrec-like insectivores (*tenrecs and golden moles*) and the shrew-like insectivores (*shrews, moles, desmans, and solenodons*).

Family: **Erinaceidae**
The only family within this suborder. Comprises two subfamilies: the true or spiny hedgehogs and the moonrats (gymnures). Representatives in the family include the desert hedgehog, long-eared hedgehog, and the greater and lesser moonrats.

Genus: *Erinaceus*
One of eight genera in this family. The genus *Erinaceus* includes four Eurasian species and another three in Africa.

7 other genera

Species: *europaeus*
The European hedgehog. Among the largest of the spiny hedgehogs. Characterised by a dense covering of spines on the back, the presence of a big toe (hallux) and 36 teeth.

6 other species

The order *Insectivora* was first introduced to group together shrews, moles, and hedgehogs. It was later extended to include tenrecs, golden moles, desmans, tree shrews, and elephant shrews, and the taxonomy of the group became very confused. Recent reclassification of the elephant shrews and tree shrews into their own separate orders has made the Insectivora a more cohesive group taxonomically.

European hedgehog
Erinaceus europaeus

KEY TERMS: Mix and Match

INSTRUCTIONS: Test your vocab by matching each term to its correct definition, as identified by its preceding letter code.

ADAPTATION

BIODIVERSITY

BINOMIAL NOMENCLATURE

CONSERVATION

COMPETITION

DIVERSITY INDICES

ECOSYSTEM

ENDEMIC

EVOLUTION

EX SITU CONSERVATION

FITNESS

FOUNDER EFFECT

GAUSE'S PRINCIPLE

GENE POOL

GENETIC DRIFT

IN SITU CONSERVATION

INTERSPECIFIC COMPETITION

INTRASPECIFIC COMPETITION

KEYSTONE SPECIES

NATURAL SELECTION

NICHE

PHYLOGENY

POPULATION BOTTLENECK

SIMPSON'S INDEX

SPECIES

SPECIES RICHNESS

A The random changes in allele frequency that occur in all populations, but are more pronounced in small populations.

B An evolutionary event in which a significant proportion of a population or species is killed or otherwise prevented from reproducing.

C An interaction between organisms exploiting the same resource.

D The active management of natural populations in order to rebuild numbers and ensure species survival.

E Any heritable trait that equips an organism for survival and reproduction (fitness) in the prevailing environment.

F The sum total of all genes of all breeding individuals in a population at any one time.

G A species that is found only within one particular geographical area.

H Community of interacting organisms and the environment (both biotic and abiotic) in which they both live and interact.

I Principle that states that two species competing for exactly the same resources cannot coexist or occupy the niche.

J The evolutionary history and line of descent of a species or higher taxonomic group.

K The loss of genetic variation when a new colony is formed by a very small number of individuals from a larger population.

L A measure of the capability of an organism with a certain genotype to reproduce. Usually expressed as a figure denoting the contribution of genes to the next generation.

M Conservation method that uses ecosystem management and legislation to protect and preserve diversity within the natural environment.

N Competitive interactions that occur between members of the same species.

O Changes in gene pools over time.

P The process by which heritable traits that are favourable in the prevailing environment become more common in successive generations.

Q Competitive interactions that occur between different species.

R Species that plays a pivotal and often disproportionate role in the stability of the ecosystem.

S A term describing the variation of life at all levels of biological organisation.

T A diversity index that takes into account both the number of species and their relative abundance and quantifies diversity by expressing the probability that two randomly sampled individuals will be of the same species.

U Method of conservation that operates away from the natural environment and is particularly useful where species are critically endangered.

V The two part scientific system that is used to for identifying organisms to genus and species.

W The number of species within a habitat or community. Sometimes compared with the number of individuals in the community.

X Mathematical expressions that quantifies the degree of diversity in an ecosystem and attempts to identify environmental stresses.

Y A group of related individuals able to breed together to produce viable offspring.

Z The functional role in the environment performed by an organism and the conditions the organism requires.

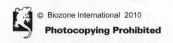

Biodiversity & Evolution

R 2

Appendix

PRACTICAL BIOLOGY AND RESEARCH SKILLS

▶ **The Truth Is Out There**

New Scientist, 26 February 2000 (Inside Science). *The philosophy of scientific method: starting with an idea, formulating a hypothesis, and following the process to theory.*

▶ **Size Does Matter**

Biol. Sci. Rev., 17 (3) February 2005, pp. 10-13. *Measuring the size of organisms and calculating magnification and scale.*

▶ **Descriptive Statistics**

Biol. Sci. Rev., 13(5) May 2001, pp. 36-37. *An account of descriptive statistics using text, tables and graphs.*

▶ **Percentages**

Biol. Sci. Rev., 17(2) Nov. 2004, pp. 28-29. *The calculation of percentage and the appropriate uses of this important transformation.*

▶ **Dealing with Data**

Biol. Sci. Rev., 12 (4) March 2000, pp. 6-8. *A short account of the best ways in which to deal with the interpretation of graphically presented data in examinations.*

▶ **Drawing Graphs**

Biol. Sci. Rev., 19(3) Feb. 2007, pp. 10-13. *A guide to creating graphs. The use of different graphs for different tasks is explained and there are a number of pertinent examples described to illustrate points.*

▶ **It's a Plot!**

Biol. Sci. Rev., 22 (2) Nov. 2009, pp. 16-19. *Using graphs to evaluate and explain data.*

▶ **Describing the Normal Distribution**

Biol. Sci. Rev., 13(2) Nov. 2000, pp. 40-41. *The normal distribution: data spread, mean, median, variance, and standard deviation.*

▶ **Experimental Animal Research**

Biol. Sci. Rev., 13(3) Jan. 2001, pp. 26-29. *The use of animals in experimental research in the UK: regulation, costs, and ethics.*

BIOLOGICAL MOLECULES

▶ **Water, Life, and Hydrogen Bonding**

Biol. Sci. Rev., 21(2) Nov. 2008, pp. 18-20. *The molecules of life and the important role of hydrogen bonding.*

▶ **Glucose & Glucose-Containing Carbohydrates**

Biol. Sci. Rev., 19(1) Sept. 2006, pp. 12-15. *The structure of glucose and its polymers.*

▶ **Designer Starches**

Biol. Sci. Rev., 19(3) Feb. 2007, pp. 18-20. *The composition of starch, and an excellent account of its properties and functions*

HEALTHY LIFESTYLE, HEALTHY HEART

▶ **The Heart**

Bio. Sci. Rev. 18(2) Nov. 2005, pp. 34-37. *The structure and physiology of the heart.*

▶ **Cunning Plumbing**

New Scientist, 6 Feb. 1999, pp. 32-37. *The arteries can actively respond to changes in blood flow, spreading the effects of mechanical stresses to avoid extremes.*

▶ **A Fair Exchange**

Biol. Sci. Rev., 13(1), Sept. 2000, pp. 2-5. Formation and reabsorption of tissue fluid (includes disorders of fluid balance).

▶ **Coronary Heart Disease**

Biol. Sci. Rev., 18(1) Sept. 2005, pp. 21-24. *An account of cardiovascular disease, including risk factors and treatments.*

▶ **Heart Stopping**

New Scientist, 11 Jan. 2003. pp. 36-39. *The atherosclerotic process may be a runaway inflammation. This article discusses its role in clot formation and the link with heart attacks and stroke.*

▶ **Skin, Scabs and Scars**

Biol. Sci. Rev., 17(3) Feb. 2005, pp. 2-6. *The roles of skin, including its role in wound healing and the processes involved in its repair when damaged.*

▶ **Atherosclerosis: The New View**

Sci. American, May 2002, pp. 28-37. *The pathological development and rupture of plaques in atherosclerosis.*

▶ **Heart Disease and Cholesterol**

Biol. Sci. Rev., 13(2) Nov. 2000, pp. 2-5. *The links between dietary fat, cholesterol level, and heart disease.*

▶ **Mending Broken Hearts**

National Geographic, 211(2), Feb. 2007, pp. 40-65. *Heart disease is becoming more prevalent; assessing susceptibility is the key to treating the disease more effectively.*

▶ **The Statin Story**

Biol. Sci. Rev., 22(3) Feb. 2010, pp. 7-9. *The development of statin drugs to treat and prevent coronary heart disease.*

▶ **Why are we so Fat?**

National Geographic, 206(2), Aug. 2004, pp. 46-61. *The occurrence and health problems associated with obesity.*

▶ **Obesity: Size Matters**

Biol. Sci. Rev., 18(4) April 2006, pp. 10-13. *Obesity and human health.*

▶ **The Good, the Fad and the Unhealthy**

New Scientist, 27 Sept. 2006, pp. 42-49. *The facts, the myths and the downright lies of nutrition.*

▶ **Feast and Famine**

Scientific American, Sept. 2007, (special issue). *Coverage of the most recent developments in health and nutrition science.*

MEMBRANES AND EXCHANGE SURFACES

▶ **Border Control**

New Scientist, 15 July 2000 (Inside Science). *The role of the plasma membrane in cell function: membrane structure, transport processes, and the role of receptors on the cell membrane.*

▶ **The Fluid-Mosaic Model for Membranes : The key points**

Biol. Sci. Rev., 22(2), Nov. 2009, pp. 20-21. *Diagrammatic revision of membrane structure and function.*

▶ **Size Does Matter**

Biol. Sci. Rev., 17 (3) February 2005, pp. 10-13. *Measuring the size of organisms and calculating magnification and scale. Surface area to volume ratio.*

▶ **Getting in and Out**

Biol. Sci. Rev., 20(3), Feb. 2008, pp. 14-16. *Diffusion: some adaptations and some common misunderstandings.*

▶ **What is Endocytosis?**

Biol. Sci. Rev., 22(3), Feb. 2010, pp. 38-41. *The mechanisms of endocytosis and the role of membrane receptors in concentrating important molecules before ingestion.*

▶ **How Biological Membranes Achieve Selective Transport**

Biol. Sci. Rev., 21(4), April 2009, pp. 32-36. *The structure of the plasma membrane and the proteins that enable the selective transport of molecules.*

Appendix

PERIODICAL REFERENCES

PROTEINS, GENES AND HEALTH

▶ **Modeling Protein Folding**

The American Biology Teacher 66(4) Apr. 2004, pp. 287-289. *How protein folding produces physical structures (teacher's reference).*

▶ **What is Tertiary Structure?**

Biol. Sci. Rev., 21(1) Sept. 2008, pp. 10-13. *How amino acid chains fold into the functional shape of a protein.*

▶ **Enzymes: Nature's Catalytic Machines**

Biol. Sci. Rev., 22(2) Nov. 2009, pp. 22-25. *Enzymes as catalysts: a very up-to-date description of enzyme specificity and binding, how enzymes work, and how they overcome the energy barriers for a reaction. Some well known enzymes are described.*

▶ **Enzymes: Fast and Flexible**

Biol. Sci. Rev., 19(1) Sept. 2006, pp. 2-5. *The structure of enzymes and how they work so efficiently at relatively low temperatures*

▶ **DNA: 50 Years of the Double Helix**

New Scientist, 15 March 2003, pp. 35-51. *A special issue on DNA: structure and function, repair, the new-found role of histones, and the functional significance of chromosome position in the nucleus.*

▶ **DNA Polymerase: Replication and Application**

Biol. Sci. Rev., 22(4) April 2010, pp. 38-41. *Illustrates the structure and replication of DNA.*

▶ **What is a Gene?**

Biol. Sci. Rev., 21(4) April 2009, pp. 10-12. *The molecular basis of genes, gene transcription, and production of a functional mRNA by removal of introns.*

▶ **What is a Mutation?**

Biol. Sci. Rev., 20(3) Feb. 2008, pp. 6-9. *The nature of mutations: causes, timing, and effects. Sickle cell disease is described.*

▶ **How do Mutations Lead to Evolution?**

New Scientist, 14 June 2003, pp. 32-39, 48-51. *An account of the five most common points of discussion regarding evolution and the mechanisms by which it occurs.*

▶ **Tertiary Structure**

Biol. Sci. Rev., 21(1), Sept. 2008, pp. 10-13. *How wrongly folded proteins can affect the way a protein functions.*

Information on the CF mutation.

▶ **Mendel's Legacy**

Biol. Sci. Rev., 18(4), April 2006, pp. 34-37. *How accurate are Mendel's laws in light of today's knowledge?*

▶ **Secrets of The Gene**

National Geographic, 196(4) Oct. 1999, pp. 42-75. *Thorough coverage of the nature of genes and the inheritance of particular genetic traits through certain populations.*

▶ **Genes Can Come True**

New Scientist, 30 Nov. 2002, pp. 30-33. *An overview of gene therapy, and a note about future directions.*

▶ **Tools You Can Trust**

New Scientist, 10 Jun. 2006, pp. 38-41. *Discusses how the development of more sophisticated gene therapy tools could allow precise repairs to be made to defective genes.*

CELLS AND MICROSCOPY

▶ **Are Viruses Alive?**

Scientific American, December 2004, pp. 77-81. *Although viruses challenge our concept of what "living" means, they are vital members of the web of life. This excellent account covers the nature of viruses, including an account of viral replication and a critical evaluation of the status of viruses in the natural world.*

▶ **Cellular Factories**

New Scientist, 23 November 1996 (Inside Science). *The role of different organelles in plant and animal cells.*

▶ **Light Microscopy**

Biol. Sci. Rev., 13(1) Sept. 2000, pp. 36-38. *An excellent account of the basis and various techniques of light microscopy.*

▶ **TEM: Transmission Electron Microscopy**

Biol. Sci. Rev., 19(4) April 2007, pp. 6-9. *An account of the techniques and applications of TEM. Includes an excellent diagram comparing features of TEM and light microscopy.*

▶ **SEM: Scanning Electron Microscopy**

Biol. Sci. Rev., 13(3) Jan. 2001, pp. 6-9. *An excellent account of the techniques and applications of SEM. Includes details of specimen preparation and recent advancements in the technology.*

▶ **The Power Behind an Electron Microscopist**

Biol. Sci. Rev., 18(1) Sept. 2005, pp.

16-20. *The use of TEMs to obtain greater resolution of finer details than is possible from optical microscopes.*

▶ **The Cell Cycle and Mitosis**

Biol. Sci. Rev., 14(4) April 2002, pp. 37-41. *Cell growth and division, stages in the cell cycle, and the complex control over different stages of mitosis.*

VARIATION AND HEREDITY

▶ **Mechanisms of Meiosis**

Biol. Sci. Rev., 15(4), April 2003, pp. 20-24. *A clear and thorough account of the events and mechanisms of meiosis.*

▶ **Mendel's Legacy**

Biol. Sci. Rev., 18(4), April 2006, pp. 34-37. *Explores the accuracy of Mendel's laws in light of today's knowledge.*

▶ **Spermatogenesis**

Biol. Sci. Rev., 15(4) April 2003, pp. 10-14. *The process and control of sperm production in humans, with a discussion of the possible reasons for male infertility.*

▶ **The Great Escape**

New Scientist, 15 Sept. 2001, (Inside Science). *How the foetus is accepted by the mother's immune system during pregnancy.*

▶ **Flower Power**

New Scientist, 9 January 1999, pp. 22-26. *Pollination and fertilisation in angiosperms, and the place of pollen competition.*

▶ **What is a Stem Cell?**

Biol. Sci. Rev., 16(2) Nov. 2003, pp. 22-23. *The nature of stem cells and their therapeutic applications.*

▶ **Cell Differentiation**

Biol. Sci. Rev., 20(4), April 2008, pp. 10-13. *How tissues arise through the control of cellular differentiation during development. The example provided is the differentiation of blood cells.*

▶ **Grown to Order**

New Scientist, 3 May, 2008, pp. 40-43. *The breakthrough in creating stem cells could be a step towards generating new tissues.*

▶ **Embryonic Stem Cells**

Biol. Sci. Rev., 22(1) Sept. 2009, pp. 28-31. *The future of embryonic stem cell research. Problems and solutions.*

▶ **What is Variation?**

Biol. Sci. Rev., 13(1) Sept. 2000, pp. 30-31. *The nature of continuous and*

Appendix

discontinuous variation. The distribution pattern of traits that show continuous variation is discussed.

▶ **The Colour Code**

New Scientist, 10 March 2002, pp. 34-37. *Researchers are uncovering the five to ten genes for skin pigmentation.*

▶ **Bring Me Sunshine**

New Scientist, 9 Aug 2003, pp. 30-33. it is widely known that *UV radiation from the sun causes skin cancer, but new research suggests moderate sun exposure can be beneficial.*

PLANTS AS RESOURCES

▶ **Designer Starches**

Biol. Sci. Rev. 19(3), Feb 2007, pp. 18-20. *The properties and composition of starch, and how it is used in many industries.*

▶ **Biofuels: How Green is Green**

Biol. Sci. Rev. 20(3), Feb 2008, pp. 10-13. *The manufacture of biofuels, the sources of material, and how environmentally friendly they really are.*

▶ **How Trees Lift Water**

Biol. Sci. Rev., 18(1), Sept. 2005, pp. 33-37. *Cohesion-tension theory and others on how trees lift water.*

▶ **High Tension**

Biol. Sci. Rev., 13(1), Sept. 2000, pp. 14-18. *Cell specialisation and transport in plants: an excellent account of the mechanisms by which plants transport water and solutes.*

▶ **Clinical Trials**

Biol. Sci. Rev. 20(1), Sept. 2007, pp. 7-9. *Outlines the process involved in clinical drug trials, including costs, process and ethical issues.*

BIODIVERSITY AND EVOLUTION

▶ **Biodiversity: Taking Stock**

National Geographic, 195(2) Feb. 1999 (entire issue). *Special issue exploring the Earth's biodiversity and what we can do to preserve it.*

▶ **Earth's Nine Lives**

New Scientist, 27 Feb. 2010, pp. 31-35. *How much can we push the Earth's support systems. This account examine the human interaction with nine critical global functions.*

▶ **Ecology and Nature Conservation**

Biol. Sci. Rev., 18(1), Sept. 2005, pp. 11-15. *The methods used to conserves and restore biodiversity in the UK. Contains some useful web links.*

▶ **What is a Species?**

Scientific American June 2008, pp. 48-55. *The science of classification; modern and traditional approaches, the value of each, and the importance of taxonomy to identifying and recognising diversity. Excellent.*

▶ **Optimality**

Biol. Sci. Rev., 17(4), April 2005, pp. 2-5. *Environmental stability and optimality of structure and function can explain evolutionary stasis in animals. Examples are described.*

▶ **The Cheetah: Losing the Race?**

Biol. Sci. Rev., 14(2) Nov. 2001, pp. 7-10. *The inbred status of cheetahs and its evolutionary consequences.*

▶ **Was Darwin Wrong?**

National Geographic, 206(5) Nov. 2004, pp. 2-35. *An excellent account of the overwhelming scientific evidence for evolution. A good starting point for reminding students that the scientific debate around evolutionary theory is associated with the mechanisms by which evolution occurs, not the fact of evolution itself.*

▶ **Skin Deep**

Scientific American, October 2002, pp. 50-57. *This article examines the evolution of skin colour in humans and presents powerful evidence for skin colour ("race") being the end result of opposing selection forces (the need for protection of folate from UV vs the need to absorb vitamin D). Clearly written and of high interest, this is a must for student discussion and a perfect vehicle for examining natural selection.*

▶ **A Passion for Order**

National Geographic, 211(6) June 2007, pp. 73-87. *The history of Carl Linnaeus and the classification of plant species.*

▶ **Uprooting the Tree of Life**

Scientific American Feb. 2000, pp. 72-77. *Using molecular techniques to redefine phylogeny and divulge the path of evolution.*

INDEX OF LATIN & GREEK ROOTS, PREFIXES, & SUFFIXES

Many biological terms have a Latin or Greek origin. Understanding the meaning of these components in a word will help you to understand and remember its meaning and predict the probable meaning of new words. Recognising some common roots, suffixes, and prefixes will make it easier to learn and understand biological terms.

The following terms are identified, together with an example illustrating their use in biology.

a(n)- without .... anoxic
affer- carrying to ....afferent
amphi- both ....amphibian
amyl- starch .... amylase
anemo- wind ....anemometer
ante- before ....antenatal
anthro- human ....anthropology
anti- against, opposite .... antibiotic
apo- separate, from .... apoenzyme
aqua- water ....aquatic
arch(ae/i)- ancient .... Archaea
arthro- joint ....arthropod
artic- jointed ....articulation
artio- even-numbered .... artiodactyl
auto- self ....autologous
avi- bird .... avian
axi- axis ....axillary
blast- germ ....blastopore
brachy- short ....brachycardia
brady- slow .... bradycardia
branch- gill ....branchial
bronch- windpipe ....bronchial
card- heart ....cardiac
centi- hundred ....centimorgan
ceph(al)- head ....cephalothorax
cerebro- brain ....cerebrospinal
chrom- color ....chromoplast
chym- juice .... chyme
cili- eyelash .... cilia
contra- opposite ....contraception
cotyl- cup .... hypocotyl
crani- skull ....cranium
crypt- hidden ....crptic
cyan- blue ....cyanobacteria
cyt- cell .... cytoplasm
dactyl- finger .... polydactylic
deci-(a) ten .... decibel, decapod
dendr- tree .... dendrogram
derm- skin ....dermal
di- two ....dihybrid
dors- back ....dorsal
dur- hard .... dura mater

Appendix

SI UNITS & MULTIPLES

echino- spiny ....echinoderm
ecto- outside ....ectoderm
effer- carrying away ....efferent
endo- inside ....endoparasite
equi- horse, equal...............equilibrium
erythr- red ....erthyrocyte
eu- well, very......................eukaryote
eury- wide ....eurythermal
ex- out of..................................explant
exo- outside ....exoskeleton
extra- beyond.................extraperitoneal
foramen- opening ..... foramen magnum
gast(e)r- stomach, pouch ..... gastropod
gymn- naked....................gymnosperm
hal- salty ....halophyte
haplo- single, simple.................hapolid
holo- complete, whole.............holozoic
hydr- water ....hydrophyte
hyper- above.......................hypertonic
hypo- beneath.......................hypotonic
infra- under ....infrared
inter- between ....interspecific
intra- within..........................intraspecific
iso- equal ....isotonic
kilo- thousand ....kilogram
lacuna- space...........................lacunae
lamella- leaf, layer...........lamellar bone
leuc- white..............................leucocyte
lip- fat.....................................lipoprotein
lith- stone ....Palaeolithic
lumen- cavity..............................lumen
lymph- clear water ....lymphatic
magni- large.........................magnification
mamma- breast.......................mammal
mat(e)ri- mother ....maternal
mega- large ....megakaryocyte
melan- black ....melanocyte
meso- middle........................Mesolithic
meta- after....................metamorphosis

micro- small ....microorganism
milli- thousand.......................millimetre
mirabile- wonderful ....rete mirabile
mono- one...........................monohybrid
morph- form ....morphology
motor- mover......................motor nerve
multi- many ....multicellular
myo- muscle ....myofibril
necro- dead ....necrosis
neo- new ....Neolithic
nephr- kidney...........................nephro
neur- nerve ....neural
noto- back, south...................notochord
oecious- house of ....monoecius
oed- swollen.............................oedema
olfact- smelling.......................olfactory
os(s/t)- bone...........................osteocyte
ovo- egg...........................ovoviviparous
palae- old.............................Palaeocene
pect(or)- chest ....pectoral fin
pent- five ....pentose sugar
per(i)- through, beyond.........peristalsis
peri- around.........................periosteum
phaeo- dark ....phaeomelanin
phag- eat ....phagosome
phyll- leaf ....sclerophyll
physio- nature.....................physiology
phyto- plant.....................phytohormone
pisc- fish ....piscivorous
plagio- oblique................. plagioclimax
pneu(mo/st)- air, lung..........pneumonia
pod- foot ....sauropod
poly- many ....polydactyly
pre- before ....premolar
pro- in front of ....Prokaryote
prot- first.............................. protandry
pseud- false...................pseudopodia
pter- wing, fern....................Pterophyta
pulmo- lung..........................pulmonary

radi- root ....radicle
ren- kidney ....renal
retic- network ,.....................reticulated
sacchar- sugar.............polysaccharide
schizo- split.................schizocoelomate
scler- hard ....sclerophyll
seba- tallow, wax ....sebaceous
semi- half ....semi-conservative
sept- seven, wall.....................septum
soma- body.........opisthosoma, somatic
sperm- seed..................spermatophyte
sphinct- closing ....sphincter
stereo- solid ....stereocilia
stom- mouth................................ stoma
strat- layer .... stratification
sub- below...............................subtidal
sucr- sugar ....sucrase
super- beyond........................superior
supra- above............supracoracoideus
sym- with.................................symbiosis
syn- with..................................synapsis
tact- touch .... tactile
tachy- fast ....tachycardia
trans- across .... transmembrane
tri- three..........................triploblastic
trich- hair ....trichome
ultra- above..........................ultraviolet
un- one ....unicellular
vas- vessel.............................. vascular
ven- vein .... venous
ventr- belly ....ventral
vern- spring............................ vernal
visc- organs of body cavity ....viscera
vitr- glass ....in vitro
xanth- yellow ....xanthophyll
xen- stranger ....xenotransplant
xer- dry .... xerophyte
xyl- wood.....................................xylem
zo- animal ....zoological

MULTIPLES

MULTIPLE	PREFIX	SYMBOL	EXAMPLE
10^9	giga	G	gigawatt (GW)
10^6	mega	M	megawatt (MW)
10^3	kilo	k	kilogram (kg)
10^2	hecto	h	hectare (ha)
10^{-1}	deci	d	decimetre (dm)
10^{-2}	centi	c	centimetre (cm)
10^{-3}	milli	m	milliimetre (mm)
10^{-6}	micro	μ	microsecond (μs)
10^{-9}	nano	n	nanometre (nm)
10^{-12}	pico	p	picosecond (ps)

INTERNATIONAL SYSTEM OF UNITS (SI)

Examples of SI derived units

DERIVED QUANTITY	NAME	SYMBOL
area	square metre	m^2
volume	cubic metre	m^3
speed, velocity	metre per second	ms^{-1}
acceleration	metre per second squared	ms^{-2}
mass density	kilogram per cubic metre	kgm^{-3}
specific volume	cubic meter per kilogram	m^3kg^{-1}
amount-of-substance concentration	mole per cubic meter	$molm^{-3}$
luminance	candela per square meter	cdm^{-2}

Index